Antibody engineeri[ng]

Institute of Cancer Research Library
Sutton
Please return this book by the last date stamped below

29. APR. 1997		
29. MAY 1997		
-3. JUL 1997		
12. SEP. 1997		
-5. NOV. 1997		
27. MAR. 1998		
-1. MAY 1998		
10. JUN. 1998		
13. MAR. 2007		

The Practical Approach Series

SERIES EDITOR

B. D. HAMES
Department of Biochemistry and Molecular Biology
University of Leeds, Leeds LS2 9JT, UK

★ **indicates new and forthcoming titles**

Affinity Chromatography
Anaerobic Microbiology
Animal Cell Culture (2nd edition)
Animal Virus Pathogenesis
Antibodies I and II
★ Antibody Engineering
★ Basic Cell Culture
Behavioural Neuroscience
Biochemical Toxicology
★ Bioenergetics
Biological Data Analysis
Biological Membranes
Biomechanics—Materials
Biomechanics—Structures and Systems
Biosensors
Carbohydrate Analysis (2nd edition)
Cell–Cell Interactions
★ Epithelial Cell Culture
The Cell Cycle
★ Cell Growth and Apoptosis
Cellular Calcium

Cellular Interactions in Development
Cellular Neurobiology
Clinical Immunology
Crystallization of Nucleic Acids and Proteins
★ Cytokines (2nd edition)
The Cytoskeleton
Diagnostic Molecular Pathology I and II
Directed Mutagenesis
★ DNA Cloning 1: Core Techniques (2nd edition)
★ DNA Cloning 2: Expression Systems (2nd edition)
★ DNA Cloning 3: Complex Genomes (2nd edition)
★ DNA Cloning 4: Mammalian Systems (2nd edition)
Electron Microscopy in Biology
Electron Microscopy in Molecular Biology
Electrophysiology
Enzyme Assays

Essential Developmental Biology
Essential Molecular Biology I and II
Experimental Neuroanatomy
★ Extracellular Matrix
Flow Cytometry (2nd edition)
★ Free Radicals
Gas Chromatography
Gel Electrophoresis of Nucleic Acids (2nd edition)
Gel Electrophoresis of Proteins (2nd edition)
★ Gene Probes 1 and 2
Gene Targeting
Gene Transcription
Glycobiology
Growth Factors
Haemopoiesis
Histocompatibility Testing
★ HIV Volumes 1 and 2
Human Cytogenetics I and II (2nd edition)
Human Genetic Disease Analysis
Immunocytochemistry
In Situ Hybridization
Iodinated Density Gradient Media
★ Ion Channels
Lipid Analysis
Lipid Modification of Proteins
Lipoprotein Analysis
Liposomes
Mammalian Cell Biotechnology
Medical Bacteriology
Medical Mycology

★ Medical Parasitology
★ Medical Virology
Microcomputers in Biology
Molecular Genetic Analysis of Populations
Molecular Genetics of Yeast
Molecular Imaging in Neuroscience
Molecular Neurobiology
Molecular Plant Pathology I and II
Molecular Virology
Monitoring Neuronal Activity
Mutagenicity Testing
★ Neural Cell Culture
Neural Transplantation
Neurochemistry
Neuronal Cell Lines
NMR of Biological Macromolecules
★ Non-isotopic Methods in Molecular Biology
Nucleic Acid Hybridization
Nucleic Acid and Protein Sequence Analysis
Oligonucleotides and Analogues
Oligonucleotide Synthesis
PCR 1
★ PCR 2
★ Peptide Antigens
Photosynthesis: Energy Transduction
Plant Cell Biology
Plant Cell Culture (2nd edition)

Plant Molecular Biology
Plasmids (2nd edition)
★ Platelets
Pollination Ecology
Postimplantation Mammalian Embryos
Preparative Centrifugation
Prostaglandins and Related Substances
Protein Blotting
Protein Engineering
Protein Function
Protein Phosphorylation
Protein Purification Applications
Protein Purification Methods
Protein Sequencing

Protein Structure
Protein Targeting
Proteolytic Enzymes
★ Pulsed Field Gel Electrophoresis
Radioisotopes in Biology
Receptor Biochemistry
Receptor–Ligand Interactions
RNA Processing I and II
Signal Transduction
Solid Phase Peptide Synthesis
Transcription Factors
Transcription and Translation
Tumour Immunobiology
Virology
Yeast

Antibody engineering
A Practical Approach

Edited by

JOHN McCAFFERTY

Cambridge Antibody Technology Ltd
The Science Park
Melbourn
Cambridgeshire SG8 6JJ

HENNIE HOOGENBOOM

CESAME at Department of Pathology
Academisch Ziekenhuis
Maastricht PO Box 5800
6202 AZ Maastricht
The Netherlands

and

DAVE CHISWELL

Cambridge Antibody Technology Ltd,
The Science Park
Melbourn
Cambridgeshire SG8 6JJ

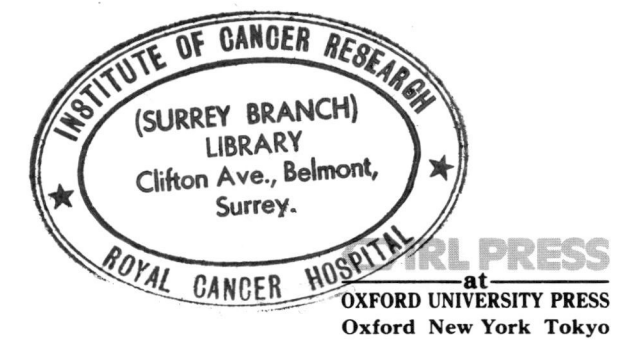

IRL PRESS
at
OXFORD UNIVERSITY PRESS
Oxford New York Tokyo

Oxford University Press, Walton Street, Oxford OX2 6DP

*Oxford   New York
Athens   Auckland   Bangkok   Bombay
Calcutta   Cape Town   Dar es Salaam   Delhi
Florence   Hong Kong   Istanbul   Karachi
Kuala Lumpur   Madras   Madrid   Melbourne
Mexico City   Nairobi   Paris   Singapore
Taipei   Tokyo   Toronto*

*and associated companies in
Berlin   Ibadan*

Oxford is a trade mark of Oxford University Press

*Published in the United States
by Oxford University Press Inc., New York*

© Oxford University Press, 1996

*All rights reserved. No part of this publication may be
reproduced, stored in a retrieval system, or transmitted, in any
form or by any means, without the prior permission in writing of Oxford
University Press. Within the UK, exceptions are allowed in respect of any
fair dealing for the purpose of research or private study, or criticism or
review, as permitted under the Copyright, Designs and Patents Act, 1988, or
in the case of reprographic reproduction in accordance with the terms of
licences issued by the Copyright Licensing Agency. Enquiries concerning
reproduction outside those terms and in other countries should be sent to
the Rights Department, Oxford University Press, at the address above.*

*This book is sold subject to the condition that it shall not,
by way of trade or otherwise, be lent, re-sold, hired out, or otherwise
circulated without the publisher's prior consent in any form of binding
or cover other than that in which it is published and without a similar
condition including this condition being imposed
on the subsequent purchaser.*

*Users of books in the Practical Approach Series are advised that prudent
laboratory safety procedures should be followed at all times. Oxford
University Press makes no representation, express or implied, in respect of
the accuracy of the material set forth in books in this series and cannot
accept any legal responsibility or liability for any errors or omissions
that may be made.*

A catalogue record for this book is available from the British Library

Library of Congress Cataloging in Publication Data
*Antibody engineering: a practical approach/edited by John
McCafferty, Hennie Hogenboom, and Dave Chiswell. – 1st ed.
(A Practical approach series ; 169)
Includes bibliographical references and index.
1. Immunoglobulins–Synthesis–Laboratory manuals.   2. Genetic
engineering–Laboratory manuals.   I. McCafferty, John, Dr.
II. Hoogenboom, Hennie.   III. Chiswell, Dave.   IV. Series.
QR186.7.A576   1996   615'.37–dc20   95-49216*
*ISBN 0 19 963593 5 (Hbk)
ISBN 0 19 963592 7 (Pbk)*

*Typeset by Footnote Graphics, Warminster, Wilts
Printed in Great Britain by Information Press, Ltd, Eynsham, Oxon.*

Preface

Recombinant DNA methods have revolutionized the isolation and production of antibodies in recent years. Initially, recombinant antibodies were engineered and expressed in mammalian cells but by the late 1980s efficient expression of antibody fragments in bacteria had been demonstrated. More recently, the isolation of new antibody specificities has been transformed by the creation of large repertoires of antibody genes and the development of methods based on bacteriophage display to isolate specific antibodies. This has been paralleled by an increased understanding of the key residues affecting antibody structure and specificity together with more quantitative analysis of the binding properties of the antibodies. In addition, the mapping and sequencing of the repertoire of human antibody germline genes has put our understanding of the changes occurring during *in vivo* affinity maturation on a firmer footing.

We believe that now is the right time to assemble in one volume, protocols which allow the researcher to isolate a new antibody, analyse its properties, format the right antibody molecule or fragment, and produce sufficient quantities to be useful.

The book is structured in two parts: the first describes the generation and analysis of antibodies and the second covers engineering and production. Chapter 1 is designed to be the source of the standard 'repertoire' of phage-display protocols. In the following chapter the 'fine art' of the system, e.g. affinity maturation, is described. Methods for introducing antibody repertoires into transgenic mice are also described in the first section, as are methods for the analysis of the characteristics of antibodies with respect to affinity and sequence.

The second section describes the engineering of natural and man-made effector functions. The conversion of rodent antibodies to human antibodies by either CDR grafting or 'guided selection' using phage display is also covered. In addition, methods to manufacture significant quantities of various antibody-based molecules in both eukaryotic and prokaryotic systems are outlined.

Traditional monoclonal antibody technology has provided a wealth of reagents for research and diagnostic applications. The methods and technologies described here can help expand the versatility of antibody-based reagents in these areas. We have endeavoured in this book to provide researchers working with antibodies in research and diagnostics with a route into the 'new technologies', as practised by the leaders in the field.

This 'route' will be of equal value to workers in the field of antibody therapeutics. The ability to generate high-affinity, high-specificity human antibodies in appropriate formats will help circumvent some of the problems

Preface

associated with earlier work using mouse monoclonal antibodies and aid in their successful use *in vivo*. As we draw to the end of the millennium we have moved closer to the 'magic bullet' first hypothesized by Paul Ehrlich at the beginning of this century.

Cambridge	J.McC
Maastricht	H.R.H
May 1996	D.J.C

Contents

List of Contributors	xvii
Abbreviations	xxi

1. Construction and use of antibody gene repertoires — 1

A.R. Pope, M. J. Embleton, and R. Mernaugh

1. Introduction	1
2. Vectors for the display of proteins on the surface of bacteriophage fd	2
3. Preparation and cloning of antibody DNA	6
Preparation of mRNA	7
cDNA preparation	8
Primary PCR	9
Assembly of scFv fragments	17
Amplification and digestion	20
Ligation and transformation	22
4. Growth and expression of phage antibodies	23
Rescue of phage	23
Growth and soluble expression	25
Purification of soluble antibody	27
5. Selection of antibody variants displayed on the surface of bacteriophage	28
6. Analysis of phage-derived antibodies by ELISA	30
7. In-cell PCR assembly	32
Introduction	32
Application	37
8. Conclusions	38
References	39

2. Affinity maturation of antibodies using phage display — 41

Kevin S. Johnson and Robert E. Hawkins

1. Introduction	41
General considerations	41

Mutagenesis	41
Selection	43
Screening	43
2. Mutagenesis	44
General considerations	44
Error-prone PCR	44
Site-directed mutagenesis	45
Mutagenesis using 'spiked' PCR primers	47
3. Selection of antibodies with altered properties	49
Selection of phage by panning	50
Solution capture on soluble antigen	51
Off-rate selection	53
4. Screening selected populations of antibodies	54
Affinity screen for phage antibody clones (K_d assay)	54
5. Sequence and fingerprint analysis	57
References	58

3. Human antibody repertoires in transgenic mice: manipulation and transfer of YACs   59

Nicholas P. Davies, Andrei V. Popov, Xiangang Zou, and Marianne Brüggemann

1. Introduction	59
2. Yeast artificial chromosomes	61
Library screening	62
Maintenance	62
3. Modification of YACs	63
Universal YAC vectors	64
Site-directed introduction	65
Mapping site-specific integration	67
Profile of single and multiple integrations	67
4. YAC transfer into embryonic stem cells	70
Spheroplast fusion	70
Picking and analysing clones	73
5. Conclusions	73
References	74

4. Measuring antibody affinity in solution   77

Lisa Djavadi-Ohaniance, Michel E. Goldberg, and Bertrand Friguet

1. General considerations	77
K_d does not directly reflect association or dissociation kinetics	77

True K_d cannot be determined when the mAb or the Ag is immobilized in a solid-phase assay	78
The determination of K_d must take into account the valency of the mAb and of the Ag molecule	78
2. Overview of methods to measure affinities in solution	78
Fluorescence	79
ELISA- and RIA-based methods	80
3. Affinity measurements in solution by competition ELISA	81
Theoretical aspects	81
Rationale	82
Requirements for the determination of K_d	82
Determination of K_d	86
Calculations	88
Determination of K_d with impure antibody	90
4. Affinity measurement in solution by an RIA-based method	91
Rationale	91
Requirements for the determination of K_d	91
Determination of K_d	94
5. Conclusions	95
References	96

5. Measuring antibody affinity using biosensors   99

Laura J. Hefta, Anna M. Wu, Michael Neumaier, and John E. Shively

1. Introduction	99
2. Theoretical aspects	101
Measuring association and dissociation rate constants	101
Limitations on measuring affinity constants	102
3. Immobilization, binding, and regeneration of the BIAcore	103
Immobilization step	104
Regeneration step	108
4. Kinetic analysis of anti-CEA antibodies	108
Direct binding assays: comparison of murine and chimeric T84.12	108
Indirect binding assays: comparison of murine and chimeric T84.66	109
Assays for engineered antibody fragments	112
Assays for anti-idiotypic antibody	115
5. Conclusions	115

Contents

Acknowledgements	116
References	116

6. Analysis of human antibody sequences   119
Gerald Walter and Ian M. Tomlinson

1. Introduction	119
2. Amplification and cloning of antibody V genes	119
Germline V segments	123
Rearranged V genes	125
3. Sequencing of immunoglobulin genes	125
Sequencing primers	126
Template preparation	126
Sequencing techniques	129
4. Analysis of antibody sequences	137
Software packages for sequence analysis	137
Editing, translating, and comparing sequences	137
Multiple alignments	138
Databases	139
Statistical analyses	140
References	144

7. Rodent to human antibodies by CDR grafting   147
Mary M. Bendig and S. Tarran Jones

1. Introduction	147
2. Cloning and sequencing mouse variable regions	147
3. Construction of a chimeric antibody	154
4. Design and construction of a reshaped human antibody	155
Analysis of the mouse variable regions	155
Design of the reshaped human antibody	155
Construction of the reshaped human antibody	157
5. Preliminary expression and analysis of the reshaped human antibodies	164
Acknowledgements	166
References	168

Contents

8. Converting rodent into human antibodies by guided selection   169
Hennie R. Hoogenboom, Deborah J. Allen, and Andrew J. Roberts

1. Introduction   169
2. Cloning, expression, and characterization of rodent scFv fragments   173
3. Construction of large chain-shuffled repertoires in guided selection   174
    The first DNA shuffle: combining murine V_H with a human V_L repertoire   175
    Construction of fully human chain-shuffled repertoires by combining the selected human V_L genes with a human V_H repertoire   181
4. Selecting half-human or completely human antibodies by display and enrichment steps   182

References   185

9. Choosing and manipulating effector functions   187
Inger Sandlie and Terje E. Michaelsen

1. Choosing effector functions   187
2. Mediation of effector functions   188
    Complement activation and lysis   189
    FcγR-mediated activities   189
3. Measuring complement activation and lysis *in vitro*   190
    The structural requirements for complement activation   190
    Comparing the IgG subclasses in complement activation and lysis   190
4. Measuring ADCC *in vitro*   195
    The structural requirements for FcγR binding   195
    Comparing the IgG subclasses in ADCC   195
5. Measuring phagocytosis and respiratory burst   197
6. Optimizing effector functions   200

Acknowledgements   201

References   201

Contents

10. Producing antibodies in *Escherichia coli*: from PCR to fermentation — 203

Andreas Plückthun, Anke Krebber, Claus Krebber, Uwe Horn, Uwe Knüpfer, Rolf Wenderoth, Lars Nieba, Karl Proba, and Dieter Riesenberg

1. Introduction — 203
2. Cloning of antibody variable domains — 204
   - Choice of antibody format — 213
3. Expression strategies — 215
   - Overview — 215
   - Secretion — 217
   - Functional antibodies from the cytoplasm — 221
4. Improving expression: the influence of the sequence on expression and folding — 225
   - Analysis and long-term solutions — 225
   - Short-term solutions — 228
5. Growth and fermentation — 229
   - Cultivation in standard shake flasks — 229
   - Cultivation in benchtop flasks to medium cell-densities — 231
   - Cultivation of *E. coli* in a 10 litre fermenter to high cell-densities — 235
6. Antibody purification — 244
   - General considerations — 244
   - Immobilized metal ion-affinity chromatography (IMAC) — 245

Acknowledgements — 248
Appendix: Abbreviations for HCDC — 249
References — 249

11. Production of single-chain Fv monomers and multimers — 253

David Filpula, Jeffrey McGuire, and Marc Whitlow

1. Introduction — 253
2. Linker designs — 254
3. scFv gene construction from hybridoma cells — 255
4. Expression of single-chain Fv in *E. coli* — 259
5. Fermentation, renaturation, and purification of an scFv protein — 261

References — 266

12. Expression of immunoglobulin genes in mammalian cells   269

Christopher Bebbington

  1. Introduction   269

  2. Transient expression of antibodies in COS cells   270
      Choice of cell line and medium   270
      Choice of expression vector   271

  3. Dihydrofolate reductase selection in Chinese hamster ovary (CHO) cells   273
      Choice of cell line and medium   273
      DHFR vector design   274
      Transfection of CHO cells   275
      Selection for amplification of DHFR vector with methotrexate   278

  4. Glutamine synthetase selection in Chinese hamster ovary (CHO) cells   279
      Choice of cell type and medium   279
      Vector design   279
      Cell transfection   280

  5. Glutamine synthetase selection in NS0 myeloma cells   282
      Choice of cells and medium   282
      GS-vector design   283
      Cell transfection   284
      Selection for GS-gene amplification in NS0 cells using MSX   287

  6. Concluding remarks   288

  References   288

13. Preparation and uses of Fab' fragments from *Escherichia coli*   291

Paul Carter, Maria L. Rodrigues, John W. Park, and Gerardo Zapata

  1. Introduction   291
      Overview   291
      High-level production of Ab fragments in *E. coli*   292
      Targeting of Fab' fragments to the periplasmic space of *E. coli*   292

  2. Recovery of Fab' fragments from *E. coli*   293
      Release of soluble Fab'–SH fragments from *E. coli*   293
      Recovery of Fab'–SH fragments from *E. coli*   293
      Endotoxin removal   294
      Determination of free thiol content   296

Contents

    3. Application of Fab' fragments from *E. coli*   297
        Construction of monospecific F(ab')$_2$ fragments   297
        Construction of bispecific F(ab')$_2$ fragments   300
        Construction of Fab'–PEG   301
        Construction of Fab'–immunoliposomes   303
        Immobilization of Fab'–SH for use in affinity purification   305

    Acknowledgements   307
    References   307

A1. *Addresses of suppliers*   309

A2. *Sequencing primers for antibody V genes*   315

A3. *Sequences of human germline V_H, V_κ, J_H, and J_κ segments*   316

Index   321

Contributors

DEBORAH J. ALLEN
Cambridge Antibody Technology Ltd, The Science Park, Melbourn, Cambridge SG8 6EJ, UK.

MARY M. BENDIG
Ludwig Institute for Cancer Research, 6th Floor, Glen House, Stag Place, London SW1E 5AJ, UK.

CHRISTOPHER BEBBINGTON
Celltech Therapeutics Ltd, 216 Bath Rd, Slough, Berkshire SL1 4EN, UK.

MARIANNE BRÜGGEMANN
Department of Development and Signalling, Babraham Institute, Babraham, Cambridge CB2 4AT, UK.

PAUL CARTER
Department of Molecular Oncology, Genentech Inc., 460 Point San Bruno Boulevard, South San Francisco, California 94080–4990, USA.

DAVE CHISWELL
Cambridge Antibody Technology Ltd, The Science Park, Melbourn, Cambridge SG8 6JJ, UK.

NICHOLAS P. DAVIES
Department of Biochemistry and Molecular Genetics, St Mary's Hospital Medical School, Norfolk Place, London W2 1PG, UK.

LISA DJAVADI-OHANIANCE
Unité de Biochimie Cellulaire, Institut Pasteur, 28 rue de Docteur Roux, 75724 Paris, cedex 15, France.

M. JIM EMBLETON
Paterson Institute for Cancer Research, Christie Hospital NHS Trust, Wilmslow Rd, Manchester M20 9BX, UK.

DAVID FILPULA
Department of Tumor Biology and Molecular Genetics, Enzon Inc., 40 Kingsbridge Rd, Piscataway, New Jersey 08854–3998, USA.

BERTRAND FRIGUET
Unité de Biochimie Cellulaire, Institut Pasteur, 28 rue de Docteur Roux, 75724 Paris, cedex 15, France.

MICHEL E. GOLDBERG
Unité de Biochimie Cellulaire, Institut Pasteur, 28 rue de Docteur Roux 75724 Paris, cedex 15, France.

Contributors

ROBERT E. HAWKINS
MRC Centre for Protein Engineering, Hills Rd, Cambridge CB2 2QH, UK.

LAURA J. HEFTA
Division of Immunology, Beckman Research Institute of the City of Hope, 1450 East Duarte Rd, Duarte, California 91010, USA.

HENNIE R. HOOGENBOOM
CESAME at Department of Pathology, Academisch Ziekenhuis Maastricht, PO Box 5800, 6202 AZ Maastricht, The Netherlands.

UWE HORN
Hans Knöll-Institut für Naturstoff-Forschung, Beutenbergstrasse 11, D-07745 Jena, Germany.

KEVIN S. JOHNSON
Cambridge Antibody Technology Ltd, The Science Park, Melbourn, Cambridge SG8 6JJ, UK.

S. TARRAN JONES
Antibody Engineering Group, MRC Collaborative Centre, 1–3 Burtonhole Lane, Mill Hill, London NW7 1AD, UK.

UWE KNÜPFER
Hans Knöll-Institut für Naturstoff-Forschung, Beutenbergstrasse 11, D-07745 Jena, Germany.

ANKE KREBBER
Biochemisches Institut, Universität Zürich, Winterthurerstrasse 190, CH-8057 Zürich, Switzerland.

CLAUS KREBBER
Biochemisches Institut, Universität Zürich, Winterthurerstrasse 190, CH-8057 Zürich, Switzerland.

JOHN McCAFFERTY
Cambridge Antibody Technology Ltd, The Science Park, Melbourn, Cambridge SG8 6JJ, UK.

JEFFREY McGUIRE
Department of Tumor Biology and Molecular Genetics, Enzon Inc., 40 Kingsbridge Rd, Piscataway, New Jersey 08854–3998, USA.

RAY MERNAUGH
Pharmacia Biotech Inc., 2202 North Bartlett Ave, Milwaukee, Wisconsin 53202, USA.

TERJE E. MICHAELSEN
Department of Vaccinology, National Institute of Public Health, Geitmyrsein 75, 0462 Oslo, Norway.

Contributors

MICHAEL NEUMAIER
Abt. für Klinisches Chemie, Universatitäts-Krankenhaus Eppendorf, Universität Hamburg, Hamburg, Germany.

LARS NIEBA
Biochemisches Institut, Universität Zürich, Winterthurerstrasse 190, CH-8057 Zürich, Switzerland.

JOHN W. PARK
Department of Molecular Biology, Genetech Inc., 460 Point San Bruno Boulevard, South San Franciso, California 94080–4990, USA.

ANDREAS PLÜCKTHUN
Biochemisches Institut, Universität Zürich, Winterthurerstrasse 190, CH-8057 Zürich, Switzerland.

ANTONY R. POPE
Cambridge Antibody Technology Ltd, The Science Park, Melbourn, Cambridge SG8 6JJ, UK.

ANDREI V. POPOV
Department of Development and Signalling, Babraham Institute, Babraham, Cambridge CB2 4AT, UK.

KARL PROBA
Biochemisches Institut, Universität Zürich, Winterthurerstrasse 190, CH-8057 Zürich, Switzerland.

DIETER RIESENBERG
Hans Knöll-Institut für Naturstoff-Forschung, Beutenbergstrasse 11, D-07745 Jena, Germany.

ANDREW J. ROBERTS
Cambridge Antibody Technology Ltd, The Science Park, Melbourn, Cambridge SG8 6JJ, UK.

MARIA L. RODRIGUES
Department of Molecular Oncology, Genentech Inc., 460 Point San Bruno Boulevard, South San Francisco, California 94080–4990, USA.

INGER SANDLIE
Department of Biology, University of Oslo, PO Box 1050, Blindern, N-0316 Oslo, Norway.

JOHN E. SHIVELY
Division of Immunology, Beckman Research Institute of the City of Hope, 1450 East Duarte Rd, Duarte, California 91010, USA.

IAN M. TOMLINSON
MRC Centre for Protein Engineering, Hills Rd, Cambridge CB2 2QH, UK.

Contributors

GERALD WALTER
Max Planck-Institut für Molekulare Genetik, Ihnestr. 73, D-14195 Berlin-Dahlem, Germany.

ROLF WENDEROTH
Hans Knöll-Institut für Naturstoff-Forschung, Beutenbergstrasse 11, D-07745 Jena, Germany.

MARC WHITLOW
Berlex Biosciences, 15049 San Pablo Ave, PO Box 4099, Richmond, California 94804, USA.

ANNA M. WU
Division of Biology, Beckman Research Institute of the City of Hope, 1450 East Duarte Rd, Duarte, California 91010, USA.

GERARDO ZAPATA
Department of Process Sciences, Genentech Inc., 460 Point San Bruno Boulevard, South San Francisco, California 94080–4990, USA.

XIANGANG ZOU
Department of Development and Signalling, Babraham Institute, Babraham, Cambridge CB2 4AT, UK.

Abbreviations

Ab	antibody
ADCC	antibody-dependent cell-mediated cytotoxicity
Ag	antigen
ARS	autonomous replicating sequences
ATCC	American tissue culture collection
bp	base pair
BSA	bovine serum albumin
CDR	complementarity-determining region
CEA	carcinoembryonic antigen
CEN	centromere
CHO	Chinese hamster ovary (cell)
CML	cell-mediated lysis
cv	column volume
DEA	diethanolamine
DEPC	diethylpyrocarbonate
DHFR	dihydrofolate reductase
DMEM	Dulbecco modified Eagle medium
DMSO	dimethylsulfoxide
dsFv	disulfide-linked Fv
DNTB	5, 5′-dithiobis (2-nitrobenzoic acid)
DTT	dithiothreitol
EDC	N-ethyl-N'-(dimethylaminopropyl) carbodiimide or 1-ethyl-3-(3-dimethylaminopropyl) carbodiimide hydrochloride
EDTA	ethylenediaminetetraacetic acid
EIA	enzyme immunoassay
ELISA	enzyme-linked immunosorbent assay
ES	embryonic stem (cell)
EtOH	ethanol
FCS	fetal calf serum
FIA	flow injection analyser
FPLC	fast protein liquid chromatography
FR	framework region
GOD	glucose oxidase
g3p	gene 3 protein
Grx	glutaredoxin
GS	glutamine synthetase
HABA	4-hydroxyazobenzene-2-carboxylic acid
HCMV	human cytomegalovirus
Hepes	N-[2-hydroxyethyl]piperazine-N'-[2-ethanesulfonic acid]
IFN-γ	gamma interferon

Abbreviations

IL	interleukin
IMAC	immobilized metal ion-affinity chromatography
IPTG	isopropyl-β-D-thio-galactopyranoside
k_{on}	kinetic rate constant—association
k_{off}	kinetic rate constant—dissociation
K_a	antibody equilibrium constant—association
K_d	antibody equilibrium constant—dissociation
lmp	low melting point
mAb	monoclonal antibody
MAC	mammalian artificial chromosome
MBS	*m*-maleimidobenzoyl-*N*-hydroxysuccinimide ester
2-ME	β-mercaptoethanol
MFI	mean fluorescence intensity
MNC	mononuclear cell
Mops	3-[*N*-morpholino]propanesulfonic acid
MSX	methionine sulfoxamine
MTX	methotrexate
MUG	methyl-umbelliferyl-β-galactopyranoside
NHS	*N*-hydroxysuccinimide
Ni–NTA	nickel–nitrilo-tri-acetic acid
NIP	4-hydroxy-3-iodo-5-nitrophenylacetic acid
NK	natural killer (cell)
NP-40	Nonidet P-40
ONPG	*o*-nitrophenyl-β-D-galactopyranoside
PBL	peripheral blood lymphocytes
PBS	phosphate-buffered saline
PCR	polymerase chain reaction
PEG	polyethylene glycol
PFGE	pulsed-field gel electrophoresis
p.f.u.	plaque-forming unit
pI	isoelectric point
PMN	polymorphonuclear leukocyte
PMSF	phenylmethylsulfonyl fluoride
PNPP	*p*-nitrophenyl phosphate
RIA	radioimmunoassay
RU	response unit
scFv	single chain Fv
SD	synthetic 'drop-out' media
SRBC	sheep red blood cell
TBS	Tris-buffered saline
TE	10 mM Tris pH 7.4, 1 mM EDTA buffer
TEL	telomere
TEMED	*N*,*N*,*N'*,*N'*-tetramethylethylenediamine

Abbreviations

TNF	tumour necrosis factor
Trx	thioredoxin
TrxB	thioredoxin reductase
X-Gal	5-bromo-4-chloro-3-indoyl-D-galactopyranoside
YAC	yeast artificial chromosome

1

Construction and use of antibody gene repertoires

A. R. POPE, M. J. EMBLETON, and R. MERNAUGH

1. Introduction

The display of antibody molecules on the surface of bacteriophage (1) provides a powerful method of selecting a specific antibody from a mixed population of antibodies together with the gene encoding it. This ability to co-select proteins and their genes has been exploited to enable the isolation of high-affinity, antigen-specific antibodies derived from an immunized mouse (2) and from unimmunized humans (3). In addition, after selecting such an antibody, a mutant antibody library can be constructed and higher affinity antibody clones selected. Alternatively, a monoclonal antibody can be subcloned from a hybridoma cell line into the bacteriophage and after mutagenesis, antibodies of improved characteristics selected (4).

Bacteriophage fd is a filamentous, single-stranded DNA phage which infects male *Escherichia coli* cells. Adsorption to the host sex pilus is mediated by the gene 3 protein (g3p) displayed at the tip of the virion. The amino-terminal domains of the three g3p molecules on each virion form knob-like structures that are responsible for binding the phage to the F-pilus whilst the C-terminal domain is anchored in the phage coat (5, 6). George Smith found that peptides could be displayed on the surface of phage by fusion to the N-terminus of g3p (7). Phage with binding activities could then be isolated from random peptide libraries after repeated rounds of growth and selection for phage with the desired binding characteristics (8–10).

The range of molecules displayed as gene 3 fusions has now been extended to include folded proteins (1). Antibodies have been displayed as functional binding molecules in the form of single chain Fv fragments (1), Fab fragments (11), and as both bivalent and bispecific dimeric Fv fragments (diabodies) (12). This has allowed the selection of phage from a mixed population according to their binding characteristics, for example single chain Fv antibody fragments with a high affinity for 2-phenyl-5-oxazolone have been selected from a library of antibodies derived from an immunized mouse (2). The enzymes alkaline phosphatase from *E. coli* (13) and *Staphylococcus* nuclease (14) have catalytic activity when displayed on bacteriophage fd. Further, phage displaying the human receptor molecules, CD4 and platelet-derived growth factor BB receptor (15), specifically bind the appropriate ligand. Bass *et al.* (16) have displayed functional human growth hormone on the surface of the closely related bacteriophage M13. Functional expression of an antibody fragment as a gene 8 protein fusion in M13 (17) has also been described, but this system is, at present, not as well characterized as that using gene 3.

In this chapter we describe in detail (see *Figure 1*) procedures for the cloning and selection of antibody genes using display on bacteriophage as g3p fusions. Many of these techniques will be directly applicable to work with other proteins.

2. Vectors for the display of proteins on the surface of bacteriophage fd

The first vectors for the display of antibodies on the surface of bacteriophage fd were derived from the vector fd-tet (18). However, phage DNA has a low efficiency of transformation when compared to plasmids such as the pUC series of vectors. Phagemid vectors (which contain both a plasmid origin of replication and a filamentous phage origin of replication) allow 100-fold higher efficiencies of transformation to be obtained compared to phage vectors. The higher efficiency of transformation as a result of the use of phagemids enables large libraries to be made. This is important when making a repertoire from an immunized mouse, and more so when making a mouse or human repertoire from a non-immunized source. Gene 3 from bacteriophage fd has therefore been inserted into phagemid vectors with restriction sites allowing insertion of foreign DNA sequences. These vectors were designed to allow the insertion of single chain Fv sequences as fusions at the N-terminus of the gene 3 protein and to be compatible with the PCR primers used by Orlandi *et al.* (19) to amplify antibody variable regions. Super infection with a helper phage, such as M13K07, results in packaging of this phagemid DNA into a phage particle, which displays the protein encoded by the insert as an N-terminal fusion with the gene 3 protein.

The phagemid pCANTAB-5 (*Figure 2*) has been derived from the phagemid pUC119 and provides unique *Sfi*I and *Not*I sites for cloning antibody genes. It allows the cloning of foreign DNA sequences (e.g. scFv fragments) as *Sfi*I/*Not*I fragments at the N-terminus of gene 3. Export of g3p fusions to the periplasm is directed using the synthetic leader sequence. A range of further developments of this vector (pCANTAB-5myc, pCANTAB-5E, and pCANTAB-6) contain an amber codon inserted at the start of the gene 3 segment, allowing expression of the inserted protein from the gene 3 fusion as a soluble fragment in a non-suppressing *E. coli* strain such as HB2151, without recloning the gene. However, in a suppressor strain such as

1: Construction and use of antibody gene repertoires

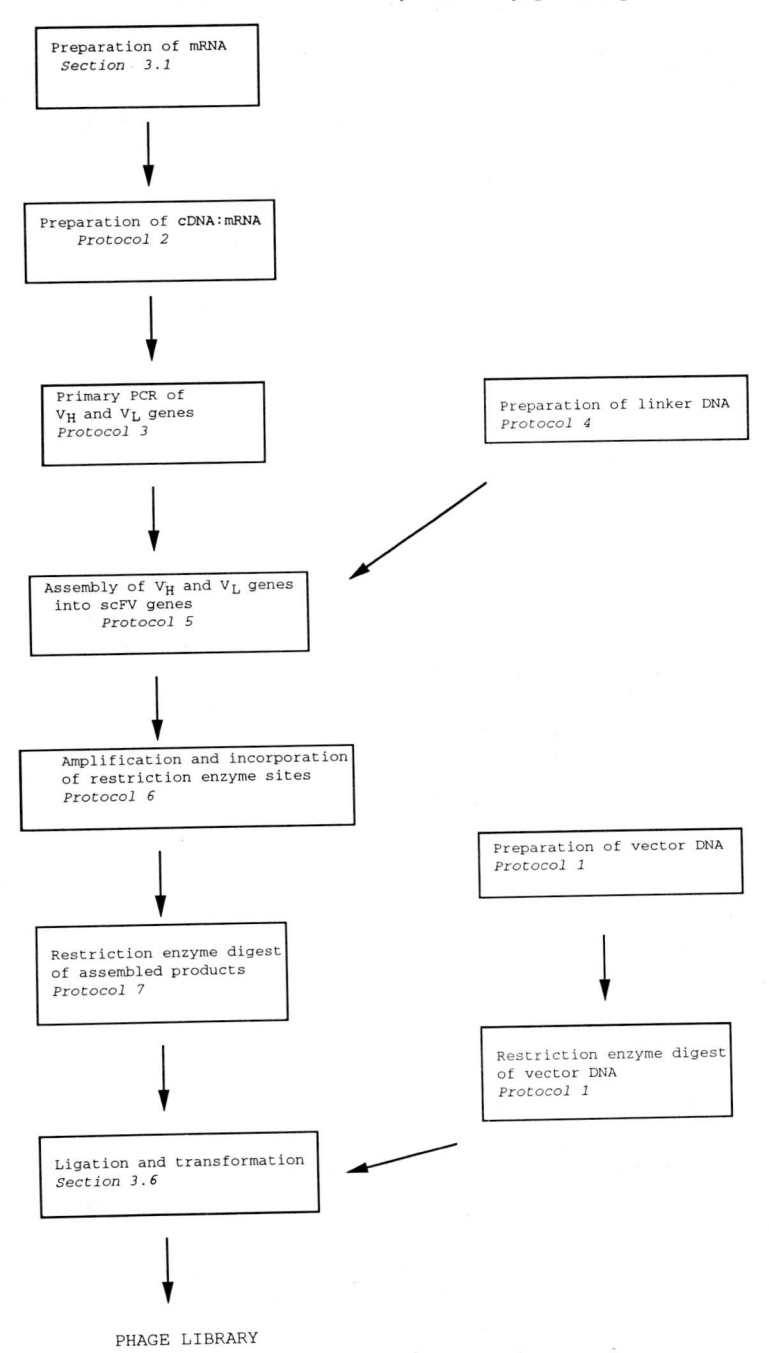

```
┌─────────────────────┐
│ Preparation of mRNA │
│     Section 3.1     │
└──────────┬──────────┘
           ↓
┌─────────────────────┐
│ Preparation of cDNA:mRNA │
│     Protocol 2      │
└──────────┬──────────┘
           ↓
┌─────────────────────┐         ┌─────────────────────────┐
│   Primary PCR of    │         │ Preparation of linker DNA│
│  V_H and V_L genes  │         │       Protocol 4        │
│     Protocol 3      │         └────────────┬────────────┘
└──────────┬──────────┘                      │
           ↓                                 ↓
┌──────────────────────────────┐
│ Assembly of V_H and V_L genes│
│      into scFV genes         │
│         Protocol 5           │
└──────────┬───────────────────┘
           ↓
┌──────────────────────────────┐
│ Amplification and incorporation│
│  of restriction enzyme sites │
│         Protocol 6           │
└──────────┬───────────────────┘
                                    ┌─────────────────────────┐
                                    │ Preparation of vector DNA│
                                    │       Protocol 1        │
                                    └────────────┬────────────┘
           ↓                                     ↓
┌──────────────────────────────┐
│   Restriction enzyme digest  │
│    of assembled products     │
│         Protocol 7           │
└──────────┬───────────────────┘
                                    ┌─────────────────────────┐
                                    │ Restriction enzyme digest│
                                    │     of vector DNA       │
                                    │       Protocol 1        │
                                    └────────────┬────────────┘
           ↓                                     │
┌──────────────────────────────┐←────────────────┘
│  Ligation and transformation │
│         Section 3.6          │
└──────────┬───────────────────┘
           ↓
       PHAGE LIBRARY
```

Figure 1. Flow diagram for the cloning of single chain Fv fragments into pCANTAB-5.

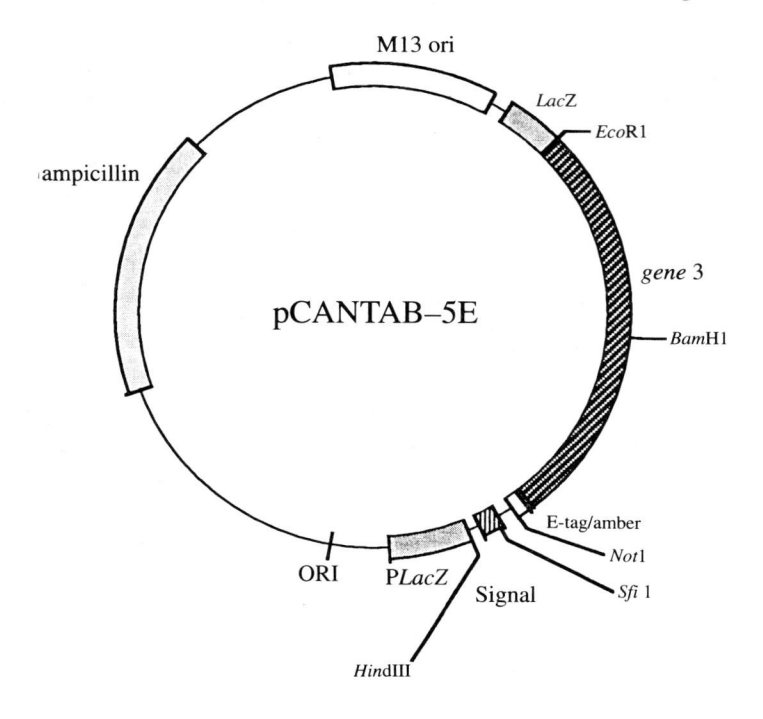

HindIII
AAGCTT TGGAGCCTTT TTTTTGGAGA TTTTCAAC GTG AAA AAA TTA TTA TTC GCA ATT
                                        V   K   K   L   L   F   A   I

signal sequence   Sfi I
CCT TTA GTT GTT CCT TTC TAT GCG GCC CAG CCG GCC ATG GCC CAG GTC CAA CTG
 P   L   V   V   P   F   Y   A   A   Q   P   A   M   A   Q   V   Q   L

                                    Not I |------------- E–tag ------
CAG GTC GAC CTC GAG ATC AAA CGG GCG GCC GCA GGT GCG CCG GTG CCG TAT CCG
 Q   V   D   L   E   I   K   R   A   A   A   G   A   P   V   P   Y   P

----- E–tag -----------| amber   fd-gene 3 sequence
GAT CCG CTG GAA CCG CGT GCC GCA TAG ACT GTT GAA AGT TGT TTA GCA AAA CCT
 N   P   L   E   P   R   A   A (*) T   V   E   S   C   L   A   K   P

Figure 2. The phagemid vector pCANTAB-5E is based on pUC119 into which gene 3 of bacteriophage fd has been inserted. The cloning sites *Sfi*I and *Not*I are used for the cloning of scFv sequences. Expression is under control of the p*lac* promoter.

1: Construction and use of antibody gene repertoires

TG1 the suppression of the amber codon is far from complete. This has the advantage that both complete gene-3/antibody fusion protein and single chain Fv antibody can be produced in the same cell with a suppressor strain. Induction of expression with IPTG will give single chain Fv antibody only with no gene-3 fusion protein. This can be used for testing the binding specificity and affinity of selected clones. Alternatively, rescue with M13KO7 helper phage will give both phage particles displaying functional antibody (which can be used for selection) and soluble antibody in the same culture.

A derivative of this vector, pCANTAB-5E (Pharmacia, cat. no. 27–9901–01) which has an E tag inserted at the C terminus of the soluble antibody has also been developed. This E tag can be detected in ELISA with the anti E monoclonal antibody which is available commercially (Pharmacia, cat. no. 27–9412–02). It can also be used for purification by affinity chromatography (Pharmacia Recombinant Phage Antibody System purification module 17–1362–01). A similar vector, pCANTAB-5myc, has a c-*myc* tag (a sequence derived from the c-*myc* oncogene) (20) inserted at the C terminus, instead of the E-tag, and can be detected and purified with the anti c-*myc* monoclonal antibody 9E10. A derivative of this, pCANTAB-6 has an additional tag of six histidines that is useful for affinity-chromatography purification (21).

Protocol 1. Preparation of vector DNA

To prepare vector DNA for cloning, RF plasmid DNA is isolated from the bacterial pellet by the plasmid alkaline lysis method (described in ref. 22). This DNA is then purified by centrifugation on a CsCl gradient. For smaller preparations a proprietary mini-prep purification system (Promega or Qiagen) can be used.

Equipment and reagents
For this protocol you will require the following basic equipment and reagents:

- Microcentrifuge
- Sterile H$_2$O
- Sterile microcentrifuge tubes (Eppendorf)
- Sterile pipette tips
- Agarose (BRL)
- Agarose gel electrophoresis tank (Pharmacia, cat. no. GNA 100)
- TAE buffer: to make a 50 × stock dissolve 242 g Tris base and 57.1 ml of glacial acetic acid in a final volume of 900 ml and then add 100 ml of 0.5 M EDTA (pH 8.0)

In addition to the basic equipment and reagents, the following will also be required:

- *Sfi*I restriction enzyme and 10 × buffer (as supplied by the manufacturer)
- *Not*I restriction enzyme and 10 × buffer (as supplied by the manufacturer)
- Chromospin 1000 (Clontech) columns or a Geneclean II DNA purification kit
- Cold (− 20 °C) ethanol
- Cold (− 20 °C) 70% ethanol
- 3 M sodium acetate buffer, pH 5.2
- Vacuum desiccator
- 65 °C waterbath
- 50 °C waterbath
- 37 °C waterbath

Method

1. Digest 10 μg of vector DNA in a volume of 200 μl with 50 units of the restriction enzyme *Sfi*I for 4 hours at 50 °C in the manufacturer's recommended buffer.

Protocol 1. *Continued*

2. Ethanol precipitate the DNA by adding a 1/10 volume of 3 M Na acetate, pH 5.2, and 2.5 × volumes of cold ethanol. Chill at −20 °C for 15 minutes. Then spin in a microcentrifuge for 5 minutes at 4 °C. Remove and discard the supernatant. Wash the pellet with 750 µl of cold (−20 °C) 70% ethanol. Spin briefly (30 seconds). Remove and discard the supernatant. Dry the pellet in a vacuum desiccator for 2–3 minutes. Do not over dry. Dissolve the pellet in the required buffer for the next step in the protocol.

3. Digest the vector DNA in a volume of 200 µl with 50 units of the restriction enzyme *Not*I for 4 hours at 37 °C in the manufacturer's recommended buffer. Cloned *Not*I enzyme should be used for the digestion. Heat inactivate at 65 °C for 20 minutes.

4. Purify the *Sfi*I–*Not*I double-cut DNA free from the resulting linker fragment. Chroma spin-1000 columns (Clontech) work well for this. Use according to the manufacturer's instructions. Use two columns and load each with 100 µl of cut DNA directly from the final restriction digest. Alternatively, a 0.7–1% low melting point agarose gel can be used to purify the cut vector and the DNA extracted using the 'Wizard PCR preps' DNA Purification System (Promega, cat. no. A7170) Geneclean II kit (BIO 101, cat. no. 3106) or Sephaglass Band Prep kit (Pharmacia, cat. no. 27-9285-01) with the final product dissolved in TE buffer ready for ligation with the DNA insert.

3. Preparation and cloning of antibody DNA

These protocols detail the production of both mouse and human antibody repertoires. These repertoires will be random combinatorial repertoires since the heavy and light chain DNA sequences are amplified separately before being assembled together and cloned. As a result, the original heavy and light chain pairings are lost. Although it is possible to find antibodies to 'self' antigens in the immune system, very often B cells producing such antibodies are eliminated by the body (e.g. by clonal selection and anergy). As a result, random combinatorial libraries make it easier to select human antibodies to human antigens.

These methods can be readily adapted for the generation of hierarchical libraries. A hierarchical library is a repertoire consisting of a single heavy chain or a restricted number of heavy chains combined with a full range of light chains (or vice versa). Such libraries are an important way of improving the affinity of an existing antibody (23) (also see Chapter 2 on affinity maturation by phage display) since they can give new antibodies with a better heavy and light chain pairing than the original.

For those researchers who wish to study the actual heavy and light chain pairings used by the immune system it is possible to preserve these by using 'in-cell PCR'. Protocols describing this new technique are described at the end of this chapter.

The following protocols have been developed for use with material from both immunized and non-immunized individuals. In addition, they can also be used to make single chain Fv clones, expressed on the surface of phage, from monoclonal antibodies. It should be noted that monoclonal antibody cell lines often contain more than a single heavy and single light chain sequence. Up to 10 different scFv clones have previously been obtained from a single monoclonal cell line (personal observation). The screening *Protocols (9, 11,* and *13)* can be used for the phage-display clones derived from a monoclonal antibody. We usually find that approximately 30% of clones are positive (by ELISA) for the antigen of interest. For some monoclonal antibodies the percentage of positive antigen binders can drop well below this and selection with the antigen of interest may be necessary.

An overall scheme for preparing libraries of antibody scFv fragments in pCANTAB-5E is shown in *Figure 1*.

A commercially available system (Pharmacia Recombinant Phage Antibody System Mouse ScFv Module, cat. no. 27-9400-01) contains the necessary reagents for cDNA preparation, primary PCR, and assembly of mouse heavy and light chain sequences preparatory to cloning and expression. An expression module (cat. no. 27-9401-01) contains vector, host cells, and helper phage for the cloning of phage antibodies and for the display and selection of recombinant phage antibodies.

3.1 Preparation of mRNA

It is possible to use total RNA made by phenol/SDS extraction of tissue (22) and this can give excellent results. However, a pure mRNA preparation prepared using an oligo (dT) column has been found to give higher yields in the PCR amplification steps that follow and is therefore generally recommended. Oligo (dT)-affinity purification systems (e.g. Pharmacia QuickPrep^r mRNA purification kit cat. no. 27-9254-01) are commercially available and produce mRNA free of RNase. Poor quality RNA preparations can cause the subsequent PCR steps to fail. In the case of repertoire constructions, even if there are no apparent problems during construction and cloning, the diversity of the repertoire could be greatly reduced and the usefulness of the final library greatly compromised if the RNA is of poor quality.

For non-immunized human repertoires, peripheral blood lymphocytes prepared from whole blood samples are used (3). Generally 500 ml of blood should contain approximately 1×10^8 cells and should give up to 100 μg of total RNA. Use 10 μg per cDNA preparation.

If using a commercial system, such as the Pharmacia QuickPrep^r mRNA kit, then it is important not to overload the oligo (dT) column with cell extract. (This can cause the column matrix to aggregate and trap contaminants reducing final yield and purity.) The white blood cells from 50 ml of blood (approximately 1×10^7 cells) will give enough mRNA from one oligo (dT) column for four cDNA reactions. The purified mRNA is ethanol precipitated and resuspended in 20 µl of RNase-free water (see *Protocol 2*, Equipment and Reagents list, for RNase-free, DEPC-treated water). For each cDNA reaction 5 µl mRNA is used. The mRNA can be safely stored in ethanol and then precipitated and resuspended prior to use.

Immunized repertoires are prepared with mRNA from the spleen of an immunized mouse. An average mouse spleen should contain approximately 1×10^8 cells. If using purified mRNA it is important not to overload the oligo (dT) column with cell extract. Therefore use 25% of the spleen and freeze the rest. This is also enough material for a total RNA preparation.

In addition, tissue culture cells from a monoclonal antibody cell line can also be used to subclone the monoclonal antibody into the phage-display format. An aliquot of 1×10^7 cells should give up to 100 µg of total RNA, 10 µg of this is used per cDNA preparation. For the Pharmacia QuickPrep^r mRNA kit, Pharmacia recommend using 5×10^6 cells per oligo (dT) column. After ethanol precipitation and resuspending in 20 µl this should be adequate for four cDNA preparations.

An alternative method for mRNA preparation is given in Chapter 6 (*Protocol 2*, step 1).

3.2 cDNA preparation

The mRNA is now used as a template to prepare cDNA. Random hexamer primers are used to prime cDNA synthesis. This generates cDNA:mRNA hybrids. Since this cDNA will be used as a PCR template there is no need to make double-stranded cDNA.

Protocol 2. Preparation of cDNA:mRNA hybrid

Equipment and reagents

In addition to basic equipment and reagents (see *Protocol 1*), the following will also be required for this protocol:

- DEPC-treated H$_2$O: prepare by adding diethylpyrocarbonate to H$_2$O (0.1% v/v), and incubate at 37°C for 2 hours. DEPC-treated H$_2$O should be autoclaved before use to inactivate the diethylpyrocarbonate
- 5 mM dNTP: equimolar mixture of dATP, dCTP, dGTP, and dTTP with a total concentration of 5 mM nucleotide (i.e. 1.25 mM each dNTP)
- 0.1 M dithiothreitol solution
- 10 × first strand buffer: 1.4 M KCl, 0.5 M Tris–HCl, pH 8.1 at 42°C, 80 mM MgCl$_2$
- Random hexamer primers (e.g. pd(N)6 from Pharmacia, cat. no. 27-2166-01)
- RNasin (a ribonuclease inhibitor) (Promega)
- Reverse transcriptase (e.g. murine reverse transcriptase (cloned) Pharmacia, cat. no. 27-0925-01)
- 100°C waterbath
- 65°C waterbath
- 42°C waterbath

1: Construction and use of antibody gene repertoires

Method

1. Set up the following reverse transcription mix:
   - H$_2$O (DEPC-treated)   19 µl
   - 5 mM dNTP   10 µl
   - 10 × first strand buffer   10 µl
   - 0.1 M dithiothreitol   10 µl
   - Random hexamer primers (10 OD$_{260}$/ml)[a]   2 µl
   - RNasin (40 U/µl)   4 µl

2. For a total RNA preparation, dilute 10 µg RNA to 40 µl final volume with DEPC-treated water. For oligo (dT) column purified mRNA preparation, add 5 µl (one quarter of the sample) to 35 µl of DEPC-treated water. Heat at 65 °C for 3 min and then place on ice for 1 min (to remove secondary structure).

3. Add 40 µl of the diluted RNA to the reverse transcription mix and then add 100 units of reverse transcriptase. Mix and then incubate at 42 °C for 1 hour.

4. Boil the reaction mix for 3 minutes, cool on ice for 1 minute and then spin in a microcentrifuge at 13 000 *g* for 5 minutes, to pellet debris. Transfer the supernatant to a new tube. Either store at −20 °C or proceed directly to *Protocol 3*.

[a] Random hexamer primers (e.g. pd(N)6 from Pharmacia) are used to prime cDNA synthesis. Specific primers complementary to the 3′ end of the CH or VL domain can also be used (2, 3), but we have found that random hexamers work better.

3.3 Primary PCR

The cDNA:mRNA hybrid is now used as a template for the 'primary' PCR amplification of the V_H and V_L domains. Specific primers have been designed for the amplification of mouse V_H and V_L genes. These primers, described in *Table 1*, have been shown to generate a diverse library when used to prepare a repertoire of antibodies from an immunized mouse (2). Specific primers have also been designed for the amplification of human V_H and V_L genes (3) and are shown in *Table 2*.

The BACK primer hybridizes to the beginning of framework 1 of the antibody sequence and the FOR primer (or in the case of the mouse kappa light chain the MJKFONX primer) binds to the end of framework 4 (contributed by the J segment during antibody rearrangement). For example, the mouse V_H genes are amplified using the primers VH1BACK and VH1-FOR2. The primers VH1BACK and VH1FOR2 are consensus primers which will cover the majority of mouse heavy chain gene families. Similarly, the primer VK2BACK is used with a mix of four J region primers (MJK1FONX, MJK2FONX, MJK4FONX, and MJK5FONX) to amplify the light chain kappa families. The primer VK2BACK is a consensus primer that will cover most of the mouse kappa families. However, there are some gaps in the sequences covered by this primer. In the case of the commercially available system from Pharmacia, a mix of five different specific VKBACK primers is used, thus covering the full range of kappa light chains.

Thus to make a mouse repertoire, two primary PCR reactions are necessary, one for the V_H genes and one for the V_L genes.

In the case of making a human repertoire the situation is more complicated since there is a great deal of sequence diversity which necessitates a large number of primers. Thus for the heavy genes there are six VHBACK primers and four JHFOR primers. For the lambda light chains (V_{lambda}), seven VLBACK and three JLFOR primers are required and for the kappa light chains (V_{kappa}), six VKBACK and five JKFOR primers. This gives a total of 31 primers. For the primary PCR amplification of the heavy genes six reactions are required, each with a separate VHBACK primer, but using the JHFOR primers as a mix. The seven lambda and six kappa light chain gene PCR amplifications are performed in a similar fashion.

After PCR amplification, check the PCR products (2 µl samples on a 2% agarose gel). Problems can be due to either the template (particularly the initial mRNA preparation) or from the PCR amplification itself. Not all thermal cyclers perform identically. The annealing temperature and possibly the ramp rate may need fine tuning. The Pharmacia Recombinant Phage Antibody System (Pharmacia, cat. no. 27-9400-01) contains a control mRNA, and cDNA prepared from this can be used as a suitable control. After PCR, the amplified products are purified ready for the next step in the process. With a human repertoire, the V_H reactions can be pooled prior to purification. The light chain PCR preparations should be processed separately as 13 distinct preparations. At the assembly stage a mix of V_H genes will be linked to each of the separate light chain preparations.

Protocol 3. Primary PCR of antibody genes

Equipment and reagents

In addition to basic equipment and reagents (see *Protocol 1*), the following will also be required for this protocol:

- Taq DNA polymerase[a] (Boehringer, under licence from Perkin Elmer/Cetus)
- 10 × Taq polymerase buffer as provided by the supplier of the Taq polymerase, or: 100 mM Tris–HCl pH 8.3, 500 mM KCl, 15 mM MgCl$_2$, and 0.01% (w/v) gelatin
- 5 mM dNTPs (see *Protocol 2*)
- Hydrochloric acid solution (0.25 M)
- DNA thermal cycler for PCR
- Mineral oil (paraffin oil) (Sigma, cat. no. M-3516)
- Ultraviolet lamp (254 nm) for irradiating contaminating DNA in amplification reaction
- 'Wizard PCR prep' DNA purification kit (Promega)

Method

1. For each PCR the following reaction mix is set up. Suitable controls (no DNA) are also required. All working solutions of primers are at concentrations of 10 pmol/µl.

1: Construction and use of antibody gene repertoires

- H_2O 35 µl
- 10 × Taq buffer   5 µl
- 5 mM dNTP   1.5 µl
- FOR primer[b] (10 pmol/µl)   2.5 µl
- BACK primer[b] (10 pmol/µl)   2.5 µl

2. Irradiate this mix with ultraviolet light at 254 nm for approximately 5 min.

3. Add 2.5 µl cDNA:mRNA hybrid (from *Protocol 2*), and 47.5 µl of the PCR mix to a 0.5 ml microcentrifuge tube. Overlay with 2 drops of mineral oil (Sigma).

4. Place on a cycling heating block preset at 94 °C.

5. Add 1 µl Taq DNA polymerase under the paraffin.

6. Amplify using 30 cycles of 94 °C for 1 min, 57 °C for 1 min, 72 °C for 2 min. Check each PCR amplification with a 2 µl sample on a 2% gel.

7. Purify on a 2% lmp (low melting point) agarose/Tris–acetate–EDTA gel.[c] Carefully excise the V_H and V_L bands (using a fresh sterile scalpel or razor blade for each) and transfer each band to a separate sterile microcentrifuge tube. The use of suitable molecular weight markers will aid selection of the correct bands. Both the V_H and V_L bands are approximately 350 bases long. (This depends on the germline sequence used and on the length of the CDR3 sequence which is highly variable. Generally, the V_H band has a slightly higher molecular weight than the V_L band.) Use 'Wizard PCR prep DNA purification system' from Promega[d] or the Sephaglass Band Prep kit (cat. no. 27–9285–01) from Pharmacia to purify each DNA band. Recover the DNA in 50 µl H_2O.

    For a human repertoire treat each one of the six kappa and the seven lambda PCRs separately. However, it is useful to pool the V_H PCRs and then concentrate them with Promega 'Wizard PCR prep' prior to gel purification. This will reduce losses of material during gel purification.

---

[a] A proof-reading Taq polymerase can help PCR amplify those monoclonal antibodies which will not amplify with standard Taq polymerase. Amplification can be prevented by somatic mutations (for example in framework 1) causing mismatches with the PCR primers.

[b] *For a mouse repertoire*
FOR and BACK primers shown in *Table 1* are:
Heavy chains: VH1-FOR2
                VH1BACK
Light chains: (MJK1FONX, MJK2FONX, MJK4FONX, MJK5FONX)
                VK2BACK

[b] *For a human repertoire*
FOR and BACK primers shown in *Table 2* are:
Heavy chains:
FOR: HuJH1–2FOR, HuJH3FOR, HuJH4–5FOR, HuJH6FOR

Protocol 3. *Continued*

BACK: HuVH1aBACK, HuVH2aBACK, HuVH3aBACK, HuVH4aBACK, HuVH5aBACK, HuVH6aBACK

Lambda chains:
FOR: HuJλ1FOR, HuJλ2–3FOR, HuJλ4–5FOR
BACK: Huλ1BACK, Huλ2BACK, Huλ3BACK, Huλ3bBACK, Huλ4BACK, Huλ5BACK, Huλ6BACK

Kappa chains:
FOR: HuJK1FOR, HuJK2FOR, HuJK3FOR, HuJK4FOR, HuJK5FOR
BACK: HuVK1aBACK, HuVK2aBACK, HuVK3aBACK, HuVK4aBACK, HuVK5aBACK, HuVK6aBACK

[c] To avoid contamination, it is essential to depurinate the electrophoresis apparatus, combs, etc. with 0.25 M HCl overnight before use.

[d] 'Wizard PCR prep' DNA purification kit is from Promega, used according to the manufacturer's instructions. After removal of wash solution (centrifugation step) incubate columns at room temperature to allow isopropanol from wash buffer to evaporate before next step.

Table 1. Oligonucleotides for mouse PCR reactions

Primary PCR oligos (restriction sites underlined):

VH1FOR-2	TGA GGA GAC GGT GAC CGT GGT CCC TTG GCC CC
VH1BACK	AGG TSM ARC TGC AGS AGT CWGG
MJK1FONX	CCG TTT GAT TTC CAG CTT GGT GCC
MJK2FONX	CCG TTT TAT TTC CAG CTT GGT CCC
MJK4FONX	CCG TTT TAT TTC CAA CTT TGT CCC
MJK5FONX	CCG TTT CAG CTC CAG CTT GGT CCC
VK2BACK	GAC ATT GAGCTC ACC CAG TCT CCA

PCR oligos to make linker:

LINKFOR	TGG AGA CTG GGT GAGCTC AAT GTC
LINKBACK	GGG ACC ACG GTC ACC GTC TCC TCA

Oligos for addition of restriction sites:

VH1BACKSfi	GTC CTC GCA ACT GCG GCC CAG CCG GCC ATG GCC CAG GTS MAR CTG CAG SAG TCW GG
JK1NOT10	GAG TCA TTC TGC GGC CGC CCG TTT GAT TTC CAG CTT GGT GCC
JK2NOT10	GAG TCA TTC TGC GGC CGC CCG TTT TAT TTC CAG CTT GGT CCC
JK4NOT10	GAG TCA TTC TGC GGC CGC CCG TTT TAT TTC CAA CTT TGT CCC
JK5NOT10	GAG TCA TTC TGC GGC CGC CCG TTT CAG CTC CAG CTT GGT CCC

Ambiguity codes

M = A or C
R = A or G
S = G or C
W = A or T

1: Construction and use of antibody gene repertoires

Primary PCR
Amplify heavy and light chains from mRNA/cDNA preparation

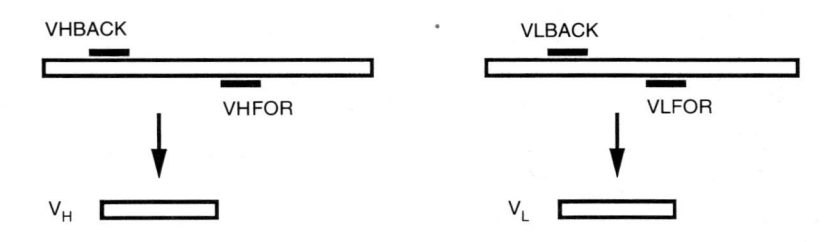

Assembly PCR
Assemble heavy and light chains with linker

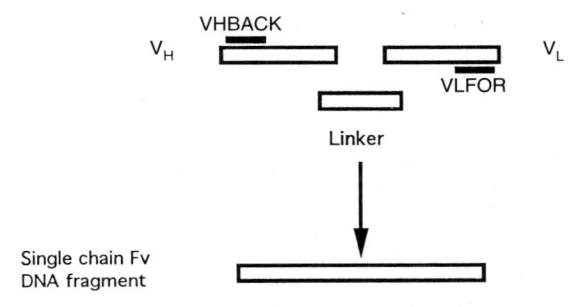

Incorporation of restriction sites
PCR amplification and incorporation of restriction sites

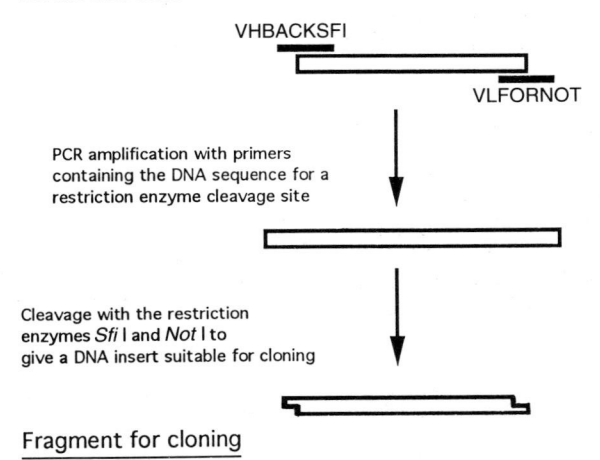

PCR amplification with primers containing the DNA sequence for a restriction enzyme cleavage site

Cleavage with the restriction enzymes *Sfi* I and *Not* I to give a DNA insert suitable for cloning

Fragment for cloning

Figure 3. Preparation of a library of inserts encoding single chain Fv fragments.

Table 2. Oligonucleotides for human PCR reactions

Primary PCR oligos:

HuJH1–2FOR	5′ TGA GGAGAC GGT GAC CAG GGT GCC
HuJH3FOR	5′ TGA AGA GAC GGT GAC CAT TGT CCC
HuJH4–5FOR	5′ TGA GGAGAC GGT GAC CAG GGT TCC
HuJH6FOR	5′ TGA GGAGAC GGT GAC CGT GGT CCC
HuVH1aBACK	5′ CAG GTG CAG CTG GTG CAG TCT GG
HuVH2aBACK	5′ CAG GTC AAC TTA AGGGAGTCT GG
HuVH3aBACK	5′ GAG GTG CAG CTG GTG GAGTCT GG
HuVH4aBACK	5′ CAG GTG CAG CTG CAG GAGTCG GG
HuVH5aBACK	5′ GAG GTG CAG CTG TTG CAG TCT GC
HuVH6aBACK	5′ CAG GTA CAG CTG CAG CAG TCA GG
HuJK1FOR	5′ ACG TTT GAT TTC CAC CTT GGT CCC
HuJK2FOR	5′ ACG TTT GAT CTC CAG CTT GGT CCC
HuJK3FOR	5′ ACG TTT GAT ATC CAC TTT GGT CCC
HuJK4FOR	5′ ACG TTT GAT CTC CAC CTT GGT CCC
HuJK5FOR	5′ ACG TTT AAT CTC CAG TCG TGT CCC
HuVK1aBACK	5′ GAC ATC CAG ATG ACC CAG TCT CC
HuVK2aBACK	5′ GAT GTT GTG ATG ACT CAG TCT CC
HuVK3aBACK	5′ GAA ATT GTG TTG ACG CAG TCT CC
HuVK4aBACK	5′ GAC ATC GTG ATG ACC CAG TCT CC
HuVK5aBACK	5′ GAA ACG ACA CTC ACG CAG TCT CC
HuVK6aBACK	5′ GAA ATT GTG CTG ACT CAG TCT CC
HuJλ1FOR	5′ ACC TAG GAC GGT GAC CTT GGT CCC
HuJλ2–3FOR	5′ ACC TAG GAC GGT CAG CTT GGT CCC
HuJλ4–5FOR	5′ ACC TAA AAC GGT GAGCTG GGT CCC
HuJλ1BACK	5′ CAG TCT GTG TTG ACG CAG CCG CC
HuJλ2BACK	5′ CAG TCT GCC CTG ACT CAG CCT GC

HuJλ3aBACK   5' TCC TAT GTG CTG ACT CAG CCA CC
HuJλ3bBACK   5' TCT TCT GAGCTG ACT CAG GAC CC
HuJλ4BACK   5' CAC GTT ATA CTG ACT CAA CCG CC
HuJλ5BACK   5' CAG GCT GTG CTC ACT CAG CCG TC
HuJλ6BACK   5' AAT TTT ATG CTG ACT CAG CCCCA

PCR oligos to make linker:

Reverse JH for scFv linker

RHuJH1–2   5' GCA CCC TGG TCA CCG TCT CCT CAG GTG G
RHuJH3   5' GGA CAA TGG TCA CCG TCT CTT CAG GTG G
RHuJH4–5   5' GAA CCC TGG TCA CCG TCT CCT CAG GTG G
RHuJH6   5' GGA CCA CGGTCA CCG TCT CCT CAG GTG G

Reverse V$_\kappa$ for scFv linker

RHuVK1aBACKFv   5' GGA GAC TGG GTC ATC TGG ATG TCC GAT CCG CC
RHuVK2aBACKFv   5' GGA GAC TGA GTC ATC ACA ACA TCC GAT CCG CC
RHuVK3aBACKFv   5' GGA GAC TGC GTC AAC ACA ATT TCC GAT CCG CC
RHuVK4aBACKFv   5' GGA GAC TGG GTC ATC ACG ATG TCC GAT CCG CC
RHuVK5aBACKFv   5' GGA GAC TGC GTG AGT GTC GTT TCC GAT CCG CC
RHuVK6aBACKFv   5' GGA GAC TGA GTC AGC ACA ATT TCC GAT CCG CC

Reverse V$_\lambda$ for scFv linker

RHuVλ1BACKFv   5' GGC GGCTGC GTC AAC ACA GAC TGC GAT CCG CCA CCG CCA GAG
RHuVλ2BACKFv   5' GCA GGCTGA GTC AGA GCA GAC TGC GAT CCG CCA CCG CCA GAG
RHuVλ3aBACKFv   5' GGT GGCTGA GTC AGC ACA TAG GAC GAT CCG CCA CCG CCA GAG
RHuVλ3bBACKFv   5' GGG TCC TGA GTC AGC TCA GAA GAC GAT CCG CCA CCG CCA GAG
RHuVλ4BACKFv   5' GGC GGT TGA GTC AGT ATA ACG TGC GAT CCG CCA CCG CCA GAG
RHuVλ5BACKFv   5' GAC GGCTGA GTC AGC ACA GAC TGC GAT CCG CCA CCG CCA GAG
RHuVλ6BACKFv   5' TGG GGCTGA GTC AGC ATA AAA TTC GAT CCG CCA CCG CCA GAG

Table 2. Continued

Oligos for addition of restriction sites:

Human V$_H$ Back Primers

HuVH1aBACKSfi	5′ GTC CTC GCA ACT GCG GCC CAG CCG GCC ATG GCC CAG GTG CAG CTG GTG CAG TCT GG
HuVH2aBACKSfi	5′ GTC CTC GCA ACT GCG GCC CAG CCG GCC ATG GCC CAG GTC AAC TTA AGG GAG TCT GG
HuVH3aBACKSfi	5′ GTC CTC GCA ACT GCG GCC CAG CCG GCC ATG GCC GAG GTG CAG CTG GTG GAG TCT GG
HuVH4aBACKSfi	5′ GTC CTC GCA ACT GCG GCC CAG CCG GCC ATG GCC CAG GTG CAG CTG CAG GAG TCG GG
HuVH5aBACKSfi	5′ GTC CTC GCA ACT GCG GCC CAG CCG GCC ATG GCC CAG GTG CAG CTG TTG CAG TCT GC
HuVH6aBACKSfi	5′ GTC CTC GCA ACT GCG GCC CAG CCG GCC ATG GCC CAG GTA CAG CTG CAG CAG TCA GG

Human J$_\kappa$ Forward Primers

HuJK1FORNOT	5′ GAG TCA TTC TCG ACT TGC GGC CGC ACG TTT GAT TTC CAC CTT GGT CCC
HuJK2FORNOT	5′ GAG TCA TTC TCG ACT TGC GGC CGC ACG TTT GAT CTC CAG CTT GGT CCC
HuJK3FORNOT	5′ GAG TCA TTC TCG ACT TGC GGC CGC ACG TTT GAT ATC CAC TTT GGT CCC
HuJK4FORNOT	5′ GAG TCA TTC TCG ACT TGC GGC CGC ACG TTT GAT CTC CAC CTT GGT CCC
HuJK5FORNOT	5′ GAG TCA TTC TCG ACT TGC GGC CGC ACG TTT AAT CTC CAG TCG TGT CCC

Human J$_\lambda$ Forward Primers

HuJλ1FORNOT	5′ GAG TCA TTC TCG ACT TGC GGC CGC ACC TAG GAC GGT GAC CTT GGT CCC
HuJλ2–3FORNOT	5′ GAG TCA TTC TCG ACT TGC GGC CGC ACC TAG GAC GGT CAG CTT GGT CCC
HuJλ4–5FORNOT	5′ GAG TCA TTC TCG ACT TGC GGC CGC ACC TAA AAC GGT GAG CTG GGT CCC

Ambiguity codes
M = A or C
R = A or G
S = G or C
W = A or T

1: Construction and use of antibody gene repertoires

3.4 Assembly of scFv fragments

The heavy and light chain domains are now linked together to give a complete single chain Fv antibody sequence. For this assembly reaction a DNA linker fragment is required. This is generated with *Protocol 4* and used to link the separately amplified V_H and V_L domains. This linker sequence creates a 15 amino acid sequence $(Gly_4Ser)_3$ between the heavy and light chain variable domains. *Figure 3* shows how the products of the primary PCR (*Protocol 3*) and the linker (*Protocol 4*) are assembled (*Protocol 5*).

The linker was originally prepared by oligonucleotide synthesis to generate the sequence: 5′ GGCACCACGGTCACCGTCTCCTCAGGTGGAGGCGGTTCAGGCGGAGGTGGCTCTGGCGGTGGCGGATCGGACATCGAGCTCACTCAGTCTCCA 3′ and this was used to construct fd scFvD1.3, a scFv construct of the lysozyme binding antibody D1.3. Having once made this construct and inserted it into a vector, it is more convenient to prepare additional linker material by PCR amplification. Thus for the construction of a mouse repertoire or for creating a phage-display scFv from a monoclonal antibody the primers LINKFOR and LINKBACK are used to make a linker. One end of this PCR product overlaps with the end of the heavy chain (the J region) and the other end with the start of the light chain sequence. Combining the PCR amplified V_H and V_L domains with the linker fragment in a PCR reaction will therefore give an assembled scFv sequence.

A commercially available linker (Pharmacia, cat. no. 27–1588–01) is available. This can be used directly in the assembly reaction for mouse scFvs and can also be used as a template to make either mouse or human linker fragments by PCR. Use 1 μl of the Pharmacia linker as template in *Protocol 4*.

For the construction of a mouse repertoire or for creating a phage-display scFv from a monoclonal antibody only a single linker preparation is required to join the V_H gene J sequence with the VKBACK sequence. In the case of a human repertoire there are seven different Vlambda BACK sequences and six Vkappa BACK sequences. Therefore 13 different linkers are required. For each linker PCR preparation use an equimolar mix of the four RHuJH primers with one of the 13 different reverse light chain primers (RHuVk1aBACKFv or RHuVk2aBACKFv, etc. or RHuVλ1BACKFv or RHuVλ2BACKFv, etc.).

Protocol 4. Preparation of linker[a]

Equipment and reagents

In addition to basic equipment and reagents (see *Protocol 1*), the following will also be required for this protocol:

- Taq DNA polymerase
- 10 × Taq polymerase buffer (see *Protocol 3*)
- 5 mM dNTPs (see *Protocol 2*)
- DNA thermal cycler for PCR
- Mineral oil (Sigma, cat. no. M-3516)
- SPIN-X columns (Costar, USA)
- Loading dye without Bromophenol Blue

All working solutions of primers are at concentrations of 10 pmol/μl.

Protocol 4. Continued

Method

1. Set up linker reaction mix. Each linker reaction provides sufficient linker for approximately three assembly reactions:
   - H_2O 36.8 µl
   - 10 × Taq buffer   5.0 µl
   - 5 mM dNTP   2.0 µl
   - FOR primer (10 pmol/µl)[a]   2.5 µl
   - BACK primer (10 pmol/µl)[a]   2.5 µl
   - linker template   1.0 µl
   - Taq DNA polymerase   0.2 µl

2. Cover with mineral oil and place on a cycling heating block at 94 °C. Amplify the linker DNA by PCR for 25 cycles of 94 °C for 1 min, 65 °C for 1 min, 72 °C for 2 min.

3. Incubate at 60 °C for a further 5 min in the heating block.

4. Purify on a 2% low melting point agarose/TAE gel (see *Protocol 1* for TAE buffer) (use loading dye without Bromophenol Blue since the linker DNA sequence required is 93 bp long). Excise the correct (93 bp) gel fragment. Place in the upper chamber of a SPIN-X column and centrifuge in a microcentrifuge for 5 minutes at 13 000 *g*. As an alternative to using SPIN-X columns the gel band can be purified using the 'Wizard PCR prep' DNA purification kit from Promega.

5. Ethanol precipitate the DNA (as in *Protocol 1*, step 2) and carefully dry the pellet before resuspending in 5 µl H_2O.

[a] The sequences of LINKFOR and LINKBACK are shown in *Table 1*. The sequences of RHuJH primers, the RHuVkBACKFv primers and the RHuVλBACKFv primers are shown in *Table 2*.

Protocol 5. Assembly of single chain Fv antibody fragments

Equipment and reagents

In addition to basic equipment and reagents (see *Protocol 1*), the following will also be required for this protocol:

- Taq DNA polymerase
- 10 × Taq polymerase buffer (see *Protocol 3*)
- 5 mM dNTPs (see *Protocol 2*)
- DNA thermal cycler for PCR
- Mineral oil (Sigma, cat. no. M-3516)
- Wizard PCR prep DNA purification kit *or*
- Wizard DNA clean-up kit from Promega

Method

1. Estimate the quantities of V_H and V_L DNA prepared by the primary PCR reactions (*Protocol 3*) and the quantity of linker DNA preparation(s) prepared in *Protocol 4* using agarose gel electrophoresis. (Use a molecular weight marker of known concentration in an adjacent

1: Construction and use of antibody gene repertoires

track.) Estimate concentration by the intensity of ethidium bromide staining. Adjust the volumes of V_H, V_L, and linker DNA added in step 2 below to give roughly equal masses of DNA fragments added to the assembly reaction (approximately 50 ng).

2. For a mouse repertoire, set up a single reaction for assembling the products of the primary V_{kappa} PCR reaction with the products of the V_H PCR reaction The assembly reaction is performed in two stages. For each first stage reaction set up the following mixture:

    For a human repertoire, set up six reactions for assembling each of the six primary V_{kappa} PCR reactions with pooled material from the V_H PCR reactions. Also set up seven reactions for assembling each of the seven primary V_{lambda} PCR reactions with the pooled material from the V_H PCR reactions. Each assembly reaction is performed in two stages. For each first stage reaction set up the following mixture:

    - V_H DNA[a] (from *Protocol 3*)   x μl (see step 1 above)
    - V_L DNA[a] (from *Protocol 3*)   y μl (see step 1 above)
    - Linker DNA (from *Protocol 4*)   z μl (see step 1 above)
    - 10 × Taq buffer   2.5 μl
    - 5 mM dNTP (*Protocol 2*)   2.0 μl
    - sterile H_2O up to 25.0 μl
    - Taq DNA polymerase (1 U/μl)   1.0 μl

3. Place on cycling heating block and incubate for 20 cycles of 94 °C for 1.5 min and 65 °C for 3 min.

4. To each reaction add 25 μl of the following mixture for the second stage.

    - 10 × Taq buffer   2.5 μl
    - 5 mM dNTP   2.0 μl
    - VHBACK[b] (10 pmol/μl)   5.0 μl
    - VLFOR[c] (10 pmol/μl)   5.0 μl
    - sterile H_2O up to 25.0 μl
    - Taq DNA polymerase (1 U/μl)   1.0 μl

5. Amplify the DNA using 30 cycles of 94 °C for 1 min, 50 °C for 1 min, and 72 °C for 2 min, with a final extension step at 72 °C for 10 min.

6. Electrophorese the product of each reaction on a 1.4% low melting point agarose/Tris–acetate–EDTA gel. Excise the band corresponding to the assembled product in each case (about 720 bp). Purify the DNA from the band with a 'Wizard DNA clean-up system' or 'Wizard PCR prep' DNA purification kit (Promega).

[a] For the generation of hierarchical libraries, the V_H and V_L DNA are derived from different sources. Either the V_H or V_L DNA is obtained by PCR amplification of DNA from a single antibody fragment clone or group of clones, by the procedure detailed in *Protocol 3*, where this domain is required to be kept constant. It is important that this DNA is purified as in *Protocol 3*,

Protocol 5. *Continued*

step 7 to avoid contamination with the original partner chain. The DNA encoding the complementary domain is derived as in *Protocol 3* for the primary PCR.

[b] For a mouse repertoire, VHBACK refers to the primer VH1BACK. For a human repertoire, VHBACK refers to an equimolar mix of HuVH1aBACK, HuVH2aBACK, HuVH3aBACK, HuVH4aBACK, HuVH5aBACK, and HuVH6aBACK.

[c] VLFOR is the corresponding light chain forward primer that was used in the primary PCR:
i.e. *for a mouse repertoire*, an equimolar mix of MJK1FONX, MJK2FONX, MJK4FONX, and MJK5FONX.
i.e. *for a human repertoire*, an equimolar mix of HuJK1FOR, HuJK2FOR, HuJK3FOR, HuJK4FOR, and HuJK5FOR for the V_H and V_K assemblies and an equimolar mix of HuJλ1FOR, HuJλ2–3FOR and HuJλ4–5FOR, for the V_H and V_λ assemblies.

Note. The assembly step can sometimes be difficult. A variation on this assembly protocol has been described by McCafferty *et al.* (21) which can also give good results. For the assembly use equimolar amounts of heavy chain, light chain, and linker. If problems do occur then repeat the assembly using a range of V_H and V_L concentrations (21). It is very important to have good quality heavy and light chain PCR products for the assembly. Repeating the primary PCR with particular attention to the purification step may solve the problem. The annealing temperature and the ramp rate are also important and may need to be fine-tuned if a different thermal cycler is used. We recommend the Perkin Elmer DNA thermal cycler.

3.5 Amplification and digestion

A further amplification is performed using primers which incorporate the restriction sites for the enzymes *Sfi*I and *Not*I (*Protocol 6*). The amplified DNA is then digested with *Sfi*I and *Not*I (*Protocol 7*) to allow cloning into *Sfi*I and *Not*I digested pCANTAB-5 or one of the other pCANTAB series of vectors.

Several identical reactions are usually set up so that there is adequate material, after restriction enzyme digestion and gel purification, to give a large enough repertoire. For a repertoire from an immunized mouse 1×10^6 clones are usually regarded as adequate, whereas for a non-immunized human repertoire at least 1×10^7 clones are required. The larger the library the better the chance of selecting a high-affinity antibody. In addition, for a human repertoire, two sets of reactions are necessary. Pool the material from the six kappa assemblies for a kappa amplification. Similarly, pool the seven lambda assemblies for a lambda amplification. Usually four identical kappa amplifications and four identical lambda amplifications will give enough DNA, but this will depend on the yield and purity of the amplification step.

A suitable control (no DNA) should also be included.

Protocol 6. Amplification and incorporation of restriction enzyme sites

Equipment and reagents

In addition to basic equipment and reagents (see *Protocol 1*), the following will also be required for this protocol:

- Taq DNA polymerase
- 10 × Taq polymerase buffer (see *Protocol 3*)
- 5 mM dNTPs (see *Protocol 2*)
- DNA thermal cycler for PCR
- Mineral oil (Sigma, cat. no. M-3516)

1: Construction and use of antibody gene repertoires

Method

1. Add the following reagents to a 500 µl tube:
   - Assembled product from *Protocol 5*   5.0 µl
   - 10 × Taq polymerase buffer   5.0 µl
   - 5 mM dNTP   2.0 µl
   - VHBACKSFI[a] (10 pmol/µl)   2.5 µl
   - VLFORNOT[a] (10 pmol/µl)   2.5 µl
   - sterile H_2O up to 50.0 µl
   - Taq DNA polymerase (2.5 U/µl)   0.5 µl

2. Cover with mineral oil and place on the block preset at 94°C. Amplify DNA by PCR with 30 cycles of 94°C for 1 min, 55°C for 1 min, 72°C for 2 min, followed by an incubation for 10 min at 72°C.

3. Analyse 5 µl of the products by electrophoresis on a 1.4% agarose–Tris–acetate–EDTA gel. If a strong band is seen at approximately 720 bp the products are pooled and digested with *Sfi*I and *Not*I (*Protocol 7*). Some undigested material should always be retained so that, if necessary, this can be reamplified to give more material for the cloning steps.

[a] *For a mouse repertoire*, VHBACKSFI is the primer VH1BACKSfi which incorporates an *Sfi*I restriction site and VLFORNOT is an equimolar mix of the primers JK1NOT10, JK2NOT10, JK4NOT10, and JK5NOT10 (*Table 1*) which incorporate *Not*I restriction sites.
For a human repertoire, two PCR reactions are required, one for the kappa assembled material and one for the lambda assembled material. **VHBACKSFI** is an equimolar mix of the primers VH1aBACKSfi, VH2aBACKSfi, VH3aBACKSfi, VH4aBACKSfi, VH5aBACKSfi, and VH6aBACKSfi, which incorporate *Sfi*I restriction sites.
For the kappa genes reaction, **VLFORNOT** which incorporate a *Not*I restriction site is an equimolar mix of the primers HuJK1FORNOT, HuJK2FORNOT, HuJK3FORNOT, HuJK4FORNOT, and HuJK5FORNOT
For the lambda reaction, **VLFORNOT** is an equimolar mix of the primers HuJλ1FORNOT, HuJλ2–3FORNOT, and HuJλ4–5FORNOT.

Protocol 7. Restriction enzyme digestion of assembled products

Equipment and reagents

In addition to basic equipment and reagents (see *Protocol 1*), the following will also be required for this protocol:

- *Sfi*I restriction enzyme and 10 × buffer
- *Not*I restriction enzyme and 10 × buffer
- Geneclean II DNA purification kit or 'Wizard PCR prep' DNA purification kit *or* Pharmacia Sephaglass Band Prep Kit
- Cold (– 20°C) ethanol
- Cold (– 20°C) 70% ethanol
- 3 M sodium acetate buffer, pH 5.2
- TE buffer: 10 mM Tris–HCl, 1 mM EDTA, pH 8.0
- Phenol (saturated with TE buffer)
- Vacuum desiccator
- 50°C waterbath
- 37°C waterbath

Method

1. Add 200 µl of phenol (TE buffer saturated) to each of the DNA products from the PCR reactions in *Protocol 6*. Mix well. Incubate at room

Protocol 7. *Continued*

temperature for 10 minutes. Mix well. Centrifuge for 5 minutes in a microcentrifuge at 13 000 *g*. Transfer the upper aqueous layer to a fresh tube. Add 200 µl of TE buffer to the phenol in the first tube. Mix well. Centrifuge for 5 minutes at 13 000 *g*. Carefully remove the upper aqueous layer and combine with the first aqueous extract.

2. Ethanol precipitate the DNA contained in the aqueous extract. Wash the pellet twice with 70% EtOH. Dry the DNA pellet but do not over dry. Resuspend in 80 µl H_2O.

3. Digest the DNA product overnight with *Not*I at 37 °C using the reaction mixture below:
   - DNA   80 µl
   - *Not*I buffer × 10   10 µl
   - *Not*I (10 U/µl)   10 µl

4. Ethanol precipitate the *Not*I digested DNA (as in *Protocol 1*, step 2). Resuspend pellet in 80 µl H_2O.

5. Digest with *Sfi*I using the reaction mixture below:
   - DNA (digested with *Not*I)   80 µl
   - *Sfi*I buffer × 10   10 µl
   - *Sfi*I (10 U/µl)   10 µl

   Incubate at 50 °C for 4 h

6. Purify on 1.5% low melting point agarose–TAE (*Protocol 1*) gel. Excise the *Not*I/*Sfi*I digested DNA band and purify the DNA using Geneclean (from BIO 101) following the manufacturer's instructions and recover the DNA in 20 µl of H_2O. Alternatively, use Pharmacia Sephaglass Band Prep kit (cat. no. 27-9285-01) or 'Wizard PCR prep' DNA purification kit (from Promega). Ethanol precipitation can be used but it is very easy to lose the small DNA pellet at this stage.

3.6 Ligation and transformation

Ligation of DNA (e.g. using a DNA ligation kit, Amersham International) and electroporation of *E. coli* cells are performed using standard procedures. Ligate 50 ng of insert with 250 ng of cut vector (*Protocol 1*) per ligation. (The Amersham ligation kit, in some instances, can approximately double the yield of transformants over standard ligation protocols.) Transformation of chemically competent cells can be used instead of electroporation but will give much lower yields of transformants. This is acceptable if constructing a phage-display antibody from a monoclonal antibody but will severely compromise the size of a phage repertoire. A repertoire of 1×10^7 or better can be obtained from 1 µg of insert with 5 µg of cut vector. Chapter 8 gives detailed protocols for ligation (*Protocol 3*) and electrotransformation (*Protocol 4*).

1: Construction and use of antibody gene repertoires

4. Growth and expression of phage antibodies

This section describes the rescue and growth of phagemid as phage particles displaying an antibody as a gene 3 fusion protein on the phage surface. Preparations of phage particles in this way gives material that can be used for antibody selection and can also be used in an ELISA assay for screening and for characterizing the selected antibody. In addition, protocols for soluble antibody expression are described.

4.1 Rescue of phage

The most widely used helper phage for superinfection is M13K07 (25) which will preferentially package phagemid DNA. With phagemid pCANTAB-5 or pCANTAB-6 a yield of around 10^{11} phage per ml is obtained.

Protocol 8. Rescue and PEG precipitation of phagemids by superinfection with helper phage[a]

Equipment and reagents

- Sterile pipette tips
- M13K07 helper phage
- 2TY media: 16 g tryptone, 10 g yeast extract (Difco), 5 g NaCl per litre, adjust pH to 7.0 with NaOH and sterilize by autoclaving.
- Ampicillin 50 mg/ml in H_2O (filter-sterilized)
- Kanamycin 25 mg/ml in H_2O (filter-sterilized)
- 2TYAG: 2TY containing 2% (w/v) filter-sterilized glucose and 100 µg/ml ampicillin
- 2TY/K/A: 2TY (NO GLUCOSE) with kanamycin (25 µg/ml) and 100 µg/ml ampicillin
- 2TY plates: 2TY medium with 15 g/litre Bacto agar (Difco)
- Glucose 20% (w/v) filter-sterilized
- 37°C shaking incubator
- 30°C shaking incubator
- Sterile 50 ml polypropylene centrifuge tubes (Falcon, cat. no. 2070)
- Sterile 250 ml flasks
- Benchtop centrifuge
- PEG/NaCl: 20% polyethylene glycol 8000, 2.5 M NaCl
- PBS
- TE buffer: 10 mM Tris, 1 mM EDTA, pH 8.0
- Skimmed milk powder

A. *Rescue*

1. Grow cells containing the phagemid in 2TYAG at 30°C overnight.
2. Add 250 µl of overnight cell culture to 25 ml 2TYAG in a 250 ml flask and grow at 37°C with fast shaking for 1–2 hours (to approximately A_{600} = 0.5 to 1.0). Alternatively, inoculate directly from a fresh colony or from a glycerol stock and grow to approximately A_{600} = 0.5 to 1.0. **Note**: never store clones on agar plates as we have found that some scFv clones will not express recombinant protein if stored in this way. Only retransforming into fresh cells will restore expression!)
3. Add M13K07 (or other helper phage) to a final concentration of 10^9 p.f.u./ml.
4. Incubate cells for 30 minutes without shaking at 37°C. Then grow cells for 30 minutes with moderate shaking (200 r.p.m.) at 37°C.

23

Protocol 8. *Continued*

5. Transfer cells to a polypropylene centrifuge tube. Centrifuge at 2500 *g* for 5 min (benchtop centrifuge) and resuspend the bacterial pellet in 25 ml 2TY/K/A. Transfer to a fresh 250 ml flask.

6. Grow overnight with fast shaking (300 r.p.m.) at 30 °C.

7. Centrifuge at 9000 *g* for 15 min at 4 °C. Transfer supernatant to fresh tube.

8. For ELISA, good signals can be obtained using this supernatant directly (see *Protocol 13*). For phage antibody selection (*Protocol 12*) the phage supernatant can either be used directly or, if required, concentrated by PEG precipitation.

B. *PEG precipitation*

1. Add 1/5 volume PEG/NaCl (20% polyethylene glycol 8000, 2.5 M NaCl). Incubate on ice for 1 h. Centrifuge at 9000 *g* for 15 min at 4 °C.

2. Retain phage pellet. Remove as much supernatant as possible by decanting and allowing tube to drain on to a paper tissue.

3. Resuspend the pellet in 2 ml PBS or TE buffer. Spin at 13 000 r.p.m. for 5 min in a microcentrifuge to remove bacterial debris. Transfer supernatant to fresh tube.

4. For phage selection by panning add 2 ml 4% skimmed milk solution in PBS. This gives a 2.5 × concentration. Adjust the initial volume of phage, as necessary, if a higher concentration is required.

[a] This protocol is for 25 ml cultures, the volumes can be adjusted proportionately for different scale preparations. It is essential that the cultures are vigorously aerated. If the volumes are increased larger vessels are needed (e.g. 25 ml cultures in a 250 ml flask or a 500 ml culture in a 2 litre flask). We also routinely grow clones in 100 μl cultures (*Protocol 9*).

Protocol 9. Rescue and growth of phagemids in microtitre plate format

This protocol is used for screening large numbers of clones for antigen-binding clones.

Equipment and reagents

- As for *Protocol 8*
- Sterile H$_2$O
- Sterile microcentrifuge tubes (Eppendorf)
- Benchtop centrifuge
- Sterile polypropylene 96-well microtitre plate (Greiner, cat. no. 650201) or 'Cell wells' polystyrene 96-well microtitre plate (Corning, cat. no. 25850)

1: Construction and use of antibody gene repertoires

Method

1. Pick colonies into 100 µl 2TYAG in a sterile microtitre plate. Use a polypropylene 96-well microtitre plate (Greiner) (autoclave before use) or a Corning 'Cell Wells' sterile polystyrene 96-well microtitre plate. Generally polypropylene is preferable for phage growth. Grow clones overnight at 30°C at 150 r.p.m. This is the master plate.

2. Prepare a replica 96-well microtitre plate with 100 µl 2TYAG in each well. Inoculate with 10 µl of cells from the master plate. Then add 40 µl of 50% glycerol to each well of the master plate and store it at −70°C. Grow the second (replica) plate at 30°C at 300 r.p.m. for approximately 5–6 hours to give an A_{600} = 0.5 to 1.0.

3. Add 20 µl of M13K07 (or other helper phage) in 2TYAG (at 10^{10} p.f.u./ml) to each well of the replica plate. Incubate the plate for 30 minutes without shaking at 37°C. Then incubate for 30 minutes with moderate shaking (150 r.p.m.) at 37°C.

4. Centrifuge the plate at 1000 *g* for 5 min (benchtop centrifuge) and resuspend each bacterial pellet in 100 µl 2TY/K/A. Alternatively, a transfer innoculum of 5–10 µl using a multi-prong device (from the plate generated in step 3) can be made into a fresh microtitre plate containing 2TY/K/A.

5. Grow overnight with shaking (150 r.p.m.) at 30°C. For ELISA good signals can be obtained using the supernatant directly, diluted 1:1 with PBS containing 4% skimmed milk powder (see *Protocol 11*).

4.2 Growth and soluble expression

For a detailed study of scFv antibodies derived from phage antibody repertoires it is necessary to express them in soluble form. The pCANTAB-5E, pCANTAB-5myc, pCANTAB-6 vectors (but not pCANTAB-5) have an amber codon between the scFv antibody sequence and gene 3. Growth in a non-suppressor strain such as HB2151 will allow soluble expression of scFv protein. However, in a suppressor strain such as TG1 the suppression of the amber codon is not complete and both complete gene3/antibody fusion protein and soluble antibody protein are made in the same cell.

After the addition of IPTG the expression of antibody is induced. Soluble antibody is transported to the periplasm of the cell. From there it can leak into the growth media. Soluble antibody can then be harvested either from the growth culture or from the periplasmic space. This latter is useful for large-scale production since it reduces the volume of crude extract prior to affinity column purification.

Protocol 10. Soluble expression of antibody[a]

Equipment and reagents

- Sterile pipette tips
- 2TY media (see *Protocol 8*)
- 2TYA: 2TY with 100 µg/ml ampicillin
- 2TYAG: 2TYA with 2% (w/v) filter-sterilized glucose added
- Ampicillin 50 mg/ml in H_2O (filter-sterilized)
- 37 °C shaking incubator
- 30 °C shaking incubator
- Sterile 50 ml polypropylene centrifuge tubes (Falcon, cat. no 2070)
- Sterile 250 ml flasks
- Benchtop centrifuge
- IPTG (isopropyl-β-D-thio-galactopyranoside) 100 mM (23.8 mg/ml) filter-sterilized

Method

1. Grow cells containing the phagemid in 2TYAG at 30 °C overnight or to mid-log phase (A_{600} = 0.5 to 1.0) if inoculation is performed on the same day as rescue.

2. Add 50 µl of cell culture to 5 ml 2TYAG in a 50 ml sterile polypropylene centrifuge tube and grow at 30 °C with fast shaking for 3–4 hours (to A_{600} = 1.0).

3. Centrifuge at 3500 *g* for 5 min (benchtop centrifuge) and resuspend the bacterial pellet in 5 ml 2TYA (2TY with 100 µg/ml ampicillin and NO glucose) and IPTG at 1 mM. (For large-scale preparations this step may be omitted if the cells are grown in 2TYA containing 0.1% glucose. At A_{600} of 1.0 add IPTG to 1 mM. **NB**. The best conditions for induction may vary between different clones.)

4. Antibody may be harvested from the culture supernatant or from the periplasm. If harvesting from the culture supernatant, grow overnight with fast shaking (350 r.p.m.) at 30 °C. For ELISA good signals can be obtained using the supernatant directly, diluted 1:1 with PBS containing 4% skimmed milk powder (see *Protocol 13*).

   If harvesting antibody from the periplasm grow for 3 hours at 30 °C after induction with IPTG. Centrifuge at 3500 *g* for 5 min (benchtop centrifuge) and resuspend the bacterial pellet in 1/10 original culture volume of PBS + 1 mM EDTA. Incubate on ice for 15 minutes. Then centrifuge at 3500 *g* for 10 minutes at 4 °C (benchtop centrifuge). Carefully remove and retain supernatant containing soluble antibody.

[a] This protocol is for 5 ml cultures; the volumes can be adjusted proportionately for different scale preparations. It is essential that the cultures are vigorously aerated. If the volumes are increased larger vessels are needed (e.g. 25 ml cultures in a 250 ml flask or a 500 ml culture in a 2 litre flask). We also routinely grow clones in 100 µl cultures in microtitre wells (*Protocol 11*) for screening large numbers of clones for antigen-binding activity.

Protocol 11. Soluble expression of antibody in microtitre plate format

Equipment and reagents

- As for *Protocol 10*
- Sterile microcentrifuge tubes (Eppendorf)
- 2TYG: 2TY with filter-sterilized glucose added to 2% (w/v)
- 37°C shaking incubator
- 30°C shaking incubator
- Sterile polystyrene 96-well microtitre plate (Corning 'Cell Wells', cat. no. 25850) *or* polypropylene 96-well microtitre plate (Greiner, cat. no. 650201)
- Glycerol (50% v/v)
- Benchtop centrifuge

Method

1. Pick colonies into 100 µl 2TYAG in a sterile microtitre plate. Use a sterile polystyrene 96-well microtitre plate (e.g. Corning) or a polypropylene 96-well microtitre plate (Greiner) (autoclave before use). Generally polystyrene is preferable for soluble expression, giving better yields than polypropylene. Grow clones overnight at 30°C at 150 r.p.m. This is the master plate.

2. Prepare a second (replica) 96-well microtitre plate with 100 µl 2TYAG in each well. Inoculate with 10 µl of cells from the master plate. Then add 40 µl of 50% glycerol to each well of the master plate and store at −70°C. Grow the second (replica) plate at 30°C at 150 r.p.m. for approximately 5–6 hours to give an A_{600} of 1.0.

3. Centrifuge at 1000 g for 5 min (benchtop centrifuge) and resuspend the bacterial pellet in each well with 100 µl of 2TYA containing 1 mM IPTG.

4. Grow overnight with shaking (150 r.p.m.) at 30°C. Centrifuge at 1000 g for 10 min (benchtop centrifuge). For ELISA good signals can be obtained using the supernatant directly, diluted 1:1 with PBS containing 4% skimmed milk powder (see *Protocol 13*).

4.3 Purification of soluble antibody

Several methods are available for purifying soluble antibody. If the vector has a polyhistidine tag then metal chelate affinity purification can be used (21). Ni–NTA (nitrilo-tri-acetic acid) agarose is available from Qiagen and is used according to the manufacturer's recommendations (in the elution buffer it is important to readjust the pH of PBS or saline when imidazole is added). Affinity chromatography with the antigen of interest cross-linked to a Sepharose column (CNBr–Sepharose) is also commonly used. In the case of small haptens, these can be cross-linked to a BSA–Sepharose column. Elution of bound antibody with 250 mM glycine buffer, pH 2.5, followed by neutralization and dialysis is usually suitable. Antibody prepared in this way is normally stable when frozen at −20°C.

Following initial purification it may be necessary to size separate monomeric scFv from dimerized and aggregated material (e.g. with Pharmacia FPLC Superdex 75 small-scale column). This is important for kinetic studies of affinity with the Pharmacia Biacore.

5. Selection of antibody variants displayed on the surface of bacteriophage

Antibody variants expressed on the surface of bacteriophage have been selected on the basis of their affinity for antigen using chromatography, panning, or adsorption to cells. Elution from affinity matrices has been achieved by specific elution using the antigen (or a related compound) or non-specific elution using, for example, 100 mM triethylamine. Washing procedures remove non-specifically bound phage. The phage binds to and is eluted from the surface according to the affinity or the nature of the binding interaction. Eluted phage are then used to infect male *E. coli* cells expressing the F pilus, allowing recovery of phage-encoding antibodies with the desired binding characteristics.

Protocol 12. Selection of antibodies displayed on the surface of bacteriophage

Equipment and reagents

- Sterile H_2O
- Sterile microcentrifuge tubes (Eppendorf)
- Sterile pipette tips
- 1 M Tris–HCl, pH 7.5
- 2TY media (see *Protocol 8*)
- Ampicillin 50 mg/ml in H_2O (filter-sterilized)
- 2TYAG: 2TY with 2% (w/v) filter-sterilized glucose and 100 μg/ml ampicillin
- 2TYAG agar plates: 2TYAG with 1.5% agar (sterilized by autoclaving)
- PBS buffer: 10 mM phosphate buffer, pH 7.2–7.4, with 150 mM NaCl
- Nunc Immuno tube Maxisorp 75 × 12 (cat. no. 4–44202) *or* Falcon 3001 Petri dishes (35 × 10 mm)
- PBS/Tween: PBS buffer with 0.1% Tween-20 (Sigma, cat. no P-2287)
- Sterile glycerol
- 37°C shaking incubator
- 30°C incubator
- Sterile polypropylene centrifuge tubes (Falcon, cat. no. 2070)
- Benchtop centrifuge
- Mid-log TG1 cells

A. *Binding*

1. Antigen is coated on to a large surface area of plastic (for example, Nunc Maxisorp tubes. Similar plastic surfaces such as Falcon 3001 Petri dishes 35 × 10 mm can also be used).

   Coat with 1 ml of 10 μg/ml antigen in PBS. This is analogous to coating an ELISA plate. This is the best coating buffer for many antigens (e.g. BSA or hapten conjugated BSA). For some antigens such as lysozyme which do not coat plastic well in PBS, use 50 mM $NaHCO_3$, pH 9.6. Coating conditions will vary from antigen to antigen. Coating

1: Construction and use of antibody gene repertoires

is overnight (12 + hours). The incubation temperature is antigen dependent. Room temperature or 30 °C will give good coating, but for many antigens where correct folding and presentation of epitopes is required then coating at 4 °C, to reduce denaturation, is preferred. If a suitable control monoclonal or polyclonal serum is available for the antigen then coating conditions for panning can be determined by ELISA before starting selection.

2. After coating overnight, rinse the coated Maxisorp tube three times with PBS and then block with 2% (w/v) skimmed milk powder in PBS for 2 h at 37 °C. (Completely fill the tube, or cover the surface if using a Petri dish.)
3. Rinse the Maxisorp tube four times with PBS.
4. Prepare phage as in *Protocol 8*. This should give 5×10^{10} to 1×10^{11} per ml. For the first round of selection it may be advantageous to use concentrated phage (prepared as in *Protocol 8*). For subsequent rounds of selection supernatants are routinely used directly, without any need for PEG precipitation and concentration. Add an equal volume of 4% (w/v) skimmed milk powder in PBS buffer and use directly in the selection. Add 1 ml of this phage preparation to the panning tube.
5. Incubate for 1 hour. Incubation can be either stationary or carried out on a turntable mixer with the tube (safely capped) turning slowly end over end. (We normally use the former.) The incubation can be done at room temperature. However, selection at 37 °C may help prevent the selection of those antibodies that are less stable.

B. *Washing and eluting*
1. Rinse the Maxisorp tube 20 times with PBS/Tween. This is carried out by filling the tube using a wash bottle and immediately tipping out.
2. Rinse the Maxisorp tube 20 times with PBS.
3. Add 1 ml 100 mM triethylamine to the Maxisorp tube. Incubate (stationary) at room temperature for 10 minutes.

C. *Infection*
1. Remove eluted phage. Add to 500 μl of 1 M Tris–HCl buffer, pH 7.5, in a 15 ml polypropylene centrifuge tube (Falcon). Add half of this (750 μl) to 5 ml of a mid-log culture of *E. coli* TG1 cells. (Store the remaining 750 μl of phage at 4 °C.) Shake the cells slowly at 37 °C for 1 hour.[a] (If 50 ml of 2 × TY in a 250 ml flask is inoculated with a fresh colony from a minimal media plate at the start of the day (step 2) it will be at approximately mid-log by this stage. It can be stored on ice when it reaches mid-log.)

Protocol 12. *Continued*

2. Centrifuge cells at 3000 *g* for 5 minutes. Carefully discard all of the supernatant and resuspend cells in 500 μl 2TYG. Spread-plate the cells on to 2 × TYAG agar plates (100 μl per plate). Grow overnight at 30 °C.
3. After overnight growth add 1.5 ml of 2 × TYAGG (2 × TYAG with 15% glycerol) per plate. Resuspend cells with a glass spreader. Pool the cells from each plate and aliquot into Eppendorf tubes. If a further round of selection is required then inoculate 25 ml of 2 × TYAG with 50 μl of these cells and grow to A_{600} = 1.0 (*Protocol 8A*, step 2) and proceed with phagemid rescue. The remainder of the scraped cells are stored at −70 °C.

[a] At this point take a small aliquot of the cells and after serial dilution, plate out on to 2 × TYAG agar plates. Grow overnight at 30 °C. This will measure the output of that round of selection. This can be an indication of the success or otherwise of the selection process. Often a rise in the relative number of phage eluted at each round of selection will occur. However, it is not always the case and not too much emphasis should be placed on this. Nevertheless, the colonies from this plating step are of real use for screening the progeny from each round of selection and should be screened for positive binders. Grow and screen as described in *Protocols 8* and *9*.

6. Analysis of phage-derived antibodies by ELISA

Phage antibody clones can be assayed directly for the ability to bind specific antigens by immunoassay techniques such as ELISA. Detection of phage antibodies with antiserum raised in sheep against bacteriophage fd (Pharmacia, cat. no. 27-9402-01) can be used for most antigens, giving very sensitive ELISA assays.

Alternatively, soluble antibody preparations can be used in the ELISA and detected with a monoclonal antibody such as the antibody to the E tag (Pharmacia, cat. no. 27-9412-01) which will bind to the E tag present on antibody clones grown in the vector pCANTAB-5E. Similarly, solubly expressed antibody from clones in the vector pCANTAB-5myc or pCANTAB-6 can be detected with the monoclonal antibody 9E10 which will bind to the c-*myc* tag.

This protocol (*Protocol 13*) should be readily adaptable to the assay of any phage antibody by substituting an appropriate antigen.

Protocol 13. Screening for binders by ELISA

Equipment and reagents

- PBS buffer: 10 mM phosphate buffer, pH 7.2–7.4, with 150 mM NaCl
- PBS/Tween: PBS buffer with 0.1% Tween-20 (Sigma, cat. no P-2287).
- ELISA plates (Falcon, cat. no. 3912)
- ELISA plate reader
- ABTS (2,2′-azinobis-(3-ethylbenzthiazoline-6-sulfonic acid); Sigma, cat. no. A-9941) Make up ABTS solution just before use; dissolve each 10 mg tablet in 20 ml of citrate buffer (made up from 10.8 ml of 50 mM citric acid and 9.2 ml 50 mM trisodium citrate stock)

1: Construction and use of antibody gene repertoires

- Anti-fd serum and peroxidase-conjugated rabbit anti-goat immunoglobulin (Sigma, cat. no.) *or* Anti-E-tag antibody (Pharmacia, cat. no. 27-9412-01) and peroxidase-conjugated goat anti-mouse immunoglobulin (Sigma, cat. no.) *or* 9E10 antibody (Cambrige Research Biochemicals) and peroxidase-conjugated anti-mouse immunoglobulin. The 9E10 monoclonal is available from the ATCC (CRL1729, named MYC1-9E10.2)
- Hydrogen peroxide

Method

1. Coat ELISA plate with 10 µg/ml antigen in PBS. This is the best coating buffer for many antigens (e.g. BSA or hapten-conjugated BSA). For some antigens such as lysozyme which do not coat plastic well in PBS use 50 mM $NaHCO_3$, pH 9.6. Coating conditions will vary from antigen to antigen. Coating is overnight (12 + hours). The incubation temperature is antigen-dependent. Room temperature or 30°C will give good coating, but for many antigens where correct folding and presentation of epitopes is required then coating at 4°C is preferred. The normal antigen concentration for coating is 10 µg/ml. With some antigens higher concentrations give improved results (e.g. for ELISA with lysozyme, which coats plastic poorly, we use 1 mg/ml).

2. After coating overnight, rinse the wells three times with PBS and block with 300 µl per well of 2% (w/v) skimmed milk powder in PBS for 2 h at 37°C.

3. Prepare phage supernatants for ELISA. Phage supernatants are routinely used directly, without any need for PEG precipitation and concentration. Add an equal volume of 4% (w/v) skimmed milk powder in PBS and use directly in the assay. Similarly, use soluble antibody from IPTG-induced cultures directly (with an equal volume of 4% (w/v) skimmed milk powder in PBS).

4. Rinse wells three times with PBS and transfer 100 µl supernatants into the wells. Incubate for 1.5 h at room temperature.

5. Wash wells three times for 2 min each time with each of PBS/0.1% Tween-20 and PBS (to remove detergent).

6. For phage supernatants add 100 µl of anti-fd serum raised in sheep (1/1000 dilution) to each well. Incubate for 1 h at room temperature. Proceed to step 7. Alternatively, use peroxidase-conjugated anti-fd (Pharmacia recombinant phage detection system cat. no. 27-9402-01) following the manufacturer's recommended dilution factor. Incubate for 1 h at room temperature. Then proceed directly to step 9. For soluble antibody from culture supernatants of phagemids with an E-tag, add 100 µl of anti E-tag antibody (Pharmacia cat. no. 27-9412-01) (following the manufacturer's recommended dilution factor) to each well. Incubate for 1 h at room temperature.

> **Protocol 13.** *Continued*
>
> For soluble antibody produced from the culture supernatants of phagemids with a *myc* tag, add 100 μl of 9E10 monoclonal antibody (1/1000 dilution) to each well. Incubate for 1 h at room temperature.
>
> 7. Wash as in step 5.
> 8. For phage supernatants add 100 μl peroxidase-conjugated rabbit anti-goat immunoglobulin (1/5000 dilution) to each well. Incubate for 1 h at room temperature. (Anti-goat or anti-sheep antibodies can be used interchangeably.) For soluble antibody from culture supernatants add 100 μl peroxidase-conjugated goat anti-mouse immunoglobulin (1/5000 dilution) to each well. Incubate for 1 h at room temperature.
> 9. Wash as in step 5.
> 10. Add 200 μl ABTS containing 1 μl of H_2O_2 per 10 ml to each well and develop until the absorbance at 405 nm is suitable (A_{405} = 0.2 to 1.0). Read absorbance in a plate reader.

Specific primers are used to PCR a sample of DNA derived by picking a colony into 20 μl of distilled water and boiling for 5 min. The diversity of clones generated can be assessed by *Bst*NI digestion of the PCR amplified insert (Chapter 8, *Protocol 6*).

7. In-cell PCR assembly

7.1 Introduction

The protocols given earlier in this chapter describe how to make mouse and human antibody repertoires. In this process, the heavy and light chain DNA sequences are amplified separately before being assembled together, and so the original heavy and light chain pairings are lost. This results in a random combinatorial library. With libraries derived from immunized B-cell donors the selected antibodies have V_H domains that are able to bind to antigen in promiscuous association with a range of V_L domains. Some of these combinations may, by chance, accurately reflect the original host B-cell repertoire, but they will be in the minority and there is currently no way to recognize the original from new antigen-binding V region combinations. Previously, the only way to rescue original repertoires has been to exploit hybridoma or other B-cell immortalization technologies (26). In-cell PCR assembly is intended to amplify and link V_H and V_L genes within fixed and permeabilized B cells in such a way that the original V gene combination in each cell is retained. When applied to a population of B cells, this should allow the expression of antibody fragments representative of the host B-cell repertoire. Essentially, the process consists of intracellular first-strand cDNA synthesis,

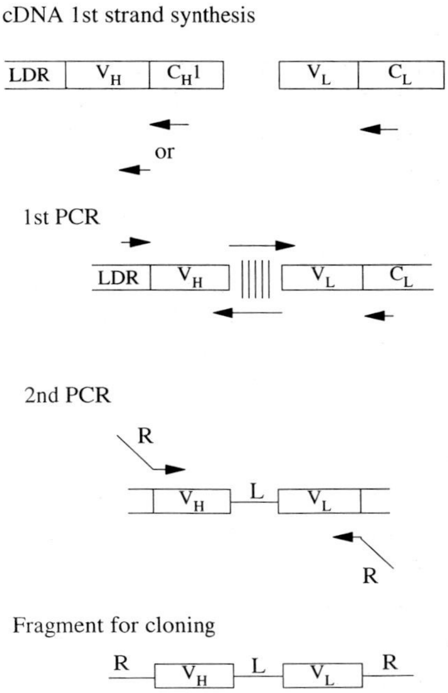

1: Construction and use of antibody gene repertoires

Figure 4. Priming sites are represented by arrows, the position of linker sequences by L, and restriction sites by R. Linkage sites are indicated by parallel vertical lines. LDR = V_H leader region. cDNA 1st strand synthesis can be accomplished using forward primer mixers corresponding to C_L and C_H1 or J_H genes. The 'outer' 1st PCR primer mixes should prime in the V_H leader regions and the C_L genes. (The latter can be the same primers as used for 1st strand synthesis.) 1st PCR linker primer mixes should prime in V_L framework 1 regions and J_H regions, and must have overlapping complementary extensions encoding the polypeptide linker $(G_4S)_3$. A J_H forward linker primer mix could double for V_H cDNA 1st strand synthesis. 2nd PCR primer mixers appending restriction sites should prime in V_H framework 1 regions, and J_L regions.

followed by intracellular amplification and linkage of V_H and V_L cDNA in single-chain Fv (scFv) format by polymerase chain reaction (PCR), then intracellular amplification of the assembled genes in a second PCR. Sufficient assembled DNA is released into the supernatant to allow cloning of the assembled gene into *E. coli*.

In-cell PCR is a novel technique, still at an early stage of development. It has been shown to work by using mixtures of different hybridoma cell lines and demonstrating that the original parental V_H and V_L pairings of each cell line are retained (see Section 7.2). For use of this technique with mouse or human B cells, careful primer design is necessary (see Section 7.3). A range of 'nested' primers are required. In the protocols describing the construction of random combinatorial libraries, the DNA from each PCR amplification was gel-purified before proceeding to the next stage. For in-cell PCR, this is not possible. The use of 'nested' primers overcomes this problem and allows each stage of the process to be performed inside the cell.

Protocol 14. In-cell PCR assembly

Equipment and reagents

- Benchtop centrifuge
- Microcentrifuge
- Haemocytometer and microscope
- Rotator
- Sterile pipetters and tips
- 1 ml syringes, and 21- and 26-gauge needles
- Waterbaths or heat blocks at 42°C and 65°C
- Cycling heat block, preferably with hot lid, and compatible tubes
- 2 ml Eppendorf tubes and 500 µl microcentrifuge tubes
- Gel electrophoresis apparatus for DNA agarose gels
- Viable cell suspension; may be hybridoma or other immortalized cells (cloned or uncloned), spleen cells from immunized mice, etc.
- 10% formal saline: 0.15 M NaCl containing 10% (v/v) formalin solution (100% formalin = 40% formaldehyde), equivalent to 4% formaldehyde in final concentration
- 0.5% v/v Nonidet P-40 (NP-40) in distilled water
- PBS (see *Protocol 12*)
- PBS–glycine, pH 7.2–7.4 PBS containing 0.1 M glycine
- AMV reverse transcriptase
- RNase inhibitor (e.g. Promega RNasin)
- Sterile water
- 5 mM dNTPs (see *Protocol 2*)
- 10 × 1st strand buffer (see *Protocol 2*)
- Dithiothreitol (DTT): 100 mM solution, diluted in distilled water from a 1 M stock frozen in 0.01 M sodium acetate
- Oligonucleotide primers, diluted in water to 10 pmol/µl
- Taq polymerase
- 10 × Taq polymerase buffer (see *Protocol 3*)
- Low melting point agarose
- TAE buffer (see *Protocol 1*)
- Ethidium bromide (stock solution 10 mg/ml in water)

A. *Preparation of B cells for in-cell PCR*

1. Wash the cells three times in PBS, centrifuging at 50–100 g for 5 min between washes.

2. Count the cells in a haemocytometer and aliquot up to 10^7 cells into a 2 ml Eppendorf tube. Spin at 12 000–14 000 g in a microcentrifuge for 2.5 min.

3. Remove the supernatant and suspend the cell pellet in 1 ml ice-cold 10% formal saline, using a syringe and 21-gauge needle. To minimize cell losses, retain the syringe and needle for re-use. Incubate the tube at 4°C on a rotator, or on ice with frequent agitation for 1 hour.

4. Spin the tube in a microcentrifuge for 2.5 min and remove the supernatant. Suspend the cells in ice-cold PBS/glycine, using the syringe and needle, and repeat the centrifugation for a total of three washes.

5. Resuspend the cells in 1 ml ice-cold 0.5% NP-40 using the same syringe and needle, and incubate for 1 hour on a rotator at 4°C or on ice with frequent agitation.

1: Construction and use of antibody gene repertoires

6. Spin down and wash three times in PBS/glycine as in step 4.
7. Finally resuspend the cells in PBS/glycine, using the syringe and a 26-gauge needle.
8. Count the cells, checking that they are free of microscopic clumps. If clumps are present, re-syringe the cells. Adjust the concentration to 2×10^6 per ml. Store in aliquots at −70°C or proceed directly to part B.

B. *cDNA first-strand synthesis*
1. Prepare a '1st strand' mix as follows:
   - V_H primer   2.5 µl
   - V_L primer   2.5 µl
   - dNTP mix   5.0 µl
   - DTT   5.0 µl
   - 10 × 1st strand buffer 5.0 µl
   - Cells[a]   26 µl
2. Heat the mix at 65°C for 3 min. Allow to cool to ambient temperature for 10 min, then place on ice.
3. Add 2 µl (40 units) AMV reverse transcriptase and 2 µl (80 units) RNase inhibitor. Incubate at 42°C for 1 hour, with occasional gentle shaking or pipetting to suspend the cells.
4. Transfer the cell suspension to a 500 µl tube and spin down at 12 000–14 000 r.p.m. for 2.5 min.
5. Discard all of the supernatant using a pipette with a fine tip. When withdrawing the last few microlitres of supernatant, slightly invert the tube and drag the pipette tip along the side of the tube to the top, while applying suction. This procedure should leave a cell pellet free of residual liquid, rendering more than a single wash unnecessary. Washes are mandatory, but must be kept to a minimum to avoid undue loss of cells; stringency is thus very important.
6. Suspend the pellet in 200 µl PBS/glycine and spin again.
7. Remove all the supernatant and resuspend the cells in 100 µl PBS/glycine. The cells can be used immediately in PCR, or frozen in aliquots at −70°C.

C. *In-cell PCR assembly*
1. Prepare the 1st PCR mix in a 500 µl tube as follows:
   - V_H back primer mix[b]   2.5 µl
   - V_L forward primer mix[b]   2.5 µl
   - V_H forward linker primer mix 2.5 µl
   - V_L back linker primer mix   2.5 µl

Protocol 14. *Continued*
- dNTP mix — 2.0 µl
- Taq DNA polymerase (2.5 units) — 0.5 µl
- 10 × Taq polymerase buffer — 5.0 µl
- Cells[c] — 10.0 µl
- water — 22.5 µl
- (total volume of PCR 50 µl)

If using a hot-lid temperature cycler, do not use oil. A hot-lid temperature cycler will give optimal results and avoid problems due to oil coating the cells. If using a basic cycler, add 1 drop of mineral oil to the tube.

2. Give the tube 30 cycles as follows:

    95°C for 1 min, 65°C for 1 min, 72°C for 1 min.

    The reaction works more efficiently at stringent annealing temperatures of 65°C or more than at lower temperatures.

3. After cycling, spin down the cells, remove (but keep) all of the supernatant and suspend the cells in 200 µl PBS/glycine. If an oil overlay is used, first remove the cells from the PCR tube by inserting a pipette tip through the oil to the bottom of the tube, and withdrawing the cells from the bottom so that the supernatant washes them into the pipette tip. Do not allow oil to enter the tip. Transfer the cell suspension to a clean tube for centrifugation and resuspension.

4. Spin the cells down again, discard all of the wash supernatant and resuspend the pellet in 10 µl PBS/glycine and use immediately as template for the 2nd PCR.

5. Prepare the 2nd PCR mix in a 500 µl tube as follows:
   - V_H back primer mix[b] — 2.5 µl
   - V_L forward primer mix[b] — 2.5 µl
   - dNTP mix — 2.5 µl
   - Taq DNA polymerase (2.5 units) — 0.5 µl
   - 10 × Taq polymerase buffer — 5.0 µl
   - washed cells (from the 1st PCR) — 10.0 µl
   - water — 27.5 µl

   Add 1 drop of oil if necessary.

6. Give the tube 30 cycles as follows:

    95°C for 1 min, 65°C for 1 min, 72°C for 1 min.

7. After cycling spin down the cells as step 3, and discard them. Keep the supernatant.

8. Prepare a 1.5% low melting point agarose gel in TAE buffer, incorporating 0.5 µg/ml ethidium bromide. Electrophorese the supernatants of the 1st and 2nd PCRs, using a suitable size marker. The 1st PCR

1: Construction and use of antibody gene repertoires

should give bands around 300–350 bp and the 2nd PCR around 650 bp.

9. Excise the approximately 650 bp bands from the 2nd PCR lane(s) and purify the DNA. A number of proprietary kits are available for this purpose. (see *Protocol 3*, step 7).

10. Digest the purified DNA with appropriate restriction endonucleases and clone into a suitable expression or selection vector (plasmid, phage, or phagemid) (*Protocol 7*).

[a] Fixed and permeabilized cells (from *Protocol 12A*) thawed, spun down, and resuspended in water to a final volume of 26 µl.
[b] See Section 7.2.
[c] 10^5 template cells in 10 µl PBS/glycine (it is not necessary to wash and suspend the cells in water).

7.2 Application

The fidelity of this process has been demonstrated (27) using 1:1 and 1:9 mixtures of two hybridoma cell lines, B1-8 (28) and NQ10/12.5 (29), and mixes of primers matched to each cell line. Results for two separate experiments with 1:9 B1-8:NQ10/12.5 mixtures are shown in Table 3. The purpose of this model was solely to identify and compare V_H and V_L gene combinations in clones prepared by in-cell assembly and random combinatorial assembly (i.e. with soluble cDNA templates), the genes being identified by hybridization with internal oligonucleotide probes and by PCR screening (30). The primers were not designed to produce full-length V_H and V_L sequences in the the second PCR. For expression of full-length scFv fragments the scheme outlined below is suggested. In the examples in *Table 3*, Experiment 1 used primers exactly as described in ref. 27. For Experiment 2, the same V_H back and V_L forward primers as in Experiment 1 were used in each PCR, but the four linker primers in the 1st PCR incorporated a loxP site (31). Linkage by the loxP site appeared to be more efficient than linkage by overlap extension as described in (27) (MJE, unpublished), but is not described here in detail because it requires the cells to be incubated in *cre* recombinase between the first and second PCRs. It is understood that this enzyme is no longer commercially available.

The results of the experiment shown in *Table 3* indicate that with both forms of linkage, in-cell PCR assembly resulted in clones containing only the parental hybridoma V_H and V_L combinations, while random combinatorial assembly using identical primers resulted in crossover combinations in which B1-8 V_H was linked to NQ10/12.5 V_L.

Assembly of full-length scFv requires that the 2nd PCR primers must prime at the 5′ end of the V_H genes (framework 1) and the 3′ end of the V_L genes (J_L). They must also be fully nested with respect to the 1st PCR primer for efficient amplification of the assembled genes. These requirements can be

Table 3. Assembly of immunoglobulin V genes from 1:9 mixes of B1-8:NQ10/12.5 murine hybridoma cells

Assembly method	Expt no.	No. of V186.2 +ve clones[a]	Fraction of V186.2 +ve vc clones with:	
			B1-8 V_H + B1-8 V^b	B1-8 V_H + NQ10/12.5 Vk^b
Random combination	1	34/450	3/34	31/34
	2	29/250	5/29	24/29
In-cell PCR	1	27/450	27/27	0/27
	2	34/250	34/34	0/34

[a] Bacterial colonies were hybridized with ^{32}P-labelled V186/2 probe specific for the B1-8 V_H gene (27).
[b] V186.2 +ve replica-plated colonies were PCR-screened using 2nd PCR primers specific for B1-8 V_H (back primer) and either B1-8 V or NQ10/12.5 V_k (forward primers) (27).

achieved by using back primers in the V_H leader regions and forward primers in the C_L regions for the 1st PCR, as indicated in *Figure 4*. Primers for cDNA 1st strand synthesis should prime in the C_L and C_H1 (or J_H) regions. In addition to priming in the V_H framework 1 and J_L regions, the 2nd PCR primers should incorporate restriction sites external to the genes in order to facilitate cloning into an appropriate vector, as suggested in *Figure 4*. Although, as yet, untested for assembly and expression of murine scFv genes, the suggested scheme should be applicable for scFv cloning from spleen cells of immunized mice. A list of purpose-designed primers is not available, but their design may be based on primers shown in *Table 3* and ref. 32 and the linker strategy employed in *Table 3*, Experiment 1 (27).

8. Conclusions

Display of proteins on the surface of bacteriophage gains its power as a method by linking proteins, presented for binding to ligand, with the DNA encoding them. The phage system mimics the natural processes by which the immune system produces molecules which bind tightly to ligands, allowing large numbers of variants to be surveyed and leading to the more rapid isolation of improved protein molecules. The monoclonal antibody technique of Kohler and Milstein (33) has enabled isolation of highly specific antibodies of importance in research, diagnosis, and therapy. Phage antibody technology extends this technology. For example, a monoclonal antibody can be cloned into the phage-display system and mutagenesis or other genetic manipulation techniques used to generate a large variant population from the original antibody. Phage antibody selection can then be used to isolate those antibodies with improved affinities or specificities.

In addition, with phage-display technology, the immune system can be bypassed and human antibodies selected directly from a non-immunized source (3). The ability to make human antibodies to a wide range of targets (e.g. parasites, bacteria, viruses, and other pathogens) is of obvious benefit. In addition, the ability to make human antibodies to human ('self') antigens also has important consequences for therapeutic use, especially for cancer treatment and for the regulation of disease states.

References

1. McCafferty, J., Griffiths, A. D., Winter, G., and Chiswell, D. J. (1990). *Nature*, **348**, 552–4.
2. Clackson, T., Hoogenboom, H. R., Griffiths, A. D., and Winter, G. (1991). *Nature*, **352**, 624–8.
3. Marks, J. D., Hoogenboom, H. R., Bonnert, T. P., McCafferty, J., Griffiths, A. D., and Winter, G. (1991). *J. Mol. Biol.*, **222**, 581–97.
4. Hawkins, R. E., Russell, S. J., and Winter, G. (1992). *J. Mol. Biol.*, **226**, 889–96.
5. Crissman, J. W. and Smith, G. P. (1984). *Virology*, **132**, 445–55.
6. Glaser-Wuttke, G., Keppner, J., and Rasched, I. (1989). *Biochim. Biophys. Acta*, **985**, 239–47.
7. Parmley, S. F. and Smith, G. P. (1988). *Gene*, **73**, 305–18.
8. Scott, J. K. and Smith, G. P. (1990). *Science*, **249**, 386–90.
9. Devlin, J. J., Panganiban, L. C., and Devlin, P. E. (1990). *Science*, **249**, 404–6.
10. Cwirla, S. E., Peters, E. A., Barrett, R. W., and Dower, W. J. (1990). *Proc. Natl Acad. Sci. USA*, **87**, 6378–82.
11. Hoogenboom, H. R., Griffiths, A. D., Johnson, K. S., Chiswell, D. J., Hudson, P., and Winter, G. P. (1991). *Nucl. Acids Res.* **19**, 4133–7.
12. Holliger, P., Prospero, T., and Winter, G. (1993). *Proc. Natl Acad. Sci. USA*, **90**, 6444–8.
13. McCafferty, J., Jackson, R. H., and Chiswell, D. J. (1991). *Prot. Engineering*, **4**, 955–61
14. Chiswell, D. J. and McCafferty, J. (1992). *Trends Biotechnol.*, **10**, 80–4
15. Jackson, R. H., Hoogenboom, H. R., Winter, G., and Chiswell, D. J. (1993). *Prot. Eng.*, **6**, 114.
16. Bass, S., Greene, R., and Wells, J. A. (1990). *Proteins*, **8**, 309–14.
17. Kang, A. S., Barbas, C. F., Janda, K. D., Benkovic, S. J., and Lerner, R. A. (1991). *Proc. Natl Acad. Sci. USA*, **88**, 4363–6.
18. Zacher, A. N., Stock, C. A., Goldern, J. W., and Smith, G. P. (1980). *Gene*, **9**, 127–40.
19. Orlandi, R., Gussow, D. H., Jones, P. T., and Winter, G. (1989). *Proc. Natl Acad. Sci. USA*, **86** 3833–7.
20. Munro, S. and Pelham, H. (1986). *Cell*, **46**, 291–300.
21. McCafferty, J., Fitzgerald, K. J., Earnshaw, J., Chiswell, D. J., Link, J., Smith, R., and Kenton, J. (1994). *Appl. Biochem. Biotechnol.*, **47**, 157–73.
22. Sambrook, J., Fritsch, E. F., and Maniatis, T. (ed.) (1989). *Molecular cloning, a laboratory manual* (2nd edn). Cold Spring Harbor Laboratory Press, NY.

23. Marks, J. D., Griffiths, A. D., Malmqvist, M., Clackson, T. P., Bye, J. M., and Winter, G. (1992). *Biotechnology*, **10**, 779–83.
24. Dower, W. J., Miller, J. F., and Ragsdale, C. W. (1988). *Nucl. Acids Res.*, **16**, 6127–45.
25. Vieira, J. and Messing, J. (1987). In *Methods in enzymology*, vol. 153, pp. 3–11. Academic Press, London.
26. Winter, G. and Milstein, C. (1991). *Nature*, **349**, 293.
27. Embleton, M. J., Gorochov, G., Jones, P. T., and Winter, G. (1992). *Nucl. Acids Res.*, **20**, 3831.
28. Cumano, A. and Rajewski, K. (1985). *Eur. J. Immunol.*, **15**, 512.
29. Griffiths, C. M., Berek, C., Kaartinen, M., and Milstein, C. (1987). *Nature*, **312**, 272.
30. Chaudhary, V. K., Batra, J. K., Gallo, M. G., Willingham, M. C., FitzGerald, D. J., and Pastan, I. (1990). *Proc. Natl Acad. Sci. USA*, **87**, 1066–70.
31. Hoess, R. H., Ziese, M., and Sternberg, N. (1982). *Proc. Natl Acad. Sci. USA*, **79**, 3389.
32. Bendig, M. (1991). *Biotechnology*, **9**, 579.
33. Kohler, G. and Milstein, C. (1975). *Nature*, **256**, 52–3.

2

Affinity maturation of antibodies using phage display

KEVIN S. JOHNSON and ROBERT E. HAWKINS

1. Introduction

1.1 General considerations

Protocols described in this section will deal with alteration of the properties of existing antibodies, formatted as scFv and displayed on bacteriophage. Although the title implies a focus on improving affinities, provisions are made for alteration in off-rate and specificity. Where the aim of the 'maturation' process is a shift in specificity, this can be achieved by introducing bias into the selection process.

Moreover, it is assumed that there is little or no structural information to guide the process, which in consequence relies heavily on the selection procedures. The apparent simplicity belies the impact the selection procedure may have on the final outcome. Many intermediate protocols are described in detail elsewhere in this book, and in these cases, the reader is referred to the relevant section.

Regardless of the aim, the process involves three distinct steps: production of mutant forms of the antibody (mutagenesis), followed by selection of those with improved properties (selection), and identification of new variants (screening).

1.2 Mutagenesis

In the absence of a known 3-D structure, the inevitable question is which parts of the antibody to mutate. The complementarity-determining regions (CDRs) are not wholly responsible for the properties of the antibody, since residues distant from the combining site ('The Vernier' region, refs 1–3, see also Bendig *et al.*, Chapter 7) can influence antigen binding indirectly (see *Figure 1*). This may be by modifying the conformation of the CDR loop or a contact residue, or influencing the way heavy and light chains pack against one another. Consequently, the emphasis is on non-localized mutagenesis strategies such as chain-shuffling and error-prone PCR. Chain-shuffling is the

Framework residues that interact with CDRs

VL D1.3 amino acid sequence

                  CDR 1   CDR2   CDR3

DI<u>QMTQ</u>SPASLSASVGETVT<u>ITC</u> RASGNIHNYLA <u>WYQQKQGKSPQLLVY</u> YTTTLAD <u>GVPSRFSGSGSGTQY</u>SLKINSLQPEDFGSYYC QHFWSTPRT FGGGTKLEIK

VH D1.3 amino acid sequence

                             CDR 1   CDR2   CDR3

QVQ<u>LKESGPGLVAPSQSLSITCTVSGFSLT</u> GYGVN <u>WVRQPPGKGLEWLG</u> MIWGDGNTDYNSALKS <u>RLSISKDNSKSQVFLKMNSLHTDD</u>TARYYC<u>AR</u> ERDYRLDY <u>WGQGTTLTVSS</u>

Figure 1. Amino acid sequence of anti-lysozyme antibody D1.3; framework residues in contact with CDRs are underlined (1).

half-way house between site-directed and general mutagenesis strategies and is described in detail in Chapter 8. PCR mutagenesis methods (2) introduce substitutions at 'random' throughout the sequence.

1.3 Selection
Selection can be any method that enables separation of clones that bind from those that do not; as such there is an endless list of possible selection methodologies, and only the two most commonly used methods are included here. These are panning on antigen adsorbed on to plastic (4, 5, 6), and selection with soluble biotinylated antigen (7). Most often, between two and five rounds of selection are needed to adequately enrich a population. The actual number of rounds is almost impossible to predict in advance being dependent on the degree of enrichment afforded by the selection method and the relative affinities of the antibodies in the population. The degree of enrichment (i.e. ratio of binders to non-binders before and following selection) typically varies from 5–1000-fold at each round, and more rounds of selection are needed to resolve antibodies with similar affinities than those having widely differing affinity constants. All selection methods automatically favour variants with stronger binding, since these are best suited to survive the selection process. These antibodies will have either highest affinity or greatest avidity. The highest affinity antibodies within a population are those which, for a given antibody concentration, bind the greatest proportion of antigen at the lowest antigen concentrations.

Selection may also favour variants that form multimers, particularly when the antigen is polymeric or present at high density. In this case, a low-affinity antibody with a tendency to multimerize may be preferentially selected by virtue of a slow rate of dissociation, caused by the rate of dissociation from antigen being a composite of the off-rates of each component of the multimer. The importance of this effect cannot be emphasized too strongly. For example, whereas a maturation cycle may have resulted in selection of a clone with a 5-fold improved off-rate, these truly improved variants can be obliterated by variants with a tendency to multimerize, where the off-rate can be slowed by 1–2 orders of magnitude with no change whatsoever in the true affinity.

Whatever the purpose of the selection, it is important to check the antigen specificity of the clones and to sample clones from all rounds of selection. Non-specific or polyreactive antibodies may predominate in late rounds of selection in the absence of a high-affinity specific antibody to compete. Moreover, sampling the early rounds of selection ensures adequate diversity, since a single species will eventually dominate any selection.

1.4 Screening
A likely outcome of the mutagenesis and selection steps is a mixture of variants with differing properties, and it may be necessary to screen large numbers of

clones to identify those best suited to your purpose. We have concentrated on ELISA-based assays for specificity and affinity since these have the highest throughput. Ideally, the screening method should be as close as possible to the eventual application. If you cannot process large numbers of crude antibodies this way, then use ELISA, Western blotting, *Bst*NI fingerprinting and sequencing to identify those most likely to have appropriate characteristics.

An important part of the screening process is DNA sequencing, described in Chapter 6. For example, you may have reselected the parental sequence, and it is important to know this before investing a large effort in analysis. Another possible outcome is the selection of 'null' sequence variants, i.e. amino acid substitutions that are neither detrimental nor beneficial. Consider any change in a CDR or Vernier residue significant until proven otherwise, even if the change is relatively minor. We have several examples of conservative substitutions (e.g. Ser → Thr, Leu → Ile) that indisputably alter the properties of the antibody (unpublished data).

2. Mutagenesis

2.1 General considerations

Methods are described here for creating mutant V-gene segments ready for cloning using the methods described in Chapter 1. Whichever procedure is used, aim to create repertoires of $> 10^6$ recombinants to ensure adequate sampling of the mutant population. It is also worthwhile quality-controlling these repertoires at an early stage to ensure that they do indeed contain mutants. *Bst*NI fingerprinting is suitable for chain-shuffled repertoires, whereas sequencing 6–10 randomly picked clones is necessary to verify that the other mutagenesis procedures have worked.

Chain-shuffling is a proven method of introducing diversity into a phage repertoire (3) where a given V_H or V_L chain is recombined with a repertoire of complementary chains. The methodologies for this are not described here but are effectively described in Hoogenboom *et al.*, Chapter 8.

2.2 Error-prone PCR

The following method has been shown to introduce an average of 1–2 base substitutions per V domain as a result of manganese reducing fidelity of Taq polymerase. It is a difficult reaction to optimize for the desired number of mutations per gene, and it is suggested that five separate reactions are set up, containing 0, 0.5 mM, 1.0 mM, 1.5 mM, and 2 mM $MnCl_2$. Concentrations of $MnCl_2$ higher than 2 mM may cause components of the PCR reaction to precipitate; moreover, it may be counterproductive to introduce large numbers of mutations. Highly defective antibodies (frameshifts, etc.) may bind non-specifically and out-compete improved variants.

2: Affinity maturation of antibodies using phage display

Protocol 1. Error-prone PCR amplification

Equipment and reagents

- 10 × Taq polymerase buffer as provided by the supplier of the Taq polymerase, or: 100 mM Tris, pH 9.0 at 25°C, 500 mM KCl, 60 mM MgCl$_2$, 1% (w/v) gelatin, and 10 % Triton X-100
- 10 mM each dNTPs (10 × stock)
- 10 μM pUC19REV (see Appendix 2)
- 10 μM FDTSEQ1 (see Appendix 2)
- 50 mM MnCl$_2$, freshly-prepared
- DNA thermal cycler for PCR
- Taq DNA polymerase (Boehringer Mannhein, under licence from Perkin Elmer/Cetus)
- Mineral oil (paraffin oil) (Sigma, cat. no. M-3516)
- 'Wizard PCR prep' DNA purification kit (Promega)
- TAE buffer: (see Chapter 1, *Protocol 1*)

Method

1. Set up a PCR reaction with your scFv DNA as template:
   - 10 × PCR buffer   10 μl
   - dNTPs   to 1 mM each
   - template DNA   20 ng
   - 10 μM pUC19REV   5 μl
   - 10 μM FDTSEQ1   5 μl
   - H$_2$O   to 100 μl final volume

2. Add 1–4 μl of freshly-prepared 50 mM MnCl$_2$ to the reaction, mix and overlay with mineral oil.

3. Place the sample in the PCR block and heat to 94°C for 2 minutes before adding 1 μl of Taq polymerase (5 units/μl; Boehringer Mannheim or Cetus) underneath the oil.

4. Perform 30 cycles at 94°C (1 min), 60°C (1 min), 72°C (4 min), followed by 10 min at 72°C after the final cycle.[a]

5. Run 3–6 μl of the PCR reaction on a 1% TAE agarose gel to check that the PCR has worked. Expect to see less product (often none at all) at the highest Mn^{2+} concentrations.

6. Recover the remainder of the reaction and process the fragment as described in Chapter 1, *Protocol 3*, step 7 onwards (purification), *Protocol 7* (restriction digestion/purification). Also Chapter 8, *Protocol 3* (ligation) and Chapter 8, *Protocol 4* (transformation). Verify that the library is mutant by sequencing 6–10 randomly picked colonies (Chapter 6).

[a] The long extension time is necessary to compensate for a reduced polymerization rate in the presence of Mn^{2+} ions.

2.3 Site-directed mutagenesis

Introducing a single, specific alteration or multiple changes simultaneously can be achieved using PCR with mutagenic primers or by oligonucleotide directed mutagenesis. There are several excellent kits for directed mutagenesis that are commercially available. We recommend the use of Amersham's oligonucleotide-directed *in vitro* mutagenesis kit. Full protocols are provided, but only production of single-stranded DNA template is covered here.

With pUC119 based vectors (e.g. pCANTAB5) the DNA strand that is packaged into phage contains the 'sense' strand of the antibody gene (the coding sequence written 5' to 3', N terminus to C terminus). All oligos should therefore be 'antisense', i.e. the reverse complement of the antibody sequence.

Protocol 2. Preparation of single-stranded template DNA

Equipment and reagents
- Phenol (BRL ultrapure redistilled, cat. no. 5509UA)
- Chloroform
- Microcentrifuge
- 95–100% ethanol
- 75% ethanol
- TE buffer: 10 mM Tris–HCl pH 7.4, 1 mM EDTA
- 3 M sodium acetate (pH 6.0)

Method

1. Perform a phagemid rescue[a] from a 10 ml culture, and PEG precipitate the resulting phage as described in *Protocol 8*.

2. Redissolve the pellet in 500 μl of TE buffer and transfer to a 1.5 ml microcentrifuge tube. Add 500 μl of freshly pH adjusted phenol.[b] Vortex for 10 seconds, leave to stand for 10 minutes, then vortex for 10 seconds again before centrifuging at 13 000 g for 10 minutes at room temperature.

3. Carefully remove up to 450 μl of the upper aqueous phase to a fresh tube and add 900 μl chloroform. Vortex for 5 sec and centrifuge at 13 000 r.p.m. for 5 min at room temperature.

4. Carefully remove the aqueous layer and measure the volume. Add 1/10th volume of 3 M sodium acetate (pH 6.0), mix, then add 2.5 vol. of ethanol (95 or 100%) and mix again.

5. Place at –20 °C overnight or –70 °C for 30 min or longer. Recover single-stranded DNA by centrifuging at 13 000 g for 20–30 min at 4 °C. DNA will be smeared up the side of the tube furthest away from the centre of the rotor.

6. Aspirate the supernatant and add 1 ml 70% ethanol. Centrifuge at 13 000 g in a microcentrifuge for 2 min, aspirate the supernatant and, if available, dry the pellet under vacuum for 2–3 min.

7. Redissolve the pellet in 30 μl TE buffer and run 2 μl of this alongside molecular weight markers on a 1% agarose gel. If using λ DNA

2: Affinity maturation of antibodies using phage display

digested with *Hin*dIII. A predominant band migrating between the 2.3 kb and 4.3 kb markers should be visible. M13KO7 single-stranded DNA migrates between the 6.5 kb and 9.4 kb markers.

[a] The presence of glucose in the medium sometimes hinders phagemid production and is best omitted.
[b] Melt at 60°C then add 200 ml distilled water to 100 g bottle and mix. Store at 4°C. Before use, remove an aliquot of the lower phenol phase and add 25 μl of 2 M Tris (unbuffered) per ml of phenol, to give a final pH of 7.0–8.0.

2.4 Mutagenesis using 'spiked' PCR primers
2.4.1 V_H CDR3 spiking

Not all CDRs will be of equal importance from one antigen–antibody pair to the next. However, a theme common to many is the importance of the third CDR of the heavy chain (V_H CDR3). This is the most variable CDR and forms the centre of the antigen combining site. It is often the best place to start. In all cases where we have 'spiked' V_H CDR3, antibodies superior to the parent have been isolated.

'Spiking' is achieved by synthesizing oligonucleotides where each phosphoramidite is mixed with a small amount of the other three phosphoramidites before loading on the synthesizer. The resulting oligonucleotide (CDR3SPIKE) will have a small number of changes along its entire length. We usually synthesize spiked oligonucleotides bounded by the invariant cysteine preceding the CDR3 and ending in the fourth framework region (FR4). Those oligonucleotides with changes at the 3' end (i.e. the part encoding FR3) will amplify poorly during PCR, whereas changes in FR4 are corrected in subsequent amplification steps where an unmutated PCR primer based in FR4 is used.

The overall procedure involves PCR amplification from the parental scFv gene to generate a V_H gene whose end (encoding CDR3) has been spiked and linking this with a fragment encoding a flexible linker peptide and the V_L region. This latter fragment is generated by PCR of the scFv gene using one primer with homology to the end of the heavy chain PCR fragment (to drive the assembly reaction) and the other based in the vector sequences. Usually the parental V_L gene is used in the assembly, but this could equally be any cloned V_L gene or repertoire thereof.

Protocol 3. Synthesis of spiked oligonucleotide

Equipment and reagents
- Oligonucleotide synthesizer and reagents

Method
1. Make up 5 ml stock containing each phosphoramidite at 100 mM.

Protocol 3. *Continued*

2. Decide what error rate you want:

    error rate = number of mistakes per oligo/length of oligonucleotide.

3. Now calculate the volume of equimolar spike mix to add to each bottle:

    volume of spike mix, µl = error rate × 1.33 × 1000 × 5[a];

    e.g. to introduce an average of three errors into a 100-base oligonucleotide:

    0.03 × 6650 = 200 µl spike mix per 5 ml of phosphoramidite.

4. Synthesize the antisense of V_H CDR3 sequence and purify.

[a] 1.33 is a correction factor due to the mix containing all four phosphoramidites; 1000 converts millilitres to microlitres; 5 ml is the final volume of phosphoramidite in the bottle.

Protocol 4. PCR amplification of V_H genes and assembly into scFv

Equipment and reagents

- See *Protocol 1*
- Spiked oligonucleotide

Method

1. Set up the following reaction to amplify the V_H segment with the spiked oligonucleotide.
    - 10 × PCR buffer   10 µl
    - 4 mM dNTP   5 µl
    - 10 µM pUC19 reverse[a] (see Appendix 2)   5 µl
    - 10 µM spiked oligo[a]   5 µl
    - template DNA   20 ng
    - H_2O to 100 µl
    - Taq polymerase   5 U

2. Amplify using 25 cycles of 94 °C for 1 min, 60 °C for 1 min and 72 °C for 1 min followed by 10 minutes at 72 °C.

3. Analyse 3–5 µl of the PCR reaction on a 1.5% gel. If a discrete band of 350 bp is visible, gel-purify the remainder of the PCR reaction as described in Chapter 1, *Protocol 3*, step 7 onwards.

4. Set up the following PCR reaction to amplify the parental (or other) light chain together with the scFv linker.
    - 10 × PCR buffer   10 µl
    - 4 mM dNTP   5 µl
    - 10 µM reverse JH[b] (see Chapter 1, *Table 2*) 5 µl
    - 10 µM FDTSEQ1[b] (see Appendix 2)   5 µl

2: Affinity maturation of antibodies using phage display

- template DNA   20 ng
- H$_2$O   to 100 µl
- Taq polymerase   5 U

5. Amplify using 25 cycles of 94°C for 1 min, 60°C for 1 min, and 72°C for 1 min, followed by 10 min at 72°C.

6. Analyse 3–5 µl of the PCR reaction on a 1.5% agarose gel. If a discrete band of 400 bp is visible, gel-purify the remainder of the PCR reaction as described in Chapter 1, *Protocol 3*, step 7 onwards.

7. Assemble and pull-through V$_H$ and linker–V$_L$ segments using *Protocol 5* in Chapter 1, but omitting the linker DNA since this is already present on the light chain fragment. The primers used should be pUC19reverse and FDTSEQ1. This will generate a scFv gene with *Sfi*I and *Not*I restriction sites already in place.

8. Digest with *Sfi*I and *Not*I (Chapter 1, *Protocol 7*) prior to gel purification and cloning (Chapter 8, *Protocols 3* and *4*).

[a] The primer pUC19reverse is located in the plasmid sequences upstream of the antibody insert. The heavy chain spiked oligo is located in CDR3 and in framework 4. This pair generate a CDR3 spiked heavy chain.
[b] The human primer reverseJH is located in framework 4 and is complementary to the spiked oligo used to generate the heavy chain fragment above. (The mouse equivalent is LINKBACK, Chapter 1, *Table 1*.) The primer FDTSEQ1 is located in *gene 3*. This pair generates a fragment encoding the end of framework 4 of the heavy chain together with the linker sequence and the light chain.

2.4.2 V$_L$ CDR3 spiking

Spiking light chain CDR3 can be performed in a similar manner to that described above, though no assembly reaction is necessary since the mutated region is close to the end of the scFv gene. In this case amplify the scFv with a primer based in the vector upstream of the antibody gene and the *Sfi* cloning site (e.g. pUC19reverse) and a 'spiked' light chain primer; this oligonucleotide should extend from CDR3 into the framework region FR4 such that the spiked fragment can be 'pulled-through' with an extended FR4 based primer, appending a *Not*I site for cloning (Chapter 1, Section 3.5).

3. Selection of antibodies with altered properties

There is no substitute for selection schemes based on the unique properties of the system under investigation, and the reader is encouraged to develop inventive selection schemes to derive antibodies with the desired characteristics. Any selection method will automatically favour antibodies of higher affinity, since by definition, these will be more long-lived interactions and bind antigen at low concentrations. The main complication is avidity, caused by two or more copies of the antibody on the phage head interacting with adjacent epitopes. Most scFvs will dimerize to a certain extent and it is important to bear in mind that the selection method may favour poor antibodies with a tendency to dimerize. The form of the antigen is the most obvious way to control this. If the antigen is a hapten coupled to a carrier, derivatize at low density. If the antigen is a repeating polymer, then selection on fragments or synthetic versions containing fewer repeat units can be used. For soluble proteins, selections incorporating solution capture are recommended.

At each round of selection, infect *E. coli* TG1 cells with selected phagemids and plate out on media containing ampicillin and glucose, then grow overnight at 30 °C. The next day, scrape the plate and rescue an aliquot as described in Chapter 1, *Protocol 8*.

3.1 Selection of phage by panning

This technique is effectively a preparative ELISA. It is particularly good at discriminating between binder and non-binder, but is poor at discriminating between antibodies of similar affinity and can result in the selection of high-avidity antibodies due to high antigen density.

Protocol 5. Solid-phase selection of phage antibodies (see also Chapter 1, *Protocol 12*)

Equipment and reagents

- 75 × 12 mm Nunc-immunotubes (Maxisorp, cat. no. 4-44202)
- Phosphate buffered saline (PBS)
- PBS containing 2% skimmed milk powder (MPBS)
- PBS containing 0.1% Tween-20
- 100 mM triethylamine, freshly made
- 1.0 M Tris–HCl, pH 7.4
- Log-phase TG 1 cells
- 1.5 ml Eppendorf tubes
- 2TYAG medium: see Chapter 1, *Protocol 8*
- Agar plates made with 2TYAG medium

Method

1. To an immunotube add 1–4 ml of an appropriate concentration of antigen in the appropriate buffer.[a, b, c, d]
2. Leave to coat overnight at 4 °C.
3. Wash the tube three times with PBS (pour PBS in and pour out again immediately).
4. Block by filling tube to the brim with PBS containing 2% skimmed milk powder. Cover with Parafilm and incubate at 37 °C for 2 h.
5. Wash tube three times with PBS.
6. Add 10^{10} to 10^{13} TU phage in 4 ml of MPBS and incubate for 1 hour at room temperature.[e]
7. Wash tubes with 20 washes of PBS, 0.1% Tween-20, then repeat with 20 PBS washes. Each washing step is performed by pouring buffer in and out immediately, best achieved using a wash bottle.

2: Affinity maturation of antibodies using phage display

8. Elute phage from tube by adding 1 ml freshly-made 100 mM triethylamine and leaving for 10 minutes at room temperature.
9. Recover supernatant and immediately transfer to a 1.5 ml Eppendorf tube containing 0.5 ml 1.0 M Tris–HCl, pH 7.4, and mix.
10. Mix 750 µl of eluted phage with 10 ml of log-phase TG1 grown at 37°C (OD_{600} around 1.0). Incubate at 37°C for 30 minutes and remove an aliquot for titration on 2TYAG plates. Centrifuge the remainder at 2000 g for 10 minutes and plate on a large 2TYAG plate. Incubate all plates at 30°C overnight.
11. Store the remaining selected phagemids at 4°C.

[a] Antigens are usually coated at 1–10 µg/ml, but concentrations as high as 3 mg/ml are used in the case of certain proteins, e.g. lysozyme and coating is sometimes better in 50 mM sodium hydrogen carbonate, pH 9.6 (pH adjusted with NaOH), than in PBS. ELISAs performed with the antigen are usually a good guide to the best coating conditions.
[b] Peptides often bind poorly to plastic and it is best to couple them to a carrier such as BSA prior to immobilization. Pierce sell a range of preactivated carriers and these are highly recommended.
[c] An alternative to non-specific adsorption is to covalently couple the antigen to the support. This may avoid large conformational changes that can result with direct immobilization. Several companies sell such cross-linking reagents.
[d] Biotinylated antigen (see below) may also be immobilized to a solid surface via streptavidin to avoid conformational changes on adsorption. Streptavidin itself adsorbs poorly to plastic regardless of pH. Covalent coupling, binding to plates coated with biotinylated BSA, or commercially available streptavidin matrices are recommended.
[e] To select against an undesirable cross-reactivity, include soluble antigen at ≥ 10-fold higher concentration than that used for coating.

3.2 Solution capture on soluble antigen

Solution capture on 'tagged' antigen is a powerful method for selecting for the highest affinity antibodies in a population, since the concentration of antigen can be controlled. The highest affinity antibodies are, by definition, those that bind most antigen at the lowest antigen concentration. The principle of the method has been described in detail (7). In brief, a low concentration of antigen which is in molar excess over phage is added; for example, to select antibodies of 10 nM or higher affinity, use antigen at 10 nM or lower concentration to select 10^{11} or fewer phage (6×10^{11} phage/ml is 1 nM, 6×10^{10} phage/ml is 0.1 nM, etc.) in a 1 ml volume.

The antigen is tagged to enable recovery of bound phage, the nature of the tag depending on the antigen in question. The usual method is biotinylation; Pierce supply many reagents for biotinylation, all of which are supplied with detailed protocols and these are recommended. The example given below is for biotinylation of a peptide by coupling of an NHS ester to lysine residues. We find it is best to add just one or two biotins to each molecule of antigen to avoid destroying epitopes.

Protocol 6. Biotinylation of proteins using NHS S–S biotin

Equipment and reagents

- NHS S–S biotin (Pierce)
- 1 M NaHCO$_3$, pH 8.5
- avidin–HABA (4-hydroxyazobenzene-2-carboxylic acid)

Method

1. Calculate the reagent concentrations, e.g.

   5 mg/ml stock of peptide mol. wt 3337 is 1.5 mM = 1.5 nanomoles/µl.
   1 mg of NHS S–S biotin dissolved in 1 ml of H$_2$O = 1.65 nanomoles/µl.[a]

2. To singly biotinylate 150 nanomoles of peptide, set up the following reaction:
   - peptide stock   100 µl
   - 1 mg/ml NHS S–S biotin   90 µl
   - 1 M NaHCO$_3$, pH 8.5   50 µl
   - H$_2$O   760 µl

3. Incubate on ice for 2 h.

4. Remove unincorporated biotin from the sample using, for example, column chromatography.[b]

5. Determine the efficiency of the coupling reaction using avidin–HABA reagent as given in the Pierce booklet or by comparing in Dot blots with a known standard. Expect the reaction to be 15–30% efficient.

[a] The NHS reagent should dissolve readily in water. If not, then it may have absorbed water due to inappropriate storage and will couple at lower efficiency. Best results have been obtained if it is < 6 months-old.

[b] The sample can be used without removal of unincorporated biotin if the coupling is performed at low density, since a vast excess of streptavidin is used in capturing phage. Note that the HABA reagent cannot then be used to determine coupling efficiency.

Protocol 7. Capture of phage on soluble biotinylated antigen

Equipment and reagents

- Streptavidin magnetic beads (Dynal, cat. no. M280) and magnetic rack
- PBS/2% skimmed milk powder (MPBS)
- 50 mM DTT
- PBS/2% skimmed milk powder/0.5% Tween-20
- Log-phase TG1 cells
- 1.5 ml microcentrifuge tubes

Method

1. Mix phage and biotinylated antigen,[a,b] in 500 µl PBS/2% skimmed milk powder/0.5% Tween-20 in a 1.5 ml microcentrifuge tube and incubate for 1 h at room temperature.[c]

2. Dispense 250 μl of streptavidin magnetic beads and collect on a magnet. Aspirate the supernatant and resuspend the beads in 250 μl of PBS/2% skimmed milk powder (MPBS) and rotate end over end to block for 1 hour at room temperature.
3. Add the streptavidin magnetic beads to the phage and incubate at room temperature for 5–15 minutes.
4. Place tubes in the magnetic rack for 1–2 min to allow the beads to migrate. Aspirate the supernatant and resuspend the beads in 1 ml of MPBS.
5. Recover the beads and repeat wash steps. Use four × 1 ml MPBS washes followed by two PBS washes. Transfer the beads to a fresh tube after the third wash to avoid recovering trapped phage.
6. After the last PBS wash, resuspend an aliquot of the beads in 100 μl of 50 mM DTT, mix and leave for 5 min at room temperature to elute the phage (the biotin reagent contains a disulfide bond separating the biotin moiety from the antigen). Store the remainder of the beads in 100 μl of PBS at 4°C, in case you need to return to them.
7. Add DTT containing eluted phage to 10 ml log-phase TG1 cells (OD_{600} around 1.0), and treat as in *Protocol 5*, steps 10 and 11.

[a] As a starting point, use 10^{10}–10^{11} phage and 100 nM antigen. Reduce phage input and antigen concentration in later rounds of selection.
[b] To select against an undesirable cross-reactivity, include unlabelled antigen at ≥ 10-fold molar excess.
[c] Temperature and incubation times and conditions can be varied as appropriate to the end-use.

3.3 Off-rate selection

One parameter that often distinguishes primary from affinity-matured antibodies is the rate of dissociation from the antigen, the 'off-rate', which is often slower for high-affinity antibodies. Off-rate selections can be performed in a modification of the above method, and is described in detail in ref. 7. Phage are first equilibrated with biotinylated antigen then a molar excess of unlabelled antigen added, such that phage dissociating from the biotinylated antigen will most likely rebind to a non-biotinylated moiety and are therefore lost. Slower off-rate antibodies will remain bound to biotinylated antigen for longer and are isolated when phage are captured at later time-points. The point of maximum discrimination has to be determined experimentally, since background binding eventually reduces discrimination. Unlike the previous protocol, the actual concentrations of antigen used are not the major issue; as a guide, use biotinylated antigen at a concentration 2–3 times the estimated K_d (20–30 nM for an antibody with an estimated dissociation constant of 10 nM). Antigen should be in molar excess over phage.

> **Protocol 8.** Off-rate selection
>
> *Equipment and reagents*
> As for *Protocol 7*
>
> *Method*
> 1. Mix phage and biotinylated antigen and allow to equilibrate.
> 2. Add at least a 10-fold molar excess of unlabelled antigen.
> 3. Capture aliquots on streptavidin dynabeads at various time-points, from 5 minutes to 1 hour.
> 4. Reinfect *E. coli* as described in the previous protocol, and sample clones from different time-points in an affinity screen.

4. Screening selected populations of antibodies

Specificity and affinity should now be checked as a matter of routine. It is vitally important to check that the antibody binds specifically. If possible, grow clones in a 96-well format and test them by ELISA, either as phage or as soluble scFv (Chapter 1, *Protocol 13*). Note that selections can go too far, such that those antibodies with desirable characteristics can disappear from later rounds of selection. Some means of sampling up to hundreds of clones from each round is desirable. Consequently, the best screening procedures use a 96-well format such as ELISA, principally because many clones can be screened in parallel. Even if the end-use is different, a crude 96-well screen can be used to focus on the best candidates which can then be analysed in detail. Though the antigen used in selections needs to be as pure as possible, that used in screening need not be as pure if availability is limited.

Clones can be grown up in 96-well arrays and tested for reactivity and specificity in ELISA as described in Chapter 1, *Protocol 13*. The method below is a rapid screen for affinity based on methods described in Chapter 4, and is for ranking clones on the basis of their equilibrium dissociation constant (K_d).

4.1 Affinity screen for phage antibody clones (K_d assay)

The underlying principle of the procedure is that the higher the affinity of the antibody, the greater the proportion that will be complexed with antigen at equilibrium. In practice, aliquots of each crude antibody are mixed either with no antigen or with soluble antigen at two concentrations. The proportion of antibody remaining free at equilibrium is determined by ELISA. The highest affinity clones will manifest the greatest signal reduction with the lowest concentrations of antigen, regardless of the absolute signal. The assay

2: Affinity maturation of antibodies using phage display

should be resistant to the two most common artefacts that affect affinity measurement, namely expression level and dimerization, since these are selected against. Variations in specific activity do not affect this assay either, since it is the active fraction that is assayed. Factors affecting this screen are discussed in more detail in Chapter 4.

A number of criteria must be met before the assay can be used:

- Temperature must be constant throughout as this affects the K_d.
- Less than 10% of the available antibody must bind to the coated antigen to prevent readjustment of the equilibrium.
- The absorbance read in the ELISA must be proportional to the antibody concentration.
- The concentration of antibody employed in the assay should be less than or equal to the K_d of the antibody tested. Note that 1.0 µg/ml of an scFv is 40 nM. Yields of scFv in crude culture supernatant range from 2 nM (50 µg/litre, e.g. for a non-induced culture) upwards.

We find that it is best if you perform a standard ELISA on the clones to be tested first, not only in order to work out which are the positives, but also to categorize them into high, medium, and low signal (usually = high, medium, and low expression level). Chapter 1, *Protocols 9, 11*, and *13* describe growing clones in 96-well arrays and performing ELISAs; note that peroxidase described in *Protocol 13* is an unsuitable detection method since colour development is non-linear. Modify the primary ELISAs for development with alkaline phosphatase as described below.

Protocol 9. Determination of per cent binding and linearity of ELISA signal

Equipment and reagents

- Microtitre plate reader
- 2% skimmed milk powder in PBS (MPBS)
- Tween-20
- ELISA plates
- Microtitre tray (for growth of culture)
- 9E10, antibody by Cambridge Research Biochemicals, cell line by ATCC (CRL1729)
- 0.9% NaCl
- Alkaline phosphatase conjugated goat anti-mouse (Pierce, cat. no. 31322X)
- PBS
- *p*-Nitrophenyl phosphate tablet (Sigma, cat. no. N-2765)
- Substrate buffer: 0.1 M glycine, 1 mM MgCl$_2$, 1 mM ZnCl$_2$, pH 10.4

Method

1. Pick one or more 'representative' clone(s) and grow and induce in a microtitre tray as for a standard ELISA.
2. Coat two ELISA plates with your test antigen using your normal coating volume and conditions, and block both for at least an hour with 2% skimmed milk powder in PBS (MPBS).

Protocol 9. *Continued*

3. Harvest the culture supernatants and add Tween-20 to 0.1% final concentration. Make dilutions from the neat supernatant ranging from 1 µl to 90 µl of supernatant per 100 µl sample in the appropriate concentration of MPBS (for plates coated with 100 µl of antigen).

4. Add your test samples, in triplicate, to one of the two coated plates and allow the antibody to bind (e.g. 37 °C for 1 h).

5. After the binding step transfer the samples to the second coated plate and allow the samples to bind as before. Meanwhile wash the first plate: two quick rinses in PBS/0.1% Tween-20 followed by three × 2 minute washes in PBS/Tween-20 followed by three quick rinses in PBS with no detergent. After the last wash, tap out all the fluid and store the plate inverted on damp tissue paper.

6. Wash the second plate as above, then process both plates simultaneously thereafter.

7. Incubate the samples with 10 µg/ml 9E10 in MPBS for 1 h at room temperature and wash as in step 5.

8. Dilute alkaline phosphatase conjugated goat anti-mouse 1/5000 in MPBS and incubate for 1 h room temperature.

9. Wash (as in step 5), but use saline instead (0.9% NaCl) for the final rinse to remove the phosphate buffer—the detection step is pH dependent.

10. Dissolve one × 20 mg tablet of *p*-nitrophenyl phosphate in 20 ml of substrate buffer and add 100 µl per well.

11. Read absorbance at 405 nm. Don't worry too much about which time-point to use; the reaction remains linear whichever time you use so long as the absorbance of your most concentrated sample does not exceed 2.0.

12. Plot the antibody concentration (i.e. supernatant volume) versus *OD*, at your preferred time-point. The graph should be linear over at least part of the range. We find that the [*Ab*] versus *OD* is linear in the range 2–20 µl of culture supernatant. Less than 2 µl generally results in > 10% of the antibody binding.

13. Calculate how much of the antibody in each sample has bound by comparing the *OD*s of the same samples on the first and second plates; you can plot readings with the two plates on the same graph and determine the 10% cut-off by the difference in slope. However, the cut-off is usually obvious from simple comparison of the *OD*s.

2: Affinity maturation of antibodies using phage display

Protocol 10. K_d assay

Equipment and reagents
- ELISA plates (Falcon)
- Polypropylene microtitre trays (Greiner)
- 1% Tween-20 in PBS
- MPBS (*Protocol 5*)
- Plate reader

Method

1. Coat ELISA plates with antigen; you will need three times the number of ELISA plates as you have microtitre tray arrays.

2. Harvest the supernatants and place in polypropylene microtitre trays containing 1/10th volume of 1% Tween-20 in PBS. Mix the supernatant with the Tween buffer and keep the plates on ice while you are setting up the assay.

3. Make dilutions of the antigen in MPBS. The concentrations of antigen to use depends on the expected/target K_ds. We typically use 1 nM and 10 nM concentrations.

4. Aliquot the antigen solutions into polypropylene microtitre trays: we usually put MPBS in the first four columns (1–4), the lowest antigen concentration in the next four columns (5–8), and the highest concentration in the last four (9–12). Each full tray of clones therefore translates into three polypropylene assay plates and three ELISA plates. We find that the assay works best if the polypropylene microtitre trays are rinsed in PBS and blocked in MPBS for an hour or so at room temperature prior to use.

5. Transfer aliquots of the test samples into the antigen solutions and mix by pipetting up and down a few times.

6. Allow the binding to come to equilibrium, e.g. 37°C for 1 hour.

7. Transfer the equilibrated samples to coated ELISA plates and allow uncomplexed antibody to bind, e.g. 37°C for 1 hour.

8. Develop as for a standard ELISA and read *OD* at 405 nM.

9. Calculate the per cent reduction in signal for each of the clones. The best clones will show the greatest signal reduction at the lowest concentration of soluble antigen.

---

5. Sequence and fingerprint analysis

An analysis of the genes in the selected clones can often prove revealing. *BstN*I fingerprinting before and after selection gives a 'quick and dirty' snapshot of whether there is in fact selection for particular clone types and how many different types of clones are being selected (Chapter 8, *Protocol 6*). Sequence analysis is definitive however, and is particularly revealing when sequence alterations are correlated with altered properties. All changes mapping to CDR or Vernier residues (*Figure 1*) should be considered significant until proven otherwise, and in our experience, there is no such beast as a 'conservative substitution'. Nevertheless, some sequence alterations may have been selected by virtue of their being 'null', i.e. neither beneficial nor detrimental. In the ideal case, a particular clone type will score highest in the K_d assay and will have been observed to comprise an increasingly greater proportion of the population as the selections have progressed. Refer to Chapter 6 to put the changes observed in your antibodies into context of those observed *in vivo*.

The acid test is whether the antibody has been improved. Improvements in affinity can be measured as described in Chapter 5, with changes in kinetic constants most readily observed by surface plasmon resonance. Changes in specificity are more readily tested, though beware that apparent loss of an undesirable cross-reactivity may be the result of a drop in affinity, such that cross-reaction is now below the level of detection.

Finally, if the antibody has not been improved check that there were sufficient mutants there to begin with; characterize the selected clones immunologically and by sequence analysis; and try changing the selection conditions. In our experience some antibodies are very difficult to improve further, though these are a minority. Improved antibodies have to be there before they can be selected, but if the selection conditions are unfavourable, they may never get selected.

References

1. Padlan, E. A. (1994). *Mol. Immunol.*, **31**, 169–217.
2. Hawkins, R. E., Russell, S. J., Baier, M., and Winter, G. (1993). *J. Mol. Biol.*, **234**, 958–64.
3. Foote, J. and Winter, G. (1992). *J. Mol. Biol.*, **224**, 487–99.
4. Clackson, T., Hoogenboom, H. R., Griffiths, A. D., and Winter, G. (1991). *Nature*, **352**, 624–8.
5. Marks, J. D., Griffiths, A. D., Malmqvist, M., Clackson, T., Bye, J. M., and Winter, G. (1992). *BioTechnology*, **10**, 779–83.
6. Kang, A. S., Jones, T. M., and Burton D. R. (1991). *Proc. Natl Acad. Sci. USA*, **88**, 11120–3.
7. Hawkins, R. E., Russell, S. J., and Winter, G. (1992). *J. Mol. Biol.*, **226**, 889–96.

3

Human antibody repertoires in transgenic mice: manipulation and transfer of YACs

NICHOLAS P. DAVIES, ANDREI V. POPOV, XIANGANG ZOU,
and MARIANNE BRÜGGEMANN

1. Introduction

Therapeutic applications of an accessible human antibody repertoire are virtually limitless. Considerable interest in the techniques for producing human antibodies are apparent from clinical results which highlight their benefits (1). The problems that occur with immunogenicity, when antibodies of a foreign species are being used for therapy, unambiguously point out that the affinity and authenticity of the antibody are critical (1–3 and references therein).

Besides the generation of *in vitro* antibody responses using human lymphocytes (4), two other approaches have been established to produce antigen-specific human antibodies. The phage-display system focuses on *in vitro* selection of antigen-binding specificities (5) and the transgenic approach relies on the transfer of immunoglobulin minigene loci into the mouse germline (6, 7). The attractions of the transgenic approach are that germline configured loci are introduced which can be rearranged, switched, and mutated by the animal in an authentic fashion which, after immunization, gives rise to specific antibodies (8). One of the hurdles associated with the transgenic strategy is that the production of high-affinity antibody repertoires may rely on the introduction of relatively large gene segments or even complete loci.

For the transfer of large regions into the mouse germline several methodologies have been established; DNA-injection into oocytes (9), lipid-mediated DNA-transfection of embryonic stem (ES) cells (10, 11), and fusion of yeast spheroplasts to ES cells (12–14). An obvious advantage of the fusion technique is that DNA purifications are unnecessary. Further practical advantages of using yeast are that it is suitable for cloning of megabase (Mb) fragments in yeast artificial chromosomes (YACs)—in fact many libraries are

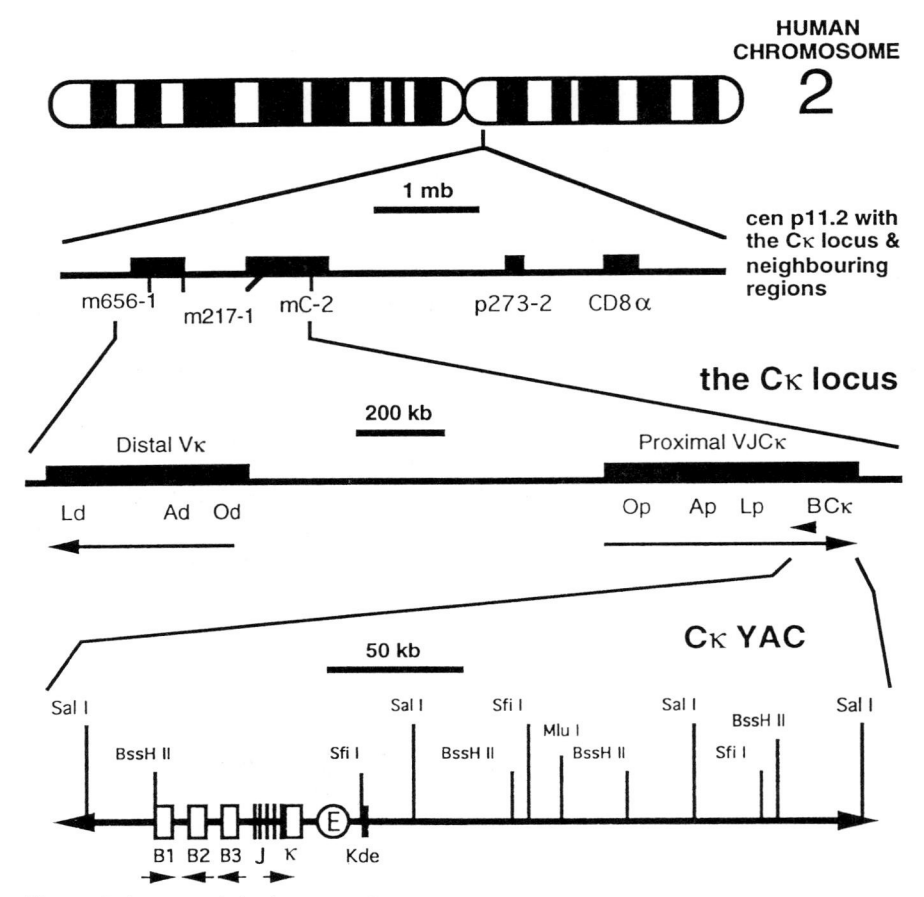

Figure 1. Layout of the human κ locus on chromosome 2p11.2 (50) showing linkage to the T-cell differentiation antigen CD8α (51). The C_κ locus is based on ref. 52. The map of the C_κ YAC is derived from ref. 13. The V_κ gene segments (B1, B2, B3), C_κ, J_κ, the 3' enhancer (E), and the K deleting element (Kde) are not drawn to scale. Arrows represent the 5'–3' transcriptional orientation.

available for screening (15–19)—and that yeast DNA can be easily manipulated by site-directed integration (20). The location of a 300 kb YAC containing the core region of the human κ light chain locus relative to its chromosomal position is illustrated in *Figure 1*. Considering that the immunoglobulin loci are in the 2 Mb size range the use of YACs offers an attractive strategy for analysing these gene loci.

In this chapter we describe experiments for the modification of YACs by site-specific integration and the transfer of manipulated YACs into ES cells by protoplast fusion. ES cells altered by such strategies have been used to generate transgenic mice, and we and others have shown that the introduced human immunoglobulin loci rearrange and express antibody repertoires (13, 21–23).

2. Yeast artificial chromosomes

The analysis of the yeast *Saccharomyces cerevisiae* revealed essential sequence elements which are crucial for yeast chromosome function (24); these are telomeres (TEL), centromeres (CEN), and autonomous replicating sequences (ARS). These sequences, in combination with a selectable marker gene (for example, *URA3*, to allow growth in uracil-deficient medium) have been combined in a plasmid cloning vector that replicates in *Escherichia coli* (15). As illustrated in *Figure 2*, a pYAC vector allows the cloning of size-fractionated DNA, at present, up to 2 Mb. More realistically, yeast artificial chromosome (YAC) libraries accommodate fragments from 200 to 1000 kb and several approaches to their construction have been previously described (15–18). The large size of the YACs and the apparent stability in the accommodating yeast strain AB1380 allowed the cloning, characterization, and

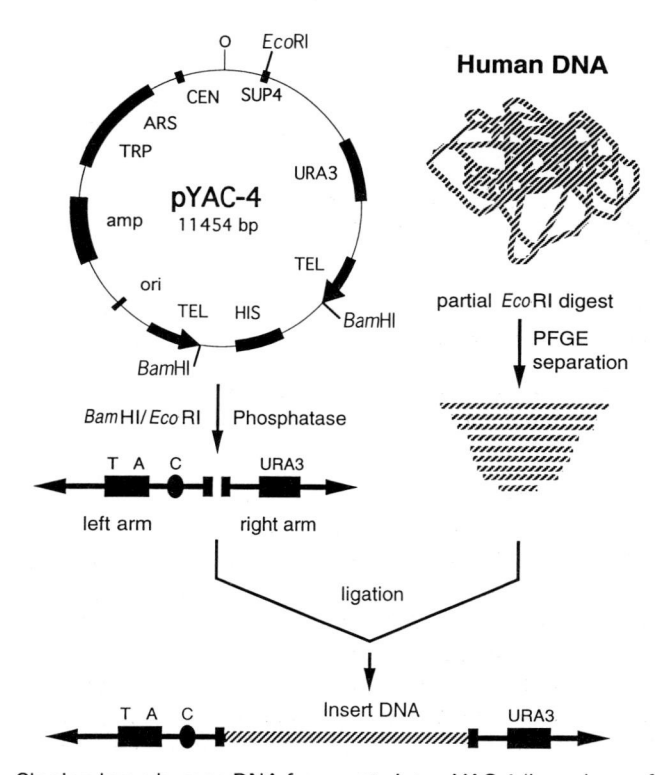

Figure 2. Cloning large human DNA fragments into pYAC-4 (based on refs 15, 53).

manipulation of large regions such as the immunoglobulin heavy and light chain loci. Essential methods for the use of YACs are described in detail in *YAC libraries: a user guide* (25).

2.1 Library screening

Various organized YAC libraries are available (26) enabling participating scientists to analyse and share biological materials effectively. We have used two different approaches for library screening; PCR and filter-hybridization. A set of specific primers is used in PCR analysis on pools of YAC clones leading towards the enrichment of the required clone(s). However, if one wishes to isolate different YACs simultaneously the filter-hybridization strategy using several oligo-labelled probes has an advantage. Access to these libraries is often generously provided with screening strategies and help readily available.

2.2 Maintenance

Yeast media and conditions are described in ref. 27, and *Protocol 1*. AB1380 is the commonly used host strain for YAC libraries and is grown under sterile conditions at 30°C on YPD plates (1% Bacto yeast extract, 2% Bacto peptone, 2% glucose, 2% Bacto agar) or in YPD liquid medium (without agar) with vigorous shaking (350 r.p.m.) and aeration. AB1380 carrying a YAC is able to grow on synthetic dextrose 'drop-out' media (SD, *Protocol 1*) lacking uracil (URA) and tryptophan (TRP). A YAC modified to carry the lysine (LYS2) gene can thus grow in SD: URA⁻, TRP⁻, LYS⁻ medium. This selection on SD plates ensures the maintenance of the modified YAC.

Protocol 1. Drop-out media for AB1380

Equipment and reagents

- Autoclave
- pH paper
- Sterile filtration units or disposable filters
- Autoclaved H$_2$O (BDH, cat. no. 44384)
- Agar (Difco, cat. no. 0145-17-0)
- Yeast nitrogen base without amino acids (Difco, cat. no. 0919-15-3)
- D-Glucose (Sigma, cat. no. G 5400)
- L-Tryptophan (BDH, cat. no. 37153); stock solution 10 mg/ml, working concentration 20 µg/ml (omit for TRP⁻ selection)
- L-Histidine monohydrochloride (BDH, cat. no. 37118); stock solution 10 mg/ml, working concentration 20 µg/ml

- L-Lysine monohydrochloride (BDH, cat. no. 37129); stock solution 10 mg/ml, working concentration 33 µg/ml (omit for LYS⁻ selection)
- L-Isoleucine (Fluka, cat. no. 58880); stock solution 10 mg/ml, working concentration 33 µg/ml[a]
- L-Threonine (BDH, cat. no. 37150); stock solution 40 mg/ml, working concentration 200 µg/ml[a]
- Uracil (BDH, cat. no. 44103); stock solution 1 mg/ml, working concentration 20 µg/ml (omit for URA⁻ selection)
- Adenine sulfate (Fluka, cat. no. 01880); stock solution 2 mg/ml, working concentration 50 µg/ml

3: Manipulation and transfer of YACs

Method

1. Mix 20 g agar (= 2%, for plates only) and 6.7 g yeast nitrogen base.
2. Autoclave in about 900 ml double-distilled H_2O.
3. After autoclaving add 20 g glucose (= 2%) sterile filtered (e.g. 50 ml of a 40% stock).
4. Add required additions (amino acids/purine/pyrimidine) from stock solutions.
5. Bring up with sterile double-distilled H_2O to the final volume of 1000 ml. The pH should be 5.8.

[a] It has been found that AB1380 requires isoleucine and threonine for growth (54) in addition to its published auxotrophies (15).

YAC carrying yeast cell lines can be maintained for a few months streaked out on drop-out media plates kept at 4°C. For long-term storage yeast cell lines can be retained in a viable state for many years by freezing, and this is the preferred method of storage as YACs can delete if successively plated on to YPD. After overnight growth (which is usually sufficient although some yeast colonies may need to be grown for several days) in 25 ml of appropriate media the cells are spun, resuspended in 850 μl YPD, and transferred into cryotubes. Fresh glycerol is added to 15%, the contents mixed, and the tubes immediately stored at –80°C. Aliquots can be taken by scratching off a small amount of frozen cells with a sterile pipette, transferring into YPD for overnight growth, and finally inoculating into the appropriate SD media to confirm the presence of the YAC.

3. Modification of YACs

Transformation of yeast with corresponding DNA sequences leads to homologous integration which allows virtually any prespecified modification. Homologous integration in yeast has been well characterized and extensively used for defined alterations (20, 28). YACs can be modified by site-specific integrations of homology sequences corresponding to the YAC left or right arm or to the cloned region, which for human DNA may involve *Alu* repeat-sequences (12, 29). Targeted integration in yeast is either performed using replacement constructs or integration vectors (20). Replacement vectors disrupt the region of homology by the insertion of exogenous DNA, whilst integration vectors recombine as a whole and duplicate a given target sequence without impairing its function. In *Figure 3* homologous integration by these two principles is illustrated. Integration vectors have several advantages; they can be rescued from the YAC (by digestion, ligation, and *E. coli* transformation) and they allow single as well as multiple integration. In order to achieve

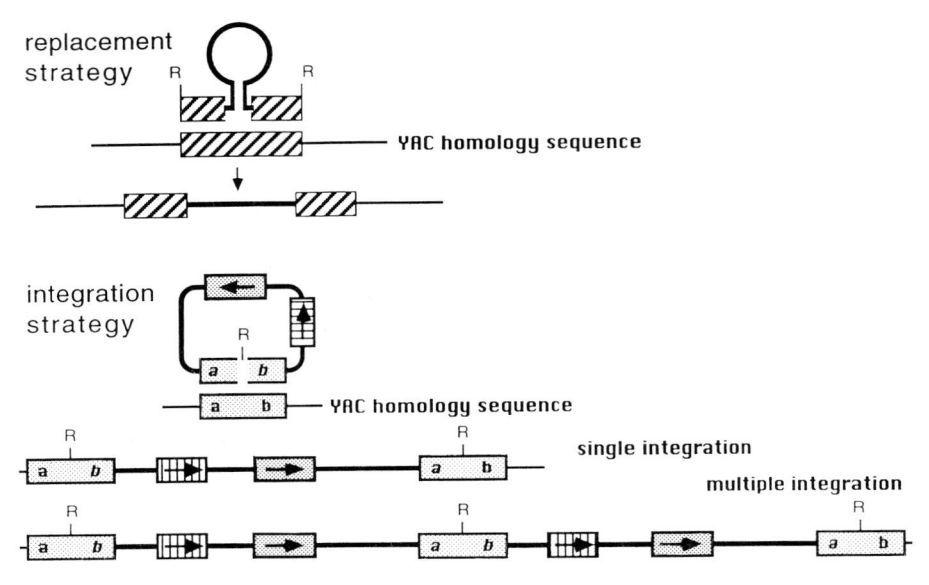

Figure 3. Site-specific integration in yeast with the use of restriction sites (R) in order to create recombinogenic homology ends.

homologous integration at the defined site it is important to linearize the vector to allow homologous ends to recombine. If both the 3' and 5' homology ends are of sufficient length this ensures a very high efficiency of site-directed and precise integration.

3.1 Universal YAC vectors

The use of a variety of insertion and deletion vectors to modify YACs, commonly described as 'retrofitting', has been reviewed recently (30). For dominant selection the neomycin resistance gene has been used successfully for many mammalian cell lines. The advantage of using the *neo* marker gene is that it does not restrict the use of available cell lines as is the case with the hypoxanthine phosphoribosyl-transferase (*HPRT*) gene which only allows selection in a deficient line (14, 31). A set of vector constructs developed in our laboratory was designed to facilitate alterations in any YAC; the left arm, right arm, or the cloned insert assuming it is of human origin (12). In addition, as illustrated in *Figure 3*, these integration vectors allow site-specific single or tandem joining when linearized in the chosen homology region. The restriction map of the four vectors, pNU, pUNA, pLNA, and pLUNA is shown in *Figure 4*. The vectors are constructed in such a way as to allow selection in yeast applying uracil (*URA*) and/or lysine (*LYS*) deficient medium, the neomycin selectable marker gene from the plasmid pMC1NeoPolyA (32) allows selection in G418 containing media of trans-

3: Manipulation and transfer of YACs

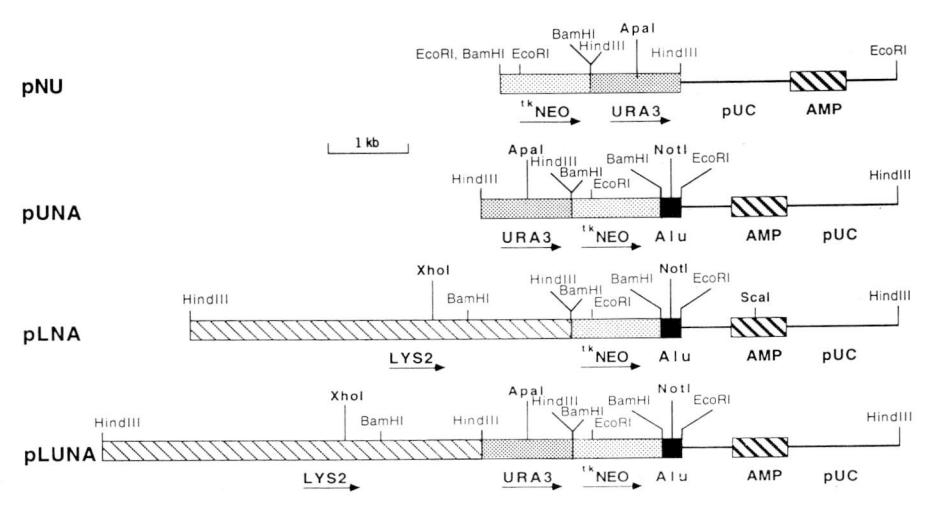

Figure 4. Cloning vectors, pNU, pUNA, pLNA, and pLUNA, used for YAC targeting. Restriction sides indicated in bold represent sites used for linearization.

formed mammalian cells after YAC uptake. Interestingly, in transfection experiments the number of G418 resistant colonies was considerably higher when using this *neo* gene driven by the herpes simplex virus thymidine kinase (*HSVTK*) promoter. Addition of the *Alu* repeat element allows integration into human DNA (12, 29), but because of the rather small region of homology there is a bias (23) yielding only a few per cent *Alu*-directed integration events after linearization with *Not*I.

To give an example; for the integration of pLUNA into the right YAC arm a plasmid miniprep is prepared (33) and 1–10 μg DNA digested with *Apa*I, a unique restriction site located in the *URA3* gene. The linearized DNA is then used for transformation.

3.2 Site-directed introduction

Using the vectors described above two essential YAC modifications have been accomplished; tandem-integration to select for YAC transfer into ES cells, and the addition of gene clusters to increase the locus (34). The latter is particularly interesting if one wishes to build up the immunoglobulin, or indeed any other locus, and the addition of V genes may help to increase the antibody repertoire. Multiple copies of the *neo* gene in the YAC appear to be essential for selection of ES cells with integrated YAC. This is based on observations (13, 35) rather than an explanation of how the expression of the *neo* gene may be influenced by yeast sequences or possible position effect. DNA integration using spheroplast transformation is described in *Protocol 2*.

Protocol 2. Yeast spheroplast transformation

Equipment and reagents

- 48°C waterbath
- 30°C shaking incubator
- Spectrophotometer
- Centrifuge (swing-out rotor and, ideally, adjustable centrifugal force)
- DMSO (Sigma, cat. no. D-2650)
- PEG 8000 (Sigma, cat. no. P-5413)
- Sorbitol (Flulka, cat. no. 85530)
- EDTA (Sigma, cat. no. ED-2SS)
- β-Mercaptoethanol (Sigma, cat. no. M-7522)
- Yeast extract (Oxoid, cat. no. L21)
- Bacto-peptone (Difco, cat. no. 0118-01-8)
- Citric acid (Sigma, cat. no. C-8532)
- 1 M Tris–HCl, pH 7.4
- Tris (Boehringer Mannheim, cat. no. 708968)
- $CaCl_2$ (Sigma, cat. no. C-3881)
- Liticase (Sigma, cat. no. L-5263)
- Zymolyase 20T (ICN, cat. no. 32-0921)
- SCE: 1 M sorbitol, 0.1 M sodium citrate, pH 5.8, 10 mM EDTA; sterile filtered
- STC: 1 M sorbitol, 10 mM Tris–HCl, pH 7.5, 10 mM EDTA; sterile filtered
- SOS: 1 M sorbitol, 0.25% yeast extract, 0.5% Bacto-peptone, 6.5 mM $CaCl_2$, 10 μg/ml uracil (*Protocol 1*); sterile filtered
- 10 ml wide-bore pipettes
- Falcon tubes, 15 ml

Method

1. Inoculate 100 ml of YPD or appropriate drop-out media with 5 ml yeast culture derived from a single colony and grow at 30°C under vigorous shaking.

2. Check the OD_{660}, when a 1 in 10 dilution is 0.25–0.45, pellet the cells at 600 *g* for 5 min at room temperature.

3. Decant the medium and wash the cells in double-distilled H_2O, then in 1 M sorbitol; each time centrifuge as before.

4. Decant and resuspend the cells in 15 ml SCE. The cell count should be ~ 1×10^9. Add fresh culture grade β-mercaptoethanol (2-ME) to 14 mM.

5. Add either 50 μl liticase (stored frozen at –20°C in aliquots at 20 mg/ml in double-distilled H_2O) or 80 μl zymolyase 20T (10 mg/ml in double-distilled H_2O, stored frozen). Mix gently and incubate at 30°C. Optimal amounts of lytic enzyme should produce 90% spheroplasts in 20 min.[a]

6. When approximately 80% of the cells are spheroplasted, immediately centrifuge the cells at 240 *g* for 10 min at 20°C.[b]

7. Remove the supernatant and carefully resuspend the cells in 15 ml STC.[c,d] Centrifuge once more and resuspend the cells as above. Spin again and finally resuspend them in 2 ml STC. At this stage DMSO may be added to 7% and cells frozen at –70°C. For transformation, frozen aliquots (≤ 1.5 ml) are defrosted rapidly, centrifuged at 200 *g* as above and resuspended in an equal volume of STC.

8. For the transformation, 1–10 μg linearized DNA (e.g. 1 μg pLUNA and for co-transformation 10 μg cosmid) and 100 μl spheroplasts are

3: Manipulation and transfer of YACs

gently mixed in a 15 ml Falcon polyethylene tube. DNA and spheroplasts should stand at room temperature for 10 min.

9. Add 1 ml of filter-sterilized 20% PEG solution in 10 mM Tris–HCl, pH 7.4, 10 mM CaCl$_2$. Mix gently, let sit for 10 min at room temperature, and centrifuge as above.

10. Carefully pipette off the PEG solution and gently resuspend the pellet in 150 μl SOS. Place at 30 °C for 40 min.

11. Add about 8 ml of recovery agar[e] kept at 48 °C, mix gently, and pour on to prewarmed selection plates. Incubate for 3–4 days at 30 °C.

[a] The extent of spheroplasting should be tested after 5, 10, 15, and 20 min. Upon adding equal amounts of 5% SDS spheroplasts burst and look like 'ghosts' under the phase-contrast microscope. A precise way to monitor spheroplasting is to check for the drop in OD at 800 nm as follows: during spheroplasting add 100 μl cells to 900 μl of water every 5 min and monitor the drop in OD_{800}. When this value is about 20% of the starting OD the spheroplasts are ready for transformation. The actual amount of zymolyase or liticase required should be titrated for every batch of enzyme.

[b] Incomplete recovery is common, but it is crucial not to centrifuge faster as the spheroplasts will burst. Instead, another spin at the same speed will usually suffice.

[c] 3×10^8 cells/ml are the optimal concentration and are stable in STC at room temperature for at least 1 h.

[d] Use of a 10 ml pipette is most useful as the wide bore prevents shearing forces on the protoplasts which could potentially burst them.

[e] Recovery agar contains 1 M sorbitol and 2.5% agar, otherwise it is prepared as described in Protocol 1.

3.3 Mapping site-specific integration

In order to analyse the integration of the exogenous DNA carrying the neomycin resistance marker, transformants are picked and grown for preparation of chromosomal DNA in agarose plugs. Fractionation by pulsed-field gel electrophoresis (PFGE), Southern blotting, and hybridization with the *neo* probe initially identifies transformed YACs.

For restriction enzyme digests the agarose plug, or a slice of it, is incubated for about 30 min with 500 μl restriction enzyme buffer. After equilibration, buffer excess is removed to leave the slice just covered and 50–150 units of the appropriate restriction enzyme are added along with 5 μl BSA (10 mg/ml) and 1 μl DTT (100 mM). The tubes are incubated for a further 30 min on ice with occasional gentle tapping to mix and to allow the enzyme to diffuse into the plug. Digests are carried out at the appropriate temperature for 1–5 h. It is important to keep the gel slice covered. This is done either by periodic centrifugation or by covering the digest with mineral oil (Sigma, cat. no. M-5904).

3.4 Profile of single and multiple integrations

Upon *Eco*RI digest and *neo* hybridization specific integration fragments can be seen in DNA from the modified yeast (12). Transformation using pLNA

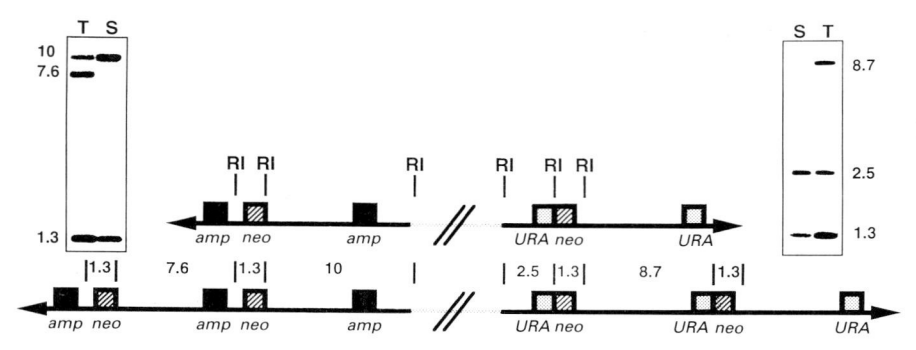

Figure 5. Southern blot profiles and integration map for YAC left and right arm modifications using universal integration vectors (12, 13). *Left hand side*: The Igκ YAC containing yeast DNA was targeted with *Sca*I linearized pLNA in the left (centromeric) arm. The hybridization profiles obtained after *Eco*RI digest (RI) and hybridization with the *neo* probe show tandem or multiple integration (T) and single integration (S). The map illustrates that single integration by homologous recombination creates upon *Eco*RI digest a 10 kb flanking region and a 1.3 kb internal *neo* band. Tandem or multiple integration, in a head to tail fashion is characterized by a 7.6 kb internal fragment, which together with the 1.3 kb band indicates integration number by signal strength. The 10 kb *neo* band does not increase in signal with multiple integration. *Right hand side*: Similarly, after transformation with *Apa*I linearized pLUNA, the Igκ YAC was digested with *Eco*RI and hybridized with the *neo* probe. Single integration by homologous recombination creates a 1.3 kb *Eco*RI internal and a 2.5 kb *Eco*RI flanking *neo* band. Tandem integration is characterized by an 8.7 kb band, which together with the 1.3 kb band indicates integration number by signal strength. The stippled lines represent the human DNA insert border. Sizes are in kb. Not drawn to scale.

digested with *Sca*I promotes site-specific integration into the *amp* gene in the left YAC arm. Following digestion with *Eco*RI this results in a 1.3 kb and 10 kb *neo* hybridizing band for single integration and a 1.3 kb, 10 kb, and 7.6 kb band characteristic for tandem integration. Similarly, using *Apa*I digested pLUNA for right YAC arm integration into the *URA3* gene results in *Eco*RI bands of 1.3 kb and 2.5 kb for single integration and a 1.3 kb, 2.5 kb, and 8.7 kb band for multiple integration. Integration analysis is shown in *Figure 5*. Such site-directed integration is universally applicable to any YACs cloned using the pYAC series and would lead to identical hybridizing fragments with the exception of the flanking fragment which is determined by the vector cloning site.

Protocol 3. Yeast chromosome mini-preparation in agarose plugs

Equipment and reagents
- Waterbath
- 30°C incubator
- 50 mM EDTA
- β-mercaptoethanol
- SCE (see *Protocol 2*)
- SCEM: SCE + 14 mM β-mercaptoethanol

3: Manipulation and transfer of YACs

- Zymolyase 20T: 10 mg/ml in double-distilled H$_2$O, see *Protocol 2*
- Low melting temperature agarose (FMC, cat. no. 50102)
- Yeast lysis solution (YLS = 1% lithium dodecyl sulfate): 100 mM EDTA, 10 mM Tris–HCl, pH 8.0,
- TE: 10 mM Tris–HCl, pH 8.0, 1mM EDTA

Method

1. Grow the yeast strain from a single colony for about 24 h in approximately 5 ml of the appropriate medium.
2. Pellet stationary phase yeast cells by centrifuging at 600 *g* for 5 min. Resuspend the pellet in 20 ml 50 mM EDTA and centrifuge again.
3. Decant the supernatant completely and resuspend the cells in 200 µl SCEM. Add 50 µl zymolyase 20T (10 mg/ml).
4. Mix with 250 µl low-melt agarose at 45°C and quickly pour into prechilled plug moulds.
5. Allow to set at 4°C for 15 min and transfer into 2 ml of SCEM. Incubate for 2 h at 37°C.
6. Remove SCEM and add 5 ml yeast lysis solution. Change solution after 2 h and incubate overnight at 37°C.
7. Wash extensively (three times at 50°C with shaking) using TE to remove all traces of YLS.

Protocol 4. Preparing and running pulsed-field gel electrophoresis

Equipment and reagents
- Pulsed-field gel electrophoresis apparatus (e.g. Pulsaphor and Gene mapper, LKB-Pharmacia)
- 0.5 × TBE: 1 × TBE is 89 mM Tris-base, 89 mM boric acid, 2 mM EDTA
- X-OMAT S Film (Kodak, cat. no. 502 3270)
- Agarose (Boehringer Mannheim, cat. no. 1388 983)
- High molecular weight markers: yeast chromosome miniprep and λ ladder (New England Biolabs, cat. nos 345 and 340)[a]

Method

1. Prepare 170 ml of 1–1.5% agarose in 0.5 × TBE. Allow to cool down to about 60°C.
2. Pour 160 ml into the gel trough and insert a standard comb for 100 µl agarose plugs.
3. Allow to set at room temperature.
4. Remove the comb and transfer the plugs into the wells using a flamed thin spatula or similar implement (a screwdriver is actually ideal). Care should be taken to avoid air bubbles trapped in the wells. A useful tip for slicing agarose plugs is to transfer them into disposable

Protocol 4. *Continued*

weighing boats whilst cutting them with a sterile scalpel blade. Another alternative is to use a disposable Petri dish placed on top of a piece of blackened X-ray film. This helps to visualize the plugs which can be almost transparent.

5. Melt the rest of the agarose and pour a few drops on top of the plugs to seal them into the slots.
6. Place the gel into the electrophoresis tank filled with precooled (see below) 0.5 × TBE and insert the electrodes.
7. Electrophoresis should be performed at 4–8 °C. The voltage, (ramped) switch times, and running time are chosen according to the manufacturer's recommendation. This is dependent on the particular size resolution required (e.g. in order to size fractionate 200–1200 kb on CHEF gel electrophoresis (Pulsaphor and Gene mapper, LKB-Pharmacia) the setting could be 200 V, 70 sec switch time for 15 h, followed by 120 sec switch time for 12 h).

[a] Yeast chromosome minipreps and commercially available λ ladder provide useful molecular weight markers.

4. YAC transfer into embryonic stem cells

Two immediate advantages come to mind with the use of YACs:
- the study of large gene families in their authentic genomic context;
- the use of the yeast system for site-directed manipulations.

The transfer of YACs into mammalian cells and animals via conventional transfection strategies is made difficult because of their large molecular size. However, the introduction of a YAC into ES cells is possible by protoplast fusion. This method facilitates the integration of one complete copy, generally without the transfer of other yeast DNA and without purification or handling of the large DNA. Other methodologies (reviewed in ref. 30 and references therein) have been used for the transfer of YACs, but problems may be encountered when transferring complete DNA molecules larger than 500 kb which we have not observed with spheroplast fusions (23). The amount of yeast DNA that can be incorporated during the fusion process is variable but this does not seem to influence the generation of germline transmitting chimeric mice (14, 22), however, many clones isolated in our laboratory do not appear to have yeast DNA integrated (13).

4.1 Spheroplast fusion

Yeast cells have a relatively thick cell wall surrounding their membrane. Before a fusion of yeast and ES cells can take place, this cell wall must be removed in order for transfer of the yeast DNA into the ES cells to occur.

3: Manipulation and transfer of YACs

This procedure, known as protoplasting, involves the use of an enzyme which breaks down components of the yeast cell wall. Intact yeast cells without a cell wall, known as protoplasts or spheroplasts, remain. We have successfully carried out the spheroplast fusion method with a large number of different ES cell lines: D3 (36), E14 (37), CCE (38), AB1 (39), HM-1 (40), and chimeric as well as germline transmission mice have been derived from them. *Protocol 5* is based on the method described in ref 41 and 42, but has been adapted for ES cells (12) and further improved.

Protocol 5. Spheroplast fusion of yeast with ES cells

Equipment and reagents

Unless otherwise stated chemicals are obtained from the suppliers mentioned in *Protocols 1, 2,* and *3*

- 37 °C CO_2 cell-culture incubator
- 30 °C shaking incubator
- ST: 1 M sorbitol, 10 mM Tris–HCl, pH 7.0; sterile filtered
- Drop-out agar plates and liquid media
- Falcon tubes, 15 and 50 ml
- Autoclaved water
- 1 M sorbitol, sterile filtered
- SCE
- β-Mercaptoethanol
- Yeast lytic enzyme (see *Protocol 2*)
- ES-cell culture medium

- 6-well tissue culture plates (ICN, cat. no. 76-058-05)
- G418 (Gibco, cat. no. 066-1811)
- Serum-free DMEM (ICN, cat. no. 12 332 54)
- Trypsin–EDTA (Sigma, cat. no. T-3924)
- PEG 1500 (Boehringer Mannheim, cat. no. 783 641; provided as a 50% solution containing 75 mM Hepes)
- PEG solution: to 1 ml 50% PEG in 75 mM Hepes add 50 μl fresh DMSO, 10 μl 0.5 M $CaCl_2$, and 10 μl 5 mM β-ME

A. Preparation of spheroplasts

1. Streak the YAC containing yeast on the appropriate drop-out plate. Grow at 30 °C until individual colonies are 1–2 mm.

2. Inoculate a single colony into 10 ml drop-out medium and grow at 30 °C with shaking at 350 r.p.m.

3. When the OD_{600} is 1–2, expand into 100 ml and continue growing until OD_{600} is 1–2.

4. Collect cells by centrifuging for 5 min at 600 *g* at room temperature in 50 ml Falcon tubes. All subsequent manipulations are performed at room temperature in the same tubes, unless otherwise stated.

5. Resuspend the cells in 20 ml double-distilled H_2O and centrifuge as before.

6. Resuspend the cells in 20 ml 1 M sorbitol and centrifuge again.

7. Finally resuspend the cells in 20 ml SCE, count the cells, and dilute in SCE to obtain 20 ml cell suspension with 7.5×10^7 cells/ml.

8. Add 46 μl β-ME. Add yeast lytic enzyme: 60–80 μl zymolyase 20T (10 mg/ml) and place tube at 30 °C.[a]

9. Monitor the extent of spheroplast formation as described in *Protocol 2*.

Protocol 5. *Continued*

   When OD_{600} is reduced to 20% as compared with the initial suspension, centrifuge cells for 5 min at 200 g at room temperature.
10. Decant the supernatant and resuspend the cells in 20 ml ST. Centrifuge again as before, resuspend in ST, count the cells, and transfer 2×10^7 cells into a 50 ml tube.
11. Centrifuge as before. Carefully aspirate the supernatant so as not to disturb the pellet.

B. *Fusion with ES cells*
1. By this time ES cells should be ready for the fusion, prepared as follows:
    (a) Remove ES cells from Petri dish by incubating with trypsin–EDTA solution (as described in ref. 43).
    (b) Resuspend the cells in a 15 ml Falcon tube containing 10 ml ES cell culture medium (described in ref. 43).
    (c) Centrifuge at 150 g, wash twice with serum-free DMEM medium and resuspend 2×10^6 cells/ml.[b]
2. Carefully layer 2×10^6 ES cells (in 1 ml) on to the spheroplast pellet and centrifuge at 150 g for 5 min.
3. Aspirate the supernatant. Gently stir the pellet and add 1 ml PEG solution dropwise.
4. Incubate at 37 °C while gently stirring, for no longer than 2 min.
5. Dilute the suspension with prewarmed (37 °C) serum-free DMEM. Dilution should be slow and dropwise (down the side of the tube). Approximately 1 ml in 1 min for the first 3 min and then quicker to a final volume of 20 ml.
6. Carefully invert once and place the tube in a 37 °C incubator for 20–30 min.
7. Centrifuge the cells at 130 g for 5 min.
8. Resuspend the cells in ES cell culture medium and plate them out in a 6-well tissue culture plate.
9. After 24 hours aspirate the supernatant and add medium containing approximately 200 µg/ml G418.[c] Change the medium every day, 5–50 colonies will be visible in 12–14 days.

[a] **Note**: other enzymes, for example liticase, can be used. The amount of enzyme used should take spheroplasting to 90% in 20–30 min (see *Protocol 2*). Treatment time with zymolyase is not very critical as successful fusions have been obtained even after a 40 min incubation, provided spheroplasting is 90% at the end of incubation.
[b] Standard ES cell techniques are described in ref. 43.
[c] The G418 concentration has to be titrated out precisely. We have found that if the concentration is too low (< 190 µg/ml) untransformed ES cells are enriched. If the concentration is too high (> 210 µg/ml) few, if any, clones appear. We strongly suggest that a negative control (using yeast containing no YAC) is always performed in parallel with this experiment.

4.2 Picking and analysing clones

After spheroplast fusion and selection in G418-containing medium, clones are visible after 12–14 days. The transformation frequency can be as high as 2×10^{-4}. Colonies are picked and expanded as described (43) and the selection in G418 is continued for at least a further 5 days. It is important to choose clones that show growth rates and morphology similar to unfused ES cells. This ensures that the selection of clones with integrated yeast DNA is a rare event (13). Initial analysis is carried out by preparing DNA using the salt–chloroform method (44) and Southern blotting. In order to identify complete integration of the YAC into one of the mouse chromosomes high molecular weight DNA digests have to be carried out. The preparation of high molecular weight mammalian DNA is similar to that described in *Protocol 2* but with the following modifications: (a) about 2×10^7 ES cells are harvested by centrifugation at 200 *g*, resuspended in 200 µl PME (PBS, 14 mM β ME, 20 mM EDTA) and kept at 37°C; (b) the suspension is gently mixed with 250 µl molten agarose (2% low-melt agarose at 45°C in PME) and quickly poured into plug moulds; (c) after incubation in YLS, plugs are washed extensively as previously described.

Digests with rare cutting enzymes are essential to verify complete integration of the large YAC into the mouse DNA. In addition, profiles of *Alu* repeats may be obtained and the co-integration of yeast DNA other than the YAC may be ruled out. This can be done by re-hybridization of the filter with total yeast DNA or yeast repetitive elements. A problem when mammalian DNA digests are compared with yeast is their retarded mobility. DNA separation on PFGE is concentration-dependent, thus separation of the yeast DNA is faster (13, 45).

5. Conclusions

Chimeric mice are obtained by blastocyst injection of the manipulated ES cells; a technology that has been described extensively (reviewed in ref. 46). Initial analysis of human antibody production in transgenic mice is carried out by serum analysis in an ELISA assay (6, 13). Rearrangement of the human heavy or light chain locus on YACs is productive and can lead to expression levels for human immunoglobulins in the µg/ml range (13, 22). Analysis of the human antibody repertoires has been assessed by PCR which highlighted junctional diversity and somatic mutation (7, 8, 13, 21–23, 47). Specific antibodies have been obtained after immunization with particularly exciting results in mice carrying transgenic human mini loci, but in which mouse antibody production was silenced (8, 21, 22, 48).

The introduction of gene loci on YACs into cells and animals aims at the production and use of an authentic human antibody repertoire. Other important questions, previously limited by the lack of YAC transfer methodologies, concern the functional analysis of gene loci. However, the use of today's YAC technology implies the use of mammalian artificial chromosomes (MACs) (49) or chromosome fragments for future analysis; this means going back to the specific human chromosome of interest as illustrated in *Figure 1*.

References

1. Waldmann, T. A. (1991). *Science*, **252**, 1656–61.
2. Waldmann, H., Hale, G., Clark, M., Cobbold, S., Benjamin, R., Friend, P., Bindon, C., Dyer, M., Qin, S., Brüggemann, M., and Tighe, H. (1988). *Prog. Allergy*, **45**, 16–30.
3. Brüggemann, M., Winter, G., Waldmann, H., and Neuberger, M. S. (1989). *J. Exp. Med.*, **170**, 2153–7.
4. Abrams, P. G., Rossio, J. L., Stevenson, H. C., and Foon, K. A. (1986). In *Methods in enzymology*, vol. 121, pp. 107–19. Academic Press, London.
5. Marks, J. D., Hoogenboom, H. R., Bonnert, T. P., McCafferty, J., Griffiths, A. D., and Winter, G. (1991). *J. Mol. Biol.*, **222**, 581–97.
6. Brüggemann, M., Caskey, H. M., Teale, C., Waldmann, H., Williams, G. T., Surani, M. A., and Neuberger, M. S. (1989). *Proc. Natl Acad. Sci. USA*, **86**, 6709–13.
7. Brüggemann, M., Spicer, C., Buluwela, L., Rosewell, I., Barton, S., Surani, M. A., and Rabbitts, T. H. (1991). *Eur. J. Immunol.*, **21**, 1323–6.
8. Wagner, S. D., Popov, A. V., Davies, S. L., Xian, J., Neuberger, M. S., and Brüggemann, M. (1994). *Eur. J. Immunol.*, **24**, 2672–81.
9. Schedl, A., Montoliu, L., Kelsey, G., and Schütz, G. (1993). *Nature*, **362**, 258–61.
10. Lamb, B. T., Sisodia, S. S., Lawler, A. M., Slunt, H. H., Kitt, C. A., Kearns, W. G., Pearson, P. L., Price, D. L., and Gearhart, J. D. (1993). *Nature Genet.*, **5**, 22–30.
11. Strauss, W. M., Dausman, J., Beard, C., Johnson, C., Lawrence, J. B., and Jaenisch, R. (1993). *Science*, **259**, 1904–7.
12. Davies, N. P., Rosewell, I. R., and Brüggemann, M. (1992). *Nucl. Acids Res.*, **20**, 2693–8.
13. Davies, N. P., Rosewell, I. R., Richardson, J. C., Cook, G. P., Neuberger, M. S., Brownstein, B. H., Norris, M. L., and Brüggemann, M. (1993). *Biotechnology*, **11**, 911–14.
14. Jakobovits, A., Moore, A. L., Green, L. L., Vergara, G. J., Maynard-Currie, C. E., Austin, H. A., and Klapholz, S. (1993). *Nature*, **362**, 255–8.
15. Burke, D. T., Carle, G. F., and Olson, M. V. (1987). *Science*, **236**, 806–12.
16. Larin, Z., Monaco, A. P., and Lehrach, H. (1991). *Proc. Natl Acad. Sci. USA*, **88**, 4123–7.
17. Albertsen, H. M., Abderrahim, H., Cann, H. M., Dausset, J., Le, P. D., and Cohen, D. (1990). *Proc. Natl Acad. Sci. USA*, **87**, 4256–60.
18. Anand, R., Villasante, A., and Tyler, S. C. (1989). *Nucl. Acids Res.*, **17**, 3425–33.
19. Bellanne, C. C., Lacroix, B., Ougen, P., Billault, A., Beaufils, S., Bertrand, S., Georges, I., Glibert, F., Gros, I., Lucotte, G., Susini, L., Codani, J-J., Gesnouin, P., Pook, S., Vaysseix, G., Lu-Kuo, J., Reid, T., Ward, D., Chumakov, I., Paslier, D., Barillot, E., and Cohen, D. (1992). *Cell*, **70**, 1059–68.

20. Rothstein, R. (1991). In *Guide to yeast genetics and molecular biology* (ed. C. Guthrie and G. R. Fink), pp. 281–301. Academic Press, London.
21. Choi, T. K., Hollenbach, P. W., Pearson, B. E., Ueda, R. M., Weddell, G. N., Kurahara, C. G., Woodhouse, C. S., Kay, R. M., and Loring, J. F. (1993). *Nature Genet.*, **4**, 117–23.
22. Green, L. L., Hardy, M. C., Maynard-Currie, C. E., Tsuda, H., Louie, D. M., Mendez, M. J., Abderrahim, H., Noguchi, M., Smith, D. H., Zeng, Y., David, N. E., Sasai, H., Garza, D., Brenner, D. G., Hales, J. F., McGuinness, R. P., Capon, D. J., Klapholz, S., and Jakobovits, A. (1994). *Nature Genet.*, **7**, 13–21.
23. Davies, N. P. (1993). PhD Thesis, Cambridge University, UK.
24. Murray, A. W. and Szostak, J. W. (1983). *Nature*, **305**, 189–93.
25. Nelson, D. L. and Brownstein, B. H. (1994). *YAC libraries: a user guide*. Oxford University Press.
26. Zehetner, G. and Lerach, H. (1994). *Nature*, **367**, 489–91.
27. Guthrie, C. and Fink, G. R. (1991) *Guide to yeast genetics and molecular biology*. Academic Press, London.
28. Orr-Weaver, T. L., Szostak, J. W., and Rothstein, R. J. (1981). *Proc. Natl Acad. Sci. USA*, **78**, 6354–8.
29. Pavan, W. J., Hieter, P., and Reeves, R. H. (1990). *Mol. Cell. Biol.*, **10**, 4163–9.
30. Huxley, C. (1994). In *Genetic engineering, principles and methods* (ed. J. Setlow), vol. 16, pp. 65–91. Plenum Press, New York.
31. Huxley, C., Hagino, Y., Schlessinger, D., and Olson, M. V. (1991). *Genomics*, **9**, 742–50.
32. Thomas, K. R. and Capecchi, M. R. (1987). *Cell*, **51**, 503–12.
33. Sambrook, J., Fritsch, E. F., and Maniatis, T. (1989). *Molecular cloning: a laboratory manual*. Cold Spring Harbor Laboratory, NY.
34. Davies, N. P. and Brüggemann, M. (1993). *Nucl. Acids Res.*, **21**, 767–8.
35. Lamb, B. T., Sisodia, S. S., Lawler, A. M., Slunt, H. H., Kitt, C. A., Kearns, W. G., Pearson, P. L., Price, D. L., and Gearhart, J. D. (1993). *Nature Genet.*, **5**, 22–30.
36. Gossler, A., Doetschman, T., Korn, R., Serfling, E., and Kemler, R. (1986). *Proc. Natl Acad. Sci. USA*, **83**, 9065–9.
37. Handyside, A. H., O'Neill, G. T., Jones, M., and Hooper, M. L. (1989). *Roux Arch. Dev. Biol.*, **198**, 48–56.
38. Robertson, E. J., Bradley, A., Kuehn, M., and Evans, M. J. (1986). *Nature*, **322**, 445–8.
39. McMahon, A. P. and Bradley, A. (1990). *Cell*, **62**, 1073–85.
40. Selfridge, J., Pow, A. M., McWhir, J., Magin, T. M., and Melton, D. W. (1992). *Somat. Cell Mol. Genet.*, **18**, 325–36.
41. Ward, M., Scott, R. J., Davey, M. R., Clothier, R. H., Cocking, E. C., and Balls, M. (1986). *Somat. Cell Mol. Genet.*, **12**, 101–9.
42. Pachnis, V., Pevny, L., Rothstein, R., and Costantini, F. (1990). *Proc. Natl Acad. Sci. USA*, **87**, 5109–13.
43. Robertson, E. J. (1987). In *Teratocarcinomas and embryonic stem cells: a practical approach* (ed. E. J. Robertson), pp. 71–112. IRL Press, Oxford.
44. Müllenbach, R., Lagoda, P. J. L., and Welter, C. (1989). *Trends Genet.*, **5**, 391.
45. Doggett, N. A., Smith, C. L., and Cantor, C. R. (1992). *Nucl. Acids Res.*, **20**, 859–64.
46. Bradley, A., Hasty, P., Davis, A., and Ramirez-Solis, R. (1992). *Biotechnology*, **10**, 534–9.

47. Lonberg, N., Taylor, L. D., Harding, F. A., Trounstine, M., Higgins, K. M., Schramm, S. R., Kuo, C.-C., Mashayekh, R., Wymore, K., McCabe, J. G., Munoz-O'Regan, D., O'Donnell, S. L., Lapachet, S. G., Bengoechea, T., Fishwild, D. M., Carmack, C. E., Kay, R. M., and Huszar, D. (1994). *Nature*, **368**, 856–9.
48. Wagner, S. D., Williams, G. T., Larson, T., Neuberger, M. S., Kitamura, D., Rajewsky, K., Xian, J., and Brüggemann, M. (1994). *Nucl. Acids Res.*, **22**, 1389–93.
49. Huxley, C., Farr, C., Gennaro, M. L., and Haaf, T. (1994). *Biotechnology*, **12**, 586–90.
50. Lautner-Rieske, A., Hameister, H., Barbi, G., and Zachau, H. G. (1993). *Genomics*, **16**, 497–502.
51. Weichhold, G. M., Huber, C., Parnes, J. R., and Zachau, H. G. (1993). *Genomics*, **6**, 512–14.
52. Weichhold, G. M., Ohnheiser, R., and Zachau, H. G. (1993). *Genomics*, **16**, 503–11.
53. Kuhn, R. M. and Ludwig, R. A. (1994). *Gene*, **141**, 125–7.
54. Spencer, F., Hugerat, Y., Simchen, G., Hurko, O., Connelly, C., and Hieter, P. (1994). *Genomics*, **22**, 118–26.

4

Measuring antibody affinity in solution

LISA DJAVADI-OHANIANCE, MICHEL E. GOLDBERG, and BERTRAND FRIGUET

1. General considerations

Determination of the affinity of a monoclonal antibody (mAb) for its antigen (Ag) is of considerable importance. It gives a quantitative indication of how strong the interaction is between the Ab and the Ag. It is the basic experimental parameter in a variety of studies, such as the analysis of mAb/Ag binding mechanism (1, 2) or the use of mAbs as conformational probes (3, 4). Hence the need for convenient and rigorous methods to determine the affinity. The affinity, K_a ($K_a = 1/K_d$), of a mAb for its antigen is defined by the Law of Mass Action as:

$$K_d = [Ag] \times [Ab]/[complex];$$

where K_d is the equilibrium dissociation constant, [*complex*] is the concentration of saturated antigen or mAb sites, [*Ag*] is the concentration of free antigenic sites on the antigen, and [*Ab*] is the concentration of free binding sites on the antibody. These concentrations are *binding site concentrations in the solution at equilibrium*. This has three important consequences, as described in Sections 1.1–1.3.

1.1 K_d does not directly reflect association or dissociation kinetics

K_d depicts an equilibrium property, and therefore does not reflect the speed at which equilibrium is reached. Yet, K_d does depend on the association and dissociation rate constants.

In many cases, the relationship between the equilibrium and rate constants is simple. When binding is a simple one-step reaction, $K_d = k_{off}/k_{on}$, where k_{on} and k_{off} are the association and dissociation rate constants. However, when a significant conformational change of either the mAb or the Ag occurs upon association, important deviations from that simple equation can be observed

(1). For such cases, measuring only k_{on} and k_{off} and using $K_d = k_{off}/k_{on}$ would provide an erroneous estimate of the affinity.

1.2 True K_d cannot be determined when the mAb or the Ag is immobilized in a solid-phase assay

Several convenient methods such as ELISA or surface plasmon resonance, with the Ag or the mAb immobilized on the titration plate or on the 'sensor chip' (e.g. gold covered with a carboxymethylated dextran hydrogel), are often used to provide affinity values. Such measurements yield real values of K_d only rarely. One reason is that immobilization often results in a partial denaturation of the protein, thus modifying its binding properties (see ref. 5 for review). Secondly, K_d is defined in solution, with both the mAb and the Ag diffusing and rotating freely in solution, but in the solid-phase assay one of the partners is immobile. This can result in estimates of K_d by solid-phase assay being orders of magnitude different from the real affinity in solution.

One should emphasize, however, that measurements of an 'apparent binding constant' by solid-phase assays may, for some purposes (such as comparative studies), be sufficient. In some instances, the binding constant obtained by solid-phase assays might even be of more significance than the K_d determined with the Ag in solution. Thus, with respect to diffusion and rotation, an antigen on the surface of a large eukaryotic cell may behave more like the Ag immobilized in a solid-phase assay than the Ag in solution, provided that precautions have been taken to avoid denaturation of the Ag upon immobilization.

1.3 The determination of K_d must take into account the valency of the mAb and of the Ag molecule

The Law of Mass Action ($K_d = [Ag] \times [Ab]/[complex]$) expresses the concentrations of reagents in terms of reactive sites, that is, antibody-binding sites (free or saturated) and antigenic sites on the antigen molecule (free or saturated). To determine the affinity from binding experiments, one must therefore determine these values. Thus, one must measure not only the antibody and antigen concentrations, but also their valencies (number of sites per molecule) and understand how this valency influences the experimental determination of the binding. This is discussed in a recent review (6).

2. Overview of methods to measure affinities in solution

Measuring K_d consists of mixing the mAb and Ag at various initial concentrations, bringing to equilibrium, then measuring the concentrations of free and saturated sites at equilibrium and analysing the binding curve. The experimental difficulty resides in distinguishing the free and bound states of either the mAb or the Ag. Several methods can be used such as:

- equilibrium dialysis for haptens and dialysable antigen
- radioimmunoassay using precipitation with salts or other agents
- filtration
- fluorescence measurements
- ELISA- or RIA-based methods.

In this chapter, we discuss the practical aspects of measurement using fluorescence, ELISA, or RIA and describe, in detail, the ELISA- and the RIA-based competition methods for measuring affinity in solution.

2.1 Fluorescence

The use of fluorescence to determine K_d requires that either the mAb or the Ag be fluorescent, and that a change in fluorescence should occur upon formation of the mAb/Ag complex. The fluorescence signal used can be either intrinsic (e.g. tryptophan residues from the mAb and, possibly, from protein antigens; some prosthetic groups such as pyridoxal phosphate, NADH, flavins, etc.) or result from prior fluorescent labelling of the Ag or the mAb with a fluorochrome. The fluorescent change observed upon association may be one of the following:

- wavelength shift
- fluorescence quenching
- fluorescence transfer
- change in fluorescence polarization.

We shall not discuss the practical aspects of these experiments because each mAb/Ag complex poses unique problems, and hence the exact procedure depends on the fluorochromes used. Thus, the wavelengths of fluorescence excitation and emission vary from mAb to mAb and from Ag to Ag, even when the same fluorochrome is studied. The procedure used also depends on the nature of the change observed (quenching, transfer, polarization, shift) and on the type of fluorimeter used. Many unlabelled mAb and Ag complexes do not give rise to a fluorescence change and individual labelling procedures must be devised for each. There is no reliable way to predict whether or not a given label will give rise to a measurable fluorescent change upon complex formation. It is therefore impossible to provide a rule of thumb as to which fluorochrome one should use and how exactly one should proceed to measure the affinity.

We shall, however, briefly discuss some of the pitfalls that should be avoided when using fluorescence for affinity measurements:

(a) Reagents used for fluorescence labelling should be highly purified because of the intense fluorescence signals sometimes produced by impurities.

(b) The sensitivity of current fluorimeters sets a lower limit of 10^{-8} to 10^{-9} M on the K_d that can be determined. Some mAbs have much higher affinities for their Ag.

(c) Labelling of antibody or antigen often gives rise to heterogeneously labelled products which require purification, since both the number of fluorochromes and their position on the molecule affect the fluorescence signal.

(d) The label may affect the binding characteristics, either by steric hindrance or through a change in conformation, thus modifying the K_d. This should *always* be checked by performing competition experiments with the unlabelled molecule.

(e) When reading the fluorescence in the fluorimeter, one should take into account the absorbance of the solution at the wavelength of excitation or emission when this absorbance is *not* negligible, i.e. above 0.05 (effect referred as to inner filter effect). Keeping these precautions in mind, fluorescence measurements should provide precise affinity values (7).

2.2 ELISA- and RIA-based methods

A method that is frequently used for estimating the affinity of a mAb for a macromolecular antigen is to measure the amount of mAb bound to an Ag-coated ELISA plate after incubation with different concentrations of mAb. In these experiments, the mAb concentration that yields half-maximal binding on the coated antigen is taken as the reciprocal of the affinity. This approach does not, however, allow determination of real affinity because equilibrium is attained at the liquid/solid interface rather than in solution (see Section 1.2). Moreover, as already discussed (see Section 1.2), coating the antigen on to the plastic by use of the conventional adsorption methods may alter its conformation (5, 8), which affects the mAb/Ag interaction and hence the affinity. Thus, although providing an estimate of the relative efficiency of the binding of a mAb to the *immobilized* antigen, such direct ELISA- or RIA-based methods fail to provide the real affinity for the *native* antigen.

Two main ELISA-based methods have been described (see refs 7 and 9) for studying the association/dissociation equilibrium *in solution* and have been recently reviewed (6). Both rely on the following principle. Mixtures of the mAb at a fixed concentration and the antigen at varying concentrations are incubated until equilibrium is reached. A solid-phase assay using Ag-coated plates is used to determine the concentration of the free (i.e. not associated with antigen) mAb at equilibrium. The methods differ in two main respects; the conditions under which equilibrium is established (in the presence or absence of the coated antigen) and the conditions under which the ELISA is performed. Thus, Friguet *et al.* (7) first incubate the mAb and the

4: Measuring antibody affinity in solution

Ag in solution for a time sufficient to reach equilibrium, and only then transfer the equilibrated solution into an ELISA plate to determine the concentration of free mAb. The amount of Ag coated in each well and the incubation time for the ELISA are such that only a small fraction (at most 5–10%) of the unsaturated mAb at equilibrium is trapped by the coated antigen. This ensures that, during the ELISA, equilibrium in solution is not significantly modified. By this means, the observed equilibrium constant corresponds to the real affinity. In the other method (see for example ref. 9) where the antibody is in contact with the soluble and immobilized antigen simultaneously, uncontrolled kinetic factors are likely to affect the relative binding of the antibody to the antigen in the liquid and solid phases, and no care is taken to avoid a shift of the equilibrium in the liquid phase when some of the free mAb becomes bound to the solid phase. This method therefore often yields underestimates of the real affinity. Thus, the method of Friguet *et al.* (7) is recommended when determination of the real affinity is required and this method is described in detail in Section 3.

3. Affinity measurements in solution by competition ELISA

3.1 Theoretical aspects

The antibody site to antigenic site association reaction can be written as follows:

$$\text{antibody} + \text{antigen} \leftrightarrow \text{complex};$$

with the concentration of antibody sites, antigen sites, and complex at equilibrium given as $[Ab]$, $[Ag]$, and $[x]$, respectively.

The concentration of antibody sites $[Ab]$ and the antigen sites $[Ag]$ at equilibrium are related to the *total* antibody sites $[Ab_t]$ and the *total* antigen sites $[Ag_t]$ by:

$$[Ab] = [Ab_t] - [x] \qquad [1]$$

$$[Ag] = [Ag_t] - [x]. \qquad [2]$$

K_d, the dissociation constant, is defined by:

$$K_d = [Ag][Ab]/[x].$$

If $[Ag_t]$ is varied while $[Ab_t]$ is kept constant:

$$K_d = [Ag]([Ab_t] - [x])/[x];$$

consequently:

$$[x]/[Ab_t] = [Ag]/([Ag] + K_d) \qquad [3]$$

Several linear plots of Equation 3 have been proposed, the most commonly used being the following:

$$\text{Scatchard equation: } [x]/[Ag] = ([Ab_t] - [x])/K_d \quad [4]$$

$$\text{Klotz equation: } [Ab_t]/[x] = K_d/[Ag] + 1. \quad [5]$$

If Ab_t and Ag_t are known, the experimental determination of K_d (or $K_a = 1/K_d$) requires precise measurement of only one of the three concentrations $[Ab]$, $[Ag]$, or $[x]$.

3.2 Rationale

The method we have developed requires two steps.

(a) In the first step, the antibody, at a constant concentration, and the antigen, at various concentrations, are incubated in *solution* until equilibrium is reached.

(b) In the second step, the concentration of the antibody that remains free at equilibrium is measured by a classical indirect ELISA in which the antigen is coated on the microtitration plate.

The state (native or partially denatured upon coating) of the coated antigen and whether or not it is recognized by the mAb differently from the soluble antigen is not important as long as the coated antigen can specifically and quantitatively trap the free antibody.

3.3 Requirements for the determination of K_d

For correct determination of the free antibody concentration at equilibrium, several requirements must be fulfilled.

(a) The absorbance obtained in the last step of the indirect ELISA, which reflects the free antibody concentration, must be proportional to the antibody concentration tested.

(b) Only a small percentage of the free antibody molecules (i.e. less than 10%) must bind to the coated antigen, to prevent any significant disruption of the equilibrium in the liquid phase.

(c) Since the dissociation constant K_d is generally dependent on the temperature, the temperature must be kept constant throughout.

To satisfy requirements (a) and (b), it is necessary to determine (using the indirect ELISA procedure described in *Protocol 1*) the total (i.e. initial) antibody concentration range that must be used, the concentration of the coated antigen, and the optimal incubation time of the antibody solutions in the coated wells.

Protocol 1. Indirect ELISA procedure

Equipment and reagents

- ELISA plate spectrophotometer or fluorimeter (Labsystems, Dynatech, SLT Lab-instruments)
- Plate washer, Handiwash type (Titertek, Dynatech)
- 96-well flat-bottom microtitration plates and plate sealers
- 5 ml glass tubes
- Repeater pipettes and tips (e.g. Eppendorf Multipette)
- Immunoconjugate: antibody directed against mouse immunoglobulins linked to alkaline phosphatase or β-galactosidase[a] (e.g. from Biosys, Southern Biotechnology Associates Inc., Promega).
- Substrate solution 1: for alkaline phosphatase.[b] Prepare this solution fresh just before use. Dissolve 20 mg of disodium p-nitrophenylphosphate (PNPP) in 10 ml DEA solution (1 M diethanolamine supplemented with 1 mM magnesium sulfate and adjusted with HCl to pH 9.8). The reaction can be stopped by adding 1 M sodium phosphate.
- Substrate solution 2: for β-galactosidase.[b] Prepare 0.4% (w/v) o-nitropheny1-β-D-galactopyranoside (ONPG) stock solution in PM_2 buffer (70 mM disodium phosphate, 30 mM monosodium phosphate, 1 mM magnesium sulfate, 0.2 mM manganese sulfate, 2 mM EDTA magnesium salt, to pH 7.0 with HCl). To prepare a working solution just before use, add 5 ml of 0.4% (w/v) ONPG stock solution to 20 ml PM_2 buffer and then add 18 µl 2-mercaptoethanol. The ONPG stock solution can be kept in the dark at 4°C until it becomes yellowish.
- Substrate solution 3: fluorogenic substrate solution for β-galactosidase.[b] Prepare a 20 mM MUG (methyl-umbelliferyl-β-galactopyranoside) stock solution by dissolving 34 mg of MUG in 5 ml dimethylformamide in a glass test tube. If necessary, warm gently by passing the tube quickly through the pilot light of a Bunsen burner until the MUG is dissolved. Keep the stock solution in the dark at −20°C. To prepare a working solution, mix 0.25 ml of the 20 mM MUG stock solution to 25 ml PM_2 buffer and add 18 µl 2-mercaptoethanol.
- Specific antibodies (at the desired concentrations in the buffer which will be used to determine the K_d in Protocol 3
- Antigen 1 µg/ml in coating buffer
- Washing buffer: PBS (pH 7.4) supplemented with 0.05% w/v Tween-20. Dissolve 8 g NaCl, 0.2 g KH_2PO_4, 2.8 g Na_2HPO_4.12 H_2O and 0.2 g KCl in distilled water to 1 litre final volume.
- Coating buffer: 50 mM carbonate buffer, pH 9.6. Dissolve 1.59 g Na_2CO_3 and 2.93 g $NaHCO_3$ in distilled water to 1 litre final volume.

Method

The assays must be done in triplicate, plus a control with a non-coated well (see legend to *Figure 1*).

1. Add 0.1 ml of a 1 µg/ml antigen solution (in coating buffer) to the well of an ELISA microtitration plate following the scheme in *Figure 1*. Do not coat the first vertical line of wells, which will be blank wells containing only the substrate solution.

2. Incubate the plate, covered with a plate sealer, for at least 3 h at room temperature or overnight at 4°C. Usually a blocking step by overcoating with BSA (0.5%) is not needed.

3. Empty the plate by turning it upside down and shaking it over a sink, then hitting it hard several times on a pile of paper towels.

4. Wash out all wells with washing buffer three times, allowing a 3 min incubation between washes at the chosen temperature.

5. After the third wash, add 0.1 ml of each antibody solution to three

Protocol 1. *Continued*

   coated wells and one non-coated well. Incubate the covered plate for 30 min at the chosen temperature.

6. Wash the wells three times, as in steps 3 and 4, at room temperature.
7. Add, to each well, 0.1 ml of the immunoconjugate previously diluted in washing buffer. Depending on the immunoconjugate supplier, use a 500- to 1000-fold dilution. Incubate for 30 min at room temperature.
8. Wash the wells three times, as in steps 3 and 4, at room temperature.
9. Add 0.1 ml of the appropriate substrate solution to each well.
10. Cover the plate with an adhesive plate sealer to avoid substrate evaporation, which can lead to erroneous results. Incubate at room temperature. If the reaction is too slow at room temperature, incubate the plate at 37 °C.
11. Follow the appearance of the product by measuring (without the plate sealer) the absorbance or fluorescence at the relevant wavelength for each substrate. Thus, for substrate solutions 1 or 2, measure the change in absorbance at 405 nm. When using substrate solution 3, measure the fluorescence at 480 nm (excitation wavelenght 355 nm).
12. If needed, stop the reaction by adding 0.05 ml of the appropriate stopping solution: 1 M sodium phosphate, pH 7, for substrate solution 1, 1.43 M Na_2CO_3 (no pH adjustment) both for substrate solution 2, and for substrate solution 3.

[a] These two enzymes are preferred to peroxidase because of their linearity in the assay response.
[b] Three substrate solutions are listed, one for use with an immunoconjugate linked to alkaline phosphatase and two for β-galactosidase linked immunoconjugates. Prepare only the appropriate substrate solution.

Protocol 2. Quantification of the amount of antibody trapped on to the coated antigen

Equipment and reagents
As for *Protocol 1*

Method

1. Prepare two coated plates as described in *Protocol 1*, step 1, using an antigen concentration of 1 μg/ml.
2. Prepare antibody solutions at different concentrations (e.g. from 10^{-8} M to 10^{-11} M) in the buffer used for the antigen, and *supplemented* with 0.02% bovine serum albumin.

4: Measuring antibody affinity in solution

3. Wash the first coated plate as described in *Protocol 1*.
4. Add 0.1 ml (per well) of each antibody solution, prepared in step 2, to three coated wells and one non-coated well of the first plate (the assays are carried out in triplicate to have enough solution to make duplicates in step 7). Incubate for 30 minutes at the temperature chosen for the affinity measurement.
5. Carefully pipette the antibody solution out of each well and pool the triplicate samples corresponding to the same antibody concentration. Follow the ELISA procedure described in *Protocol 1* from step 6 for this first plate.
6. Wash the second coated plate as in step 3.
7. Add 0.1 ml of each of the pooled antibody solutions that were recovered from the first plate to the wells. Carry out the assays in duplicate. Incubate for the *same time* and at the *same temperature* as in step 4.
8. Follow the ELISA described in *Protocol 1* from step 6 for the second plate. The enzymatic reaction must be stopped after the *same incubation time* in the two plates. Plot the absorbance read in each plate versus the initial antibody concentrations used for the first plate.

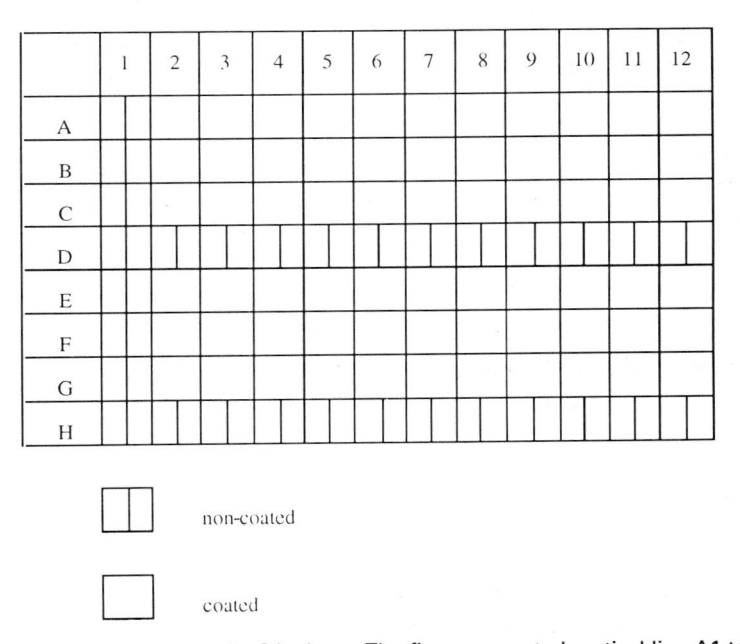

Figure 1. Scheme for coating ELISA plates. The first non-coated vertical line A1 to H1 will become the blank wells containing only the substrate solution, when measuring absorbances in the spectrophotometer. The wells used for assays (triplicate plus control in a non-coated well) are A2 to D2 for the first sample, A3 to D3 for the second one, etc.

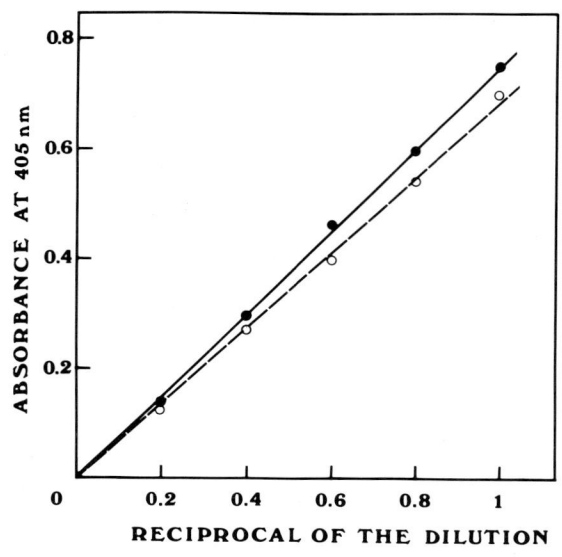

Figure 2. An example of the determination of the amount of antibody retained on a coated antigen. The mAb 46–9 directed against the β₂ subunit of *E. coli* tryptophan synthase was incubated for 1 h at 20°C in plates previously coated with 1 μg/ml β₂. The range of antibody concentrations was 10^{-10} to 7×10^{-10} M. After incubation, the liquid in the wells was transferred to a second coated plate and incubated under identical conditions (see *Protocol 2*). The absorbance at 405 nm was monitored after 2.5 h. Closed and open symbols correspond, respectively, to the absorbance values obtained with the first and the second plate.

The plot obtained from protocol 2, step 8 is linear over a limited concentration range of antibody and reaches a plateau. The constant concentration of antibody used for the determination of K_d must be chosen from the linear part of this plot. In this linear range (see *Figure 2*), the fraction of antibody retained on the coated antigen in the first plate is deduced from the ratio $(S_1 - S_2/S_1)$, where S_1 is the slope obtained with the first plate and S_2 with the second plate. In the case described in *Figure 2*, the amount of antibody retained on the coated antigen represents 6% of the total amount of antibody incubated in the plate.

If the amount of antibody retained is higher than 10% of the total amount of antibody, it is *absolutely necessary* to repeat the experiment with a lower concentration of antigen in the coating step 1 and/or to reduce the incubation time of the antibody solutions on the plate in step 4, until conditions are found under which the two slopes differ by no more than 10%.

3.4 Determination of K_d

Binding equilibrium studies require that the total concentration of antibody $[Ab_t]$ should be close to, or lower than the value of the dissociation constant

4: Measuring antibody affinity in solution

(K_d). Since the dissociation constant is a priori not known, the total antibody concentration should be chosen to be as small as possible. Choose the lowest antibody concentration that gives an absorbance of 1 for the enzymatic reaction after a reasonable time. Do not forget that this antibody concentration should be in the linear part of the plot established in *Protocol 2*.

The results are unsatisfactory if the total antibody concentration $[Ab_t]$ is higher than the K_d. The sensitivity of the immunoenzymatic assay with a chromogenic substrate limits the minimum total antibody concentration to about 10^{-10} M. Thus for an antibody with very high affinity, the experiment should be carried out with the β-galactosidase immunoconjugate and its fluorogenic substrate which increase the sensitivity of the assay and permits the use of lower antibody concentration. Taking into account these considerations determine the K_d as described in *Protocol 3*.

Protocol 3. Determination of dissociation constant (K_d)

Equipment and reagents
As for *Protocol 1*

Method

1. Prepare a coated plate as described in *Protocol 1*, step 1, with the antigen at a concentration of 1 µg/ml, or the one determined in *Protocol 2*.

2. Prepare 3 ml of the antibody solution at twice the concentration required (since it will be diluted 2-fold with the antigen) as determined in *Protocol 2*. The antibody and antigen solutions are prepared in the buffer usually required for the antigen and *supplemented* with 0.02% bovine serum albumin.

3. Prepare the antigen solutions (0.3 ml for each concentration) at 10 different concentrations (e.g. by serial dilutions from 10^{-7} to 10^{-10} M). The value of $[Ag_t]$ corresponds to the total concentration of antigenic sites (for oligomeric proteins, it is the concentration of protomers if each protomer binds one antibody molecule).

4. Prepare 11 tubes, 10 containing 0.25 ml of the antigen at different concentrations (prepared in step 3) and one containing 0.25 ml of incubation buffer (no antigen).

5. Add 0.25 ml of the antibody solution prepared in step 2, to each tube and incubate at the desired constant temperature until equilibrium is reached (e.g. overnight).

6. When equilibrium is reached wash the coated plate prepared in step 1, as described in *Protocol 1*, steps 3 and 4.

7. Add 0.1 ml (per well) from each tube to three coated wells and one

Protocol 3. *Continued*

non-coated well. Incubate at the same temperature as in step 5 and for the time determined in *Protocol 2*.

8. Follow the ELISA procedure described in *Protocol 1* from step 6. Since the antibody concentration was chosen in a range where the absorbance is proportional to the antibody concentration, the absorbance read for each well will correspond to the free antibody concentrations.

3.5 Calculations

K_d is calculated from either a Scatchard plot (using Equation 4 (Section 3.1) or Klotz plot (using Equation 5 (Section 3.1)). First, the relative absorbance readings at various antigen concentrations are used to calculate the amount of complex and free antigen at each point as follows: A0 is the absorbance measured for the well containing the total antibody concentration $[Ab_t]$ in the absence of antigen. A1, A2, ... A10 are the absorbances measured for the wells containing the total antigen concentrations $[Ag_{t1}]$, $[Ag_{t2}]$ ... $[Ag_{t10}]$.

$[Ab]$, the free antibody concentration in each well, is related to the absorbance A measured in the ELISA by the following equation:

$$[Ab] = [Ab_t] (A/A_0),$$

therefore:

$$[Ab_1] = [Ab_t] (A_1/A_0);$$
$$[Ab_2] = [Ab_t] (A_2/A_0); \ldots$$
$$[Ab_{10}] = [Ab_t] (A_{10}/A_0).$$

$[x]$, the antigen–antibody complex concentration, is related to the total antibody site concentration $[Ab_t]$ and the free antibody site concentration $[Ab]$ (see Equation 1 in Section 3.1):

$$[x] = [Ab_t] - [Ab] = [Ab_t] - [Ab_t] [A/A_0] = [Ab_t] (A_0 - A)/A_0,$$

therefore:

$$[x_1] = [Ab_t] (A_0 - A_1)/A_0;$$
$$[x_2] = [Ab_t] (A_0 - A_2)/A_0; \ldots$$
$$[x_{10}] = [Ab_t] (A_0 - A_{10})/A_0.$$

$[Ag]$, the free antigen concentration, is related to the total antigen site $[Ag_t]$ and the antibody–antigen complex concentration $[x]$ (see Equation 2 in Section 3.1):

$$[Ag] = [Ag_t] - [x],$$

4: Measuring antibody affinity in solution

therefore:

$$[Ag_1] = [Ag_{t1}] - [x_1];$$
$$[Ag_2] = [Ag_{t2}] - [x_2]; \ldots$$
$$[Ag_{10}] = [Ag_{t10}] - [x_{10}].$$

Since as indicated above $[x] = [Ab_t] (A_0 - A)/A_0$,

$$[Ag_1] = [Ag_{t1}] - [Ab_t] (A_0 - A_1)/A_0;$$
$$[Ag_2] = [Ag_{t2}] - [Ab_{t2}] (A_0 - A_2)/A_0; \ldots$$
$$[Ag_{10}] = [Ag_{t10}] - [Ab_t] (A_0 - A_{10})/A_0.$$

For the Scatchard equation calculate the values of $[x]$ and $[Ag]$ using the absorbance obtained for each triplicate.

The Scatchard equation (see Equation 4 in Section 3.1) can be also written:

$$[x]/[Ab_t] [Ag] = (1 - [x]/[Ab_t])/K_d;$$

$[x]/[Ab_t]$ (the fraction of bound antibody) is usually referred as v.

Plot $v/[Ag]$ versus v. The straight line obtained (see *Figure 3*) has a slope equal to K_a. The K_d can be calculated since $K_a = 1/K_d$.

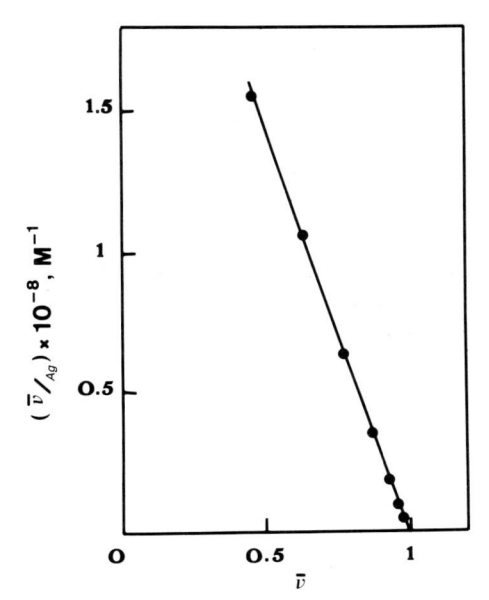

Figure 3. Scatchard plot of the binding of the holo-β_2 subunit of tryptophan synthase to mAb 46–9 measured by ELISA at 20°C. v is the fraction of bound antibody $[x]/[Ab_t]$ and $[Ag]$ the concentration of free antigen at equilibrium. The total concentration $[Ab_t]$ of antibody sites was 3×10^{-10} M. The slope of the straight line is equal to the $K_a = 1/K_d$. The value of K_d is 3.6×10^{-9} M.

For Klotz representation (see Equation 5 in Section 3.1) calculate $1/[Ag]$ and $[Ab_t]/[x]$ using the absorbance obtained for each triplicate.

Plot $[Ab_t]/[x]$ versus $1/[Ag]$. The straight line obtained has a slope equal to $K_d = (1/K_a)$.

Moreover, by using programs now readily available for any standard microcomputer, a non-linear regression method allows the extraction of K_d directly from the saturation curve in solution (e.g. plotting $(A_0 - A)/A_0$ versus $[Ag]$).

3.6 Determination of K_d with impure antibody

The high sensitivity of the ELISA permits the total antibody concentration $[Ab_t]$ to be very low compared to the total antigen concentration $[Ag_{t1}]$, $[Ag_{t2}] \ldots [Ag_{t10}]$ (e.g. $[Ag_t] = 10[Ab_t]$). In such a case, the free antigen concentration can be approximated by the total antigen concentration, since the antibody–antigen complex concentration becomes negligible compared to the total antigen concentration. For example, for the total antigen concentration $[Ag_{t1}]$ the equation:

$$[Ag_1] = [Ag_{t1}] - [Ab_t](A_0 - A_1)/A_0$$

becomes: $[Ag_1] = [Ag_{t1}]$,

if $[Ab_t]$ is sufficiently smaller than $[Ag_{t1}]$.

Using the approximation $[Ag_t] = [Ag_{t1}]$ and replacing $[x_1][Ab_t]$ by its expression given by the absorbances:

$$[x_1]/[Ab_t] = (A_0 - A_1)/A_0,$$

the Scatchard equation (see Equation 6 in Section 3.5) can be written:

$$(A_0 - A_1)/A_0 \cdot 1/[Ag_{t1}] = 1/K_d \cdot (1 - (A_0 - A_1)/A_0); \qquad [7]$$

and the Klotz equation (see Equation 5 in Section 3.1) can be written:

$$A_0/(A_0 - A_1) = K_D/[Ag_{t1}] + 1. \qquad [8]$$

Equations 7 and 8 no longer contain the total antibody concentration $[Ab_t]$. Therefore, when $[Ag_1]$ can be approximated by $[Ag_{t1}]$, Equations 7 and 8 each permit the determination of the dissociation constant K_d even if the antibody site concentration is not known, such as in ascitic fluids, impure monoclonal antibody, or non-titrated Fab (see the Klotz plot of *Figure 4*). Remember that this is true only if the antigen is in large excess over the antibody (e.g. 10-fold).

4: Measuring antibody affinity in solution

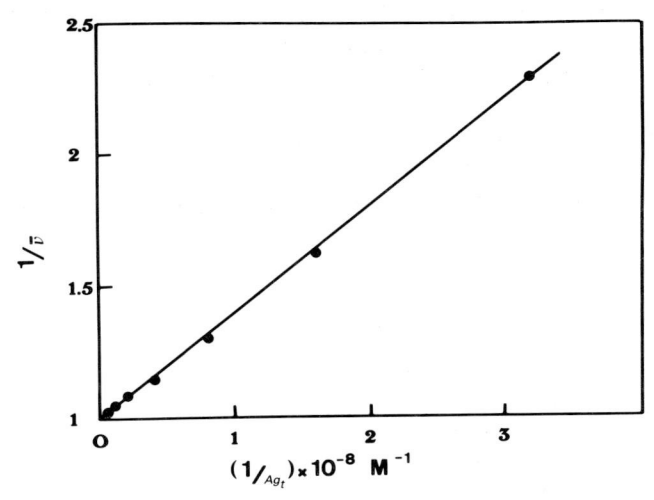

Figure 4. Klotz plot of the binding of holo β$_2$ to mAb 46–9 at 20°C measured by the ELISA with impure antibodies. v is the fraction of bound antibody [x]/[Ab$_t$] and [Ag$_t$] the total antigen concentration. The concentration of impure antibody was 10^{-4} mg/ml. The slope of the straight line is equal to the K_d. The value of K_d is 3.8 × 10^{-9} M.

4. Affinity measurement in solution by an RIA-based method

4.1 Rationale

The rationale of the competition ELISA described has been adapted to an RIA-based method where the antigen is available as a radiolabelled molecule in minute amounts and is not necessarily pure. It has been designed to measure the affinity of polypeptide chains synthesized in a cell-free system in the presence of radioactive amino acids (see ref. 10). Chemically radiolabelled antigen can also be used, providing that the conformation of the protein is not altered upon labelling. In principle, the radiolabelled antigen at a constant concentration and the antibody at various known concentrations in large excess over the antigen concentration are incubated in solution until equilibrium is reached. The concentration of the free antigen (that has not reacted with the antibody) is then determined by using immunobeads, i.e. dextran beads, (instead of the ELISA coated plate) to which the same antibody is covalently attached. The free antigen molecules are specifically trapped by the immobilized antibody and the radioactivity of the immunotrapped antigen fraction is then determined.

4.2 Requirements for the determination of K_d

The same requirements that have been underlined for determining the free antibody concentration at equilibrium in the competition ELISA must be also fulfilled for determining the free antigen concentration in this method:

(a) The concentration of antigen trapped on the immunobeads must be proportional to the concentration of antigen free in solution.

(b) Only a small percentage of the free antigen molecules in the equilibrium solution must bind to the antibody immobilized on the beads to prevent any readjustment of the concentrations at equilibrium.

These two requirements are fulfilled by comparing the total radioactivity of the antigen in solution to the radioactivity bound to the beads during the RIA (see *Protocol 4*). Assuming that the antigen is the only radioactive molecule present in the incubation mixture, a simple radioactivity measurement in a scintillation counter can be done. In a more complex situation where the antigen is present in a heterogeneous mixture of radioactive molecules, a separation step, such as gel electrophoresis, has to be carried out followed by the quantification of the amount of radioactive antigen on the gel.

Protocol 4. Coupling antibody on dextran beads

Equipment and reagents

- Coupling buffer: 0.1 M NaHCO$_3$, pH 8.3, containing 0.5 M NaCl
- Wash buffers:
  —0.1 M Tris–HCl, pH 8;
  —0.1 M sodium acetate, pH 4, containing 0.5 M NaCl;
  —0.1 M Tris–HCl, pH 8, containing 0.5 M NaCl
- Cyanogen bromide activated Sepharose 4B beads commercially available from Pharmacia have been routinely used for coupling the antibody by following the instructions given by the supplier

Method

1. Suspend 1 g of freeze-dried powder in 1 mM HCl and filter-wash with 200 ml of 1 mM HCl.

2. Make up 5 mg of immunoglobulin in 2 ml coupling buffer and mix it with 2 ml of the swollen gel prepared in step 1.

3. Rotate the mixture end-over-end for 2 h at room temperature.

4. Check that no free immunoglobulin is present in the coupling buffer after pelleting the beads and measuring the protein concentration in the supernatant.

5. Resuspend the coupled beads in 15 ml of 0.1 M Tris–HCl buffer, pH 8, to block the remaining active groups on the beads. Incubate for 16 h at 4 °C.

6. Wash the antibody-coupled Sepharose beads with three cycles of

4: Measuring antibody affinity in solution

alternating pH: each cycle consists of a wash with 0.1 M sodium acetate buffer, pH 4 (containing 0.5 M NaCl), followed by a wash with 0.1 M Tris–HCl buffer, pH 8 (containing 0.5 M NaCl).

7. Store the immunobeads at 4°C as a 50% suspension in 0.1 M Tris–HCl buffer, pH 8 (containing 0.5 M NaCl).

Protocol 5. Quantification of the amount of antigen trapped on to the immunobeads

Equipment and reagents

- Immunobeads (see *Protocol 4*)
- Microcentrifuge for 1.5 ml plastic micro test tubes
- Incubation and wash buffer: PBS or any buffer with pH and ionic strength compatible with antibody/antigen interaction and supplemented with a non-ionic detergent such as 0.1% Nonidet P-40 to minimize non-specific protein–protein interaction
- Radiolabelled antigen (see section 4.1)
- Liquid scintillation radioactivity counter
- Electrophoresis sample buffer (if required): 10% SDS, 20% 2-mercaptoethanol, 16% glycerol, 0.04% Bromophenol Blue, 4 mM EDTA, and 40 mM Tris–HCl, at pH 8.0
- Electrophoresis apparatus and power supply (if required)
- Radioactivity scanner (if required)

Method

NB: Carry out steps 1 and 2 at an appropriate fixed temperature.

1. In a series of micro test tubes, mix a constant volume (50 µl) of the immunobeads (prepared as described in *Protocol 4*) with increasing amounts of the radiolabelled antigen in a final volume of 200 µl.

2. Incubate the mixtures for a constant period of 1 min (or a little more if required to increase the sensitivity of the radioactive signal).

3. Add 850 µl of wash buffer and immediately spin at 10 000 *g* for 15 sec in a microcentrifuge. Remove the supernatant and repeat this washing procedure twice with 1 ml of wash buffer.

    If the antigen is the only radiolabelled molecule present in the antigen solution, proceed as indicated in step 4 and then go directly to step 8; otherwise, skip step 4 and proceed from steps 5 to 7.

4. Suspend the pelleted beads in incubation buffer and count the amount of radiolabelled antigen trapped by the immunobeads in a scintillation counter.

5. If the antigen is not the only radiolabelled molecule, add 10 µl SDS sample buffer to the pelleted beads to dissociate the antigen bound to the beads.

6. Run an SDS gel to separate the antigen of interest from the other radiolabelled molecules.

Protocol 5. Continued

7. Determine the radiolabelled antigen concentration by quantitatively scanning the gel (after drying the gel).
8. For each antigen concentration, compare the amount of antigen trapped by the immunobeads with the total amount of antigen initially present in the solution.

A constant fraction of the free antigen initially present in the solution must be recovered in the pelleted immunobeads. Moreover, this fraction must be low enough (not more than 10%) to ensure that the equilibrium is not significantly perturbed in the immunobeads binding step. As already pointed out for the affinity ELISA method, it is critical that this condition should be fulfilled to avoid underestimation of the affinity constant. To minimize the fraction of antigen bound to the immunobeads, several parameters can be adjusted—such as the amount of immunobeads used, the density of antibody molecule on the beads, or the incubation time of the antigen with the immunobeads.

4.3 Determination of K_d

As was the case for the antibody concentration in the ELISA (see Section 3.4), the antigen concentration for the RIA should be choosen to be close to or lower than the K_d value. Using the RIA under the conditions described in Section 4.2, the affinity of a monoclonal antibody for its antigen can be determined as described in Protocol 6.

Protocol 6. Determination of the K_d by an RIA-based method

Equipment and reagents

- Immunobeads (see *Protocol 4*)
- Microcentrifuge for 1.5 ml plastic micro test tubes
- Incubation and wash buffer: PBS or any buffer with pH and ionic strength compatible with antibody/antigen interaction and supplemented with a non-ionic detergent such as 0.1% Nonidet P-40 to minimize non-specific protein–protein interaction
- Radiolabelled antigen (see section 4.1)
- Stock solution of the monoclonal antibody
- Electrophoresis sample buffer (if required): 10% SDS, 20% 2-mercaptoethanol, 16% glycerol, 0.04% Bromophenol Blue, 4 mM EDTA, and 40 mM Tris–HCl, pH 8.0
- Electrophoresis apparatus and power supply (if required)
- Radioactivity scanner (if required)

Method

NB: Carry out steps 1 to 3 at an appropriate temperature. Run the experiment in *duplicate* at least and include control samples with Sepharose beads coupled with a non-specific antibody to evaluate non-specific binding of the antigen to the immunobeads.

1. Make up the radiolabelled antigen (1.4 ml) at twice the required concentration (as determined in *Protocol 5*).

4: Measuring antibody affinity in solution

2. Make up the antibody solutions at about seven different concentrations (0.2 ml for each concentration) ranging, for instance, from 10^{-7} to 10^{-10} M.
3. Prepare micro test tubes containing 75 µl of the antibody solution at different concentrations (as prepared in step 2) and two tubes containing 75 µl incubation buffer (zero antibody).
4. Add 75 µl of the antigen solution prepared in step 1 to each micro test tube containing the antibody solutions and also to the tubes containing the incubation buffer. Incubate at a constant temperature until equilibrium is reached. In most cases, 2 hours has been found to be enough.
5. Add 50 µl of the immunobeads suspension prepared as described in *Protocol 4*.
6. Follow the procedure of *Protocol 5* from steps 2 to 7.

If R_0 is the radioactivity obtained for the samples without antibody, R_i the radioactivity for the samples with the monoclonal antibody at a concentration M_i (M_i = concentration of antibody-binding sites), the fraction v of bound antigen is equal to:

$$v = (R_0 - R_i)/R_0.$$

This expression is obtained assuming that the non-specific binding of the radiolabelled antigen to the immunobeads is negligible, otherwise the background values obtained with the control samples have to be subtracted from both R_0 and R_i. If M_i has been kept much larger than the total concentration of antigen, M_i can be considered as practically equal to the concentration of free antibody sites. Thus, plotting $R_0/(R_0 - R_i)$ as a function of $1/M_i$ results in a straight line, the slope of which gives the K_d.

5. Conclusions

The indirect competition ELISA method described above offers many advantages, as long as it is used under conditions where equilibrium in solution is reached and not perturbed by the solid-phase immunoassay. Only minute amounts of mAb and antigen are needed. Only one of the components (antigen or mAb) needs to be purified or titrated. It provides the real saturation curve in solution. The species at equilibrium in solution are not labelled, which ensures that the observed affinity is characteristic of *native* mAb and antigen.

The rationale of the competition ELISA described in Section 2 has been adapted to an RIA-based method aimed at measuring the affinity of polypeptide chains synthesized in cell-free systems in the presence of radioactive amino acids (see Section 4 and ref. 10).

With the indirect competition ELISA or RIA, the two main problems one may encounter are related to the sensitivity of the solid-phase assay (which sets a higher limit to the affinity values that can be measured) and to the complexity of the experimental curves obtained with multivalent antibodies (which complicates the extraction of the affinity constant from the experimental data).

The lower limits to the K_d values that can be determined by ELISA using classical chromogenic substrates are about 10^{-10} M. However, commercially available fluorogenic substrates for ELISA increase the assay sensitivity by a factor of about 100, and render it possible to determine K_d of 10^{-11}–10^{-12} M (3). Higher affinities could certainly be approached by RIA-based indirect competition methods, using very high specific activity radiolabelled antigens.

The multivalency of antibodies and antigens, and sometimes the heterogeneity of the antibodies, complicate the determination of affinity. This has become the object of much attention (11, 12). The simple data analysis proposed by Friguet et al. (7) provides satisfactory results even when divalent immunoglobulins are used, as long as one extracts the affinity value only from that part of the saturation curve obtained at high saturation of the mAb by the antigen.

In summary, the competition ELISA method described above is the method of choice to determine the real equilibrium dissociation constants of mAb–Ag complexes *in solution* because of its simplicity, its modest requirements in terms of equipment, as well as quantity and purity of antigen or monoclonal antibody. Its rationale adapted to an RIA-based method permits the measurement of the affinity of monoclonal antibodies for radioactive polypeptide chains present in minute amounts in crude heterogeneous mixtures.

These methods can be transposed to the study of protein–protein interactions (see, for example, ref. 13) that do not involve antibodies.

References

1. Friguet B., Djavadi-Ohaniance, L., and Goldberg, M. E. (1989). *Res. Immunol.*, **140**, 355–76.
2. Raman C. S., Jemmerson, R., Nall, B. T., and Allen, M. J. (1992). *Biochemistry*, **31**, 10370–9.
3. Djavadi-Ohaniance, L., Friguet, B., and Goldberg, M. E. (1986). *Biochemistry*, **25**, 2502–8.
4. Goldberg, M. (1991). *TIBS*, **16**, 358–62.
5. Djavadi-Ohaniance, L. and Friguet, B. (1991). In *The immunochemistry of solid-phase immunoassay* (ed. J. E. Butler), Chapter 10, pp. 201–6. CRC Press Inc., Roca Baton, FL.
6. Goldberg, M. E. and Djavadi-Ohaniance, L. (1993). *Curr. Opin. Immunol.*, **5**, 278–81.

7. Friguet, B., Chaffotte, A. F., Djavadi-Ohaniance, L., and Goldberg, M. E. (1985). *J. Immunol. Methods*, **77**, 305–19.
8. Friguet, B., Djavadi-Ohaniance, L., and Goldberg, M. E. (1984). *Mol. Immunol.*, **21**, 673–7.
9. Rath, S., Stanley, C. M., and Steward, M. W. (1988). *J. Immunol. Methods*, **106**, 245–9.
10. Friguet, B., Fedorov, A., Serganov, A. A., Navon, A., and Goldberg, M. E. (1993). *Anal. Biochem.*, **210**, 344–50.
11. Stevens, F. J. (1987). *Mol. Immunol.*, **24**, 1055–60.
12. Winzor, D. J., Bowles, M. R., Pentel, P. R., Schoof, D. D., and Pond, S. M. (1991). *Mol. Immunol.*, **28**, 995–1001.
13. Nelson, R. M. and Long, G. L. (1992). *J. Biol. Chem.*, **267**, 8140–5.

5

Measuring antibody affinity using biosensors

LAURA J. HEFTA, ANNA M. WU, MICHAEL NEUMAIER, and JOHN E. SHIVELY

1. Introduction

Antibodies are often engineered to improve their specificity and affinity for a given antigen. One of the first goals in evaluating newly engineered antibodies is to measure their affinity constants. Recently, a new technology has become available that allows the measurement not only of antibody equilibrium constants, K_a (association) or K_d (dissociation), but also of their kinetic rate constants, k_{on} and k_{off}, where $K_d = k_{off}/k_{on}$. The biosensor-based technology permits real-time mass measurements using either surface plasmon resonance (SPR, BIAcore) or a resonant mirror (IAsys, Fisons). In general, a protein sample (ligand) is coupled to the sensor chip via a hydrogel (carboxymethyl dextran) and solutions containing different concentrations of a binding protein (ligate) are allowed to flow across the chip while monitoring binding with mass sensitive detection (association step). At the end of a binding experiment, it is possible to flow buffer across the chip and observe the dissociation step. This is usually done at saturating conditions, where rebinding of the ligate is not appreciable. Appropriate analysis of the binding and dissociation steps allows the calculation of kinetic constants k_{on} and k_{off}, and as mentioned above, K_d can be calculated from the ratio of the two kinetic constants. In the case of measuring affinity constants of antibodies, the antigen can be coupled to the sensor chip, and the kinetic constants determined by flowing different dilutions of antibody across the chip. (The reverse experiment can also be performed, namely, coupling antibody to the chip and flowing antigen across the chip.) Since the experiment requires that each antibody dilution be performed on the same chip, it is necessary to dissociate the antibody (ligate) without denaturing the antigen (ligand). This is usually done with a brief pulse of 10–100 mM HCl. If the antigen is a protein that can be obtained in purified form, it can be coupled to the chip by a variety of conjugation chemistries, selecting one which preserves its reactivity with antibody. If the antigen is a small molecular weight compound, it may either be directly coupled to the chip or via an inert carrier protein. If the antigen is not available in purified form, or is inactivated by the coupling chemistry, then this method is not appropriate for determining antibody affinity.

Descriptions of the SPR and IAsys systems have been presented in excellent reviews (1, 2). Both systems rely on the covalent coupling of the ligand to a hydrogel surface and mass analysis on ligate binding by measuring a change in the angle of reflectance of an incident light beam at the sample surface. In this chapter we will describe only results obtained from an SPR system, namely the BIAcore system (Pharmacia BIAcore). In the case of SPR, monochromatic, plane polarized light (760 nm) undergoes total internal reflectance at the surface of a prism coated with a 50 nm layer of gold. The resulting evanescent wave causes plasmon resonance in the gold film (a full discussion of the physics involved is beyond the scope of this chapter, see refs 1 and 7). The reflected light intensity has a minimum caused by energy absorption at a unique angle over a range of angles of the incident light beam. The angle at which the minimum occurs is sensitive to the mass of the material applied to the gold surface. The effect is propagated only to the extent of the wavelength of the incident light, and decays exponentially. In the BIAcore instrument the incident light wavelength is 760 nm, and the effective range of the effect is about 300 nm, of which 100 nm is occupied by the hydrogel coat (carboxymethyl dextran linked to the gold surface via hydroxyalkyl thiol). The output is measured in RUs (response units). A change in signal of 1000 RUs corresponds to a 0.1° shift in the SPR angle, with a full range of 30 000 RUs for the detector. A flow system (microfluidics) allows ligate to be passed over the coupled ligand at a variety of concentrations and flow rates. A typical binding curve may run from 20–2000 RUs. A simple schematic of the biosensor optics is shown in *Figure 1*.

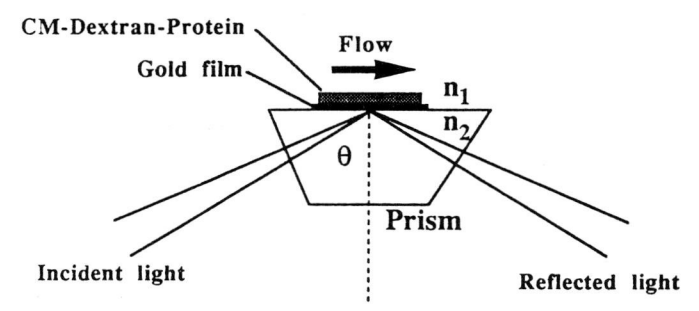

Figure 1. Schematic of biosensor optics. The ligand (protein) is coupled to carboxymethyl (CM)-dextran which is covalently linked to gold film attached to a prism. A light beam is reflected at the gold film interface (n_1/n_2) and the angle of the minimum in the reflected beam recorded. The sample (ligate) flows across the ligand film and causes a change in the angle of the reflected minimum according to the mass of ligate bound.

2. Theoretical aspects

The SPR detector can measure k_{on} and k_{off} from the simple reversible interaction:

$$Ab + Ag \underset{k_{off}}{\overset{k_{on}}{\longleftrightarrow}} AgAb; \quad [1]$$

where *Ab* is the antibody, *Ag* is the antigen. As explained above, either the antibody or the antigen can be immobilized. Arbitrarily, we have selected Ab as the flowing ligate (ligate is the BIAcore term for the substance in the flow) and *Ag* for the immobilized ligand (ligand is the BIAcore term for the substance immobilized). Using these terms, the rate of formation of product *AbAg* at time *t* is:

$$d[AbAg]/dt = k_{on}[Ab][Ag] - k_{off}[AbAg]; \quad [2]$$

and since after time *t*, $[Ag] = [Ag]_0 - [AbAg]$, where $[Ag]_0$ is the concentration of *Ag* at $t = 0$, the equation can be rewritten as:

$$d[AbAg]/dt = k_{on}[Ab]([Ag]_0 - [AbAg]) - k_{off}[AbAg]. \quad [3]$$

In SPR the signal observed, *R*, is proportional to the formation of *AbAg* complexes, and R_{max} is the total amount of binding sites of the immobilized ligand *Ag* (in RU). Since *Ab* is continuously flowing across the biosensor, [*Ab*] remains constant, *C*. Thus, Equation 2 becomes:

$$dR/dt = k_{on}C(R_{max} - R) - k_{off}R, \quad [4]$$

or

$$dR/dt = k_{on}CR_{max} - R(k_{on}C + k_{off}). \quad [5]$$

2.1 Measuring association and dissociation rate constants

Equation 5 is the basic equation used for analysis of BIAcore association kinetics. Thus, one can determine the kinetic constants without the need to measure absolute concentrations of [*AbAg*] at the surface of the biosensor. A plot of dR/dt versus *R* gives a slope, k_s, where:

$$k_s = k_{on}C + k_{off} \quad [6]$$

and a plot of k_s versus *C*, will yield the value of k_{on} from the slope. In theory, the value of k_{off} could be obtained from the y-intercept, however, in practice, more accurate values of k_{off} are obtained from observing the direct dissociation of $AbAg \rightarrow Ab + Ag$. This is performed at the highest concentration of ligate, *Ab*, allowing [*AbAg*] to reach saturation levels, and then watching *R* decrease over an extended period of time while flowing buffer containing no ligate. The rate of dissociation is given by:

$$dR/dt = -k_{off}R. \quad [7]$$

A plot of $\ln(R_1/R_n)$ versus time $(t_n - t_1)$ gives the slope of the line as k_{off}. A complete discussion of the derivation of these equations is given by Karlsson et al. (3). For both Equations 5 and 7 the usual procedure is to perform linear transforms (as described) to calculate k_{on} and k_{off}; however, O'Shannessy et al. (4) have correctly pointed out that more accurate data is obtained by using the integrated forms of the rate equations:

$$R_t = \frac{Ck_{on}R_{max}}{Ck_{on} + k_{off}}(1 - e^{-[(Ck_{on} + k_{off})t]}), \quad [5']$$

and:

$$R_t = R_a e^{-k_{off}t}. \quad [7']$$

The use of the integrated form of the equations gives values for both k_{on} and k_{off} for each data set, and minimizes errors carried over from each run (5). Fortunately, the newest software release from BIAcore allows the use of non-linear least squares analysis of the raw data in the integrated form of the rate equations. In order to be used successfully, the method employs a careful analysis of the baseline, selection of the appropriate data region, and curve fitting.

Another possibility is to determine K_a at equilibrium. If equilibrium is reached (net binding rate is 0, R_{eq} is the equilibrium value in RU), $dR/dt = 0$, and:

$$k_{on}C(R_{max} - R_{eq}) = k_{off}R_{eq}, \quad [8]$$

or

$$R_{eq}/C = k_{on}R_{max} - K_aR_{eq}; \quad [9]$$

since $K_a = k_{on}/k_{off}$. Thus, a plot of R_{eq}/C versus R_{eq} will give K_a as the slope (and $K_d = 1/K_a$). This is a linear transform of the rate equation and subject to the same errors as mentioned above. Also, in most experiments the limit on injection time does not allow equilibrium to be reached, prohibiting the routine use of this method.

2.2 Limitations on measuring affinity constants

A major limitation of kinetic analysis involves consideration of mass transport from bulk flow of ligate to the biosensor surface. It is assumed in the above analyses that sufficient ligate exists in the flow so that $[Ab]$ remains constant. At low values of $[Ab]$ this may not be the case, and the association rate will be affected. Thus, especially in the linear transform methods, it is important to analyse the upper portions of the curves, and to avoid including data from the early time points.

In the case of engineered antibodies, monovalent constructs such as Fab or single chain Fv fragments (scFv) will follow the above described kinetics; however, bivalent molecules such as whole IgGs will deviate from this analysis, primarily affecting the dissociation phase where biphasic results are usually observed (6). This is expected since the antibody (ligate) may re-associate with the antigen (ligand) by its other binding site, resulting in the phenomenon termed avidity. The net result is a 2–10-fold decrease in K_d over the equivalent monovalent fragments of the antibody. A good discussion of the analysis of bivalent antibodies is given by Altschuh *et al.* [7].

3. Immobilization, binding, and regeneration on the BIAcore

We will use engineered antibodies to CEA (carcinoembryonic antigen) to illustrate the general concepts of measuring kinetic and equilibrium constants for antigen–antibody interactions. Since the reader may not be familiar with CEA, we will briefly review its properties. CEA was first described as a colon specific tumour marker 30 years ago by Gold and Freedman (8). Since then it has been shown that CEA is present in the normal colon (9, 10) and may be expressed in a variety of cancers of epithelial origin, including breast, lung, ovary, and medullary thyroid (11). Radiolabelled anti-CEA antibodies can be used for the *in vivo* targeting of CEA positive tumours. Due to its apical or luminal expression in normal colon, antibodies to CEA do not target the normal colon. In CEA-positive tumours, the altered cellular morphology leads to antibody accessibility and excellent targeting. In order to improve the biodistributions of anti-CEA antibodies it is necessary to engineer anti-CEA antibodies of different size, affinity, and valency. Analysis of their kinetic properties on the BIAcore is a high priority in the initial evaluation of each engineered antibody.

Although CEA contains about 50% carbohydrate by weight, the majority of antibodies raised against CEA recognize protein determinants (12). The domain structure of CEA and the epitopes recognized by the antibodies studied in this report are shown in *Figure 2*. Most of our clinical studies have focused on the high-affinity antibody mT84.66 (13, 14) or its chimeric murine/human engineered version, cT84.66 (15). This antibody, and a related antibody, T84.12, recognize the A3 subdomain within the A3B3 domain. A third antibody, H6C8, recognizes the B3 subdomain (12). In order to study the domain specificity of T84.66, we have also developed an anti-idiotypic antibody to mT84.66, 6G6.C4H (16), and expressed various domains of CEA in soluble form (17).

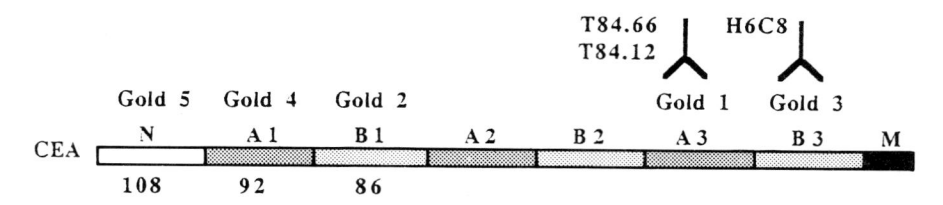

```
                                    T84.66  │  H6C8
                                    T84.12  │    │
                                           Y     Y
       Gold 5    Gold 4    Gold 2          Gold 1  Gold 3
        N         A1        B1       A2    B2    A3    B3    M
CEA  ┌────────┬─────────┬──────────┬──────┬──────┬──────┬──────┐
       108       92        86
```

Figure 2. Epitope map for CEA. The domains are labelled sequentially, N, A1, B1, etc. The number of amino acids in each domain is shown below each domain. Domains with similar designations (e.g. A1, A2, A3, or B1, B2, B3) are the same size and have > 90% sequence homology. The epitopes are labelled Gold 1–5 and their domain locations are shown. Three antibodies studied in this work are T84.66, T84.12, and H6C8, the first two interacting with Gold 1, and the third with Gold 3.

3.1 Immobilization step

SPR analysis depends on immobilizing sufficient amounts of ligand to the carboxymethyl dextran hydrogel. In the case of protein ligands, proteins usually possess sufficient surface lysine residues so that coupling to the carboxyl groups of the hydrogel can be accomplished without completely destroying the ligand's antibody binding activity. Johnson *et al.* (18) have shown that maximum protein binding to the hydrogel occurs at a pH below the protein's isoelectric point and at low ionic strength (10 mM or less) where ionic interactions between the protein and hydrogel are favoured. Coupling of protein to EDC/NHS (*N*-ethyl-*N*'-(dimethylaminopropyl) carbodiimide hydrochloride/*N*-hydroxysuccinimide) activated carboxyl groups of the hydrogel, the standard BIAcore activation chemistry, occurs best at the pH where the maximum amount of protein electrostatically interacts with the hydrogel. For most proteins, even those of low pI, the optimum pH is from 4.0–4.5, the optimum pH for EDC activation of carboxyl groups. This is true even though the optimum pH for the coupling reaction to the ε-amino groups of lysine are expected at pH 8 or greater, since the predominant effect is the concentration of the protein at the hydrogel surface at low pH. Since the BIAcore allows real-time mass sensitive detection, the electrostatic binding of protein to the hydrogel can be monitored for each ligand, and individually optimized. However, in practice, ligand is often a precious commodity, and the investigator is motivated to start with a protocol that is optimized for most proteins.

Protocol 1 is a good starting point for coupling proteins (ligand) to the hydrogel. The entire process is monitored by a sensorgram (*Figure 3*). The target value for net covalent binding to the hydrogel should be from a minimum of 1000 to a maximum of 10 000 RUs. At the lower end, there may not be sufficient range (in RUs) to cover the binding kinetics of 4–5 concentrations of ligate (antibody). At the upper end, too much ligand may be present on the biosensor, causing problems with mass transport at low ligate concentrations. The last step of the protocol includes treatment with acid or other

5: Measuring antibody affinity using biosensors

appropriate agent. This step is necessary to establish the amount of protein actually coupled, since some protein may be non-specifically adsorbed. In addition, this step will determine the stability of the immobilized ligand to the surface regeneration step that is required during the kinetic measurements. In this example, the protein concentration was 0.7 μM and the overall coupled yield was about 1700 RUs.

Protocol 1. Immobilization of protein to hydrogel using EDC/NHS

Equipment and reagents

- BIAcore (Pharmacia BIAcore)
- Protein (ligand; in this case, CEA) 0.1 to 1.0 μM, in 0.01 M sodium acetate, pH 4.0
- 0.1 M NHS (supplied by manufacturer)
- 0.4 M EDC (supplied by manufacturer)
- Ethanolamine HCl: 1 M in water, pH 8.5, adjusted with NaOH; supplied by manufacturer
- HBS: pH 7.4, 0.01 M Hepes, 0.15 M NaCl, 3.4 mM EDTA, 0.05% surfactant P-20 or Tween-20

Method

1. Activate the carboxyl groups of the hydrogel by flowing 30 μl of freshly mixed (50/50(v/v)) NHS and EDC (mixing programed by BIAcore) for 6 min at 5 μl/min. Rinse for 1 min with HBS to remove excess reagents.

2. Couple protein (ligand or CEA) for 6 min at 5 μl/min (total volume of 30 μl). Rinse for 1 min with HBS to remove excess protein.

3. Cap excess activated carboxyl groups with ethanolamine HCl for 6 min at 5 μl/min (total volume of 30 μl). Rinse with HBS for 1–2 min to remove excess reagent.

4. Treat the ligand with 0.01–0.1 M HCl (or other appropriate agent) to remove non-covalently bound ligand.

The results shown for CEA are a good example of the successful coupling of a low pI protein (CEA has a pI of 2–3 (19)). The low pI of CEA is due to its high sialic content (1–8% by weight); however, when CEA is treated with neuraminidase, there is no substantial increase in coupling efficiency. If poor coupling or poor binding activity after coupling is observed, several approaches can be attempted. The simplest approach is to increase the concentration of protein until sufficient RUs and binding activity is obtained. The second approach is to perform an indirect coupling using either biotinylated ligand bound to avidin coupled hydrogel, or ligand bound through a second antibody recognizing a distinct epitope on the antigen (see below). If this fails, the coupling chemistry can be changed. A recent review by O'Shannessy describes several modifications of the EDC/NHS chemistry leading to new functional groups on the hydrogel (20). For example, if the

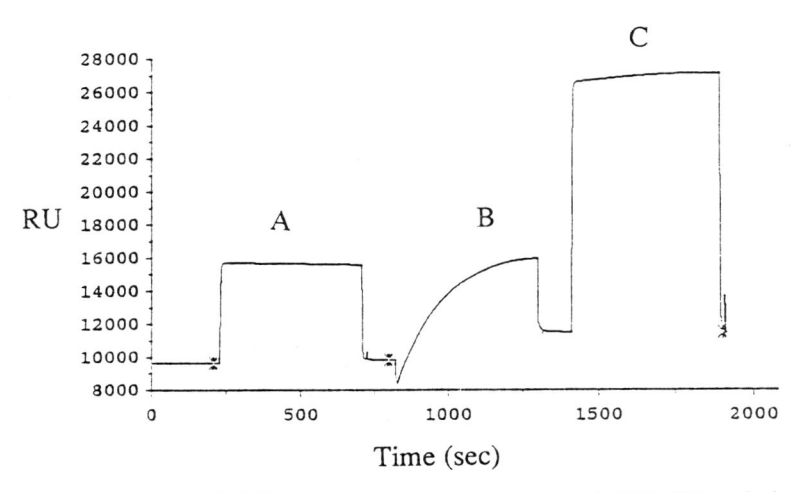

Figure 3. Immobilization of CEA on a biosensor chip. In region A (200–700 sec), the CM-dextran is activated with EDC/NHS, and washed with HBS. In region B (800–1300 sec), CEA (50 µg/ml) is reacted with the activated CM-dextran, and washed with HBS. In region C (1400–1900 sec), excess activated carboxyl groups are capped with ethanolamine, and washed with HBS. The difference in RUs from the start to the finish baseline is 1678 RUs, the amount of CEA covalently coupled to the CM-dextran.

ligand has a free cysteine, sulfhydryl coupling chemistry can be performed. This involves reaction of the EDC/NHS activated carboxyl groups of the hydrogel first with ethylenediamine, capping with ethanolamine, and reaction with sulfo-MBS (m-maleimidobenzoyl-N-hydroxysuccinimide ester). The sulfo-MBS moiety will then couple to the free sulfhydryl group in the protein. In glycoproteins like CEA, it is possible to couple via the carbohydrate moieties after periodate oxidation. The EDC/NHS activated carboxyl groups of the hydrogel are first reacted with hydrazine, capped with ethanolamine, and coupled to periodate-oxidized glycoprotein (20). Although the resulting hydrazone linkage is more stable at low pH than corresponding Schiff bases, it may be advisable to reduce the linkage to the hydrazide form using either sodium borohydride or sodium cyanoborohydride (a milder reducing reagent). Sodium borohydride may be preferred since it will also reduce the excess aldehydes generated by the periodate oxidation, and prevent the aldehydes from forming Schiff bases with amino groups on the protein. An example of this immobilization method is given in *Protocol 2*.

Protocol 2 is general and should work for a variety of glycoproteins, including IgGs that also have carbohydrate groups. The protocol has the potential to cross-link the ligand to itself, since Schiff bases can form between the protein's amino groups and the oxidized carbohydrate. In the case of CEA, the protein is rather stable to periodate and forms very little cross-linked species. A general procedure for the periodate oxidation of a glycoprotein is given in

5: Measuring antibody affinity using biosensors

Protocol 3. The amount of hydrazine coupled to the hydrogel will be the main determinant for the amount of oxidized glycoprotein coupled. Thus, if too many RUs are coupled, the concentration of hydrazine or volume (total flow) can be adjusted.

Protocol 2. Immobilization of glycoprotein to hydrogel using hydrazine

Equipment and reagents

- BIAcore (*Protocol 1*)
- Glycoprotein (freshly periodate-oxidized): 0.1–1.0 μM in 0.01 M sodium acetate, pH 4.0 (see *Protocol 3*)
- NHS (*Protocol 1*)
- EDC (*Protocol 1*)
- Hydrazine HCl, (freshly prepared): 1 M in water, pH 8.5 adjusted with NaOH
- Ethanolamine HCl (*Protocol 1*)
- Sodium borohydride: 0.1 M in pH 9.0, 0.05 M sodium carbonate/bicarbonate
- HBS (*Protocol 1*)

Method

1. Activate the carboxyl groups of the hydrogel as described in *Protocol 1*.
2. Couple hydrazine to the activated hydrogel by flowing hydrazine for 6 min at 5 μl/min (total volume of 30 μl).
3. Cap excess activated carboxyl groups with ethanolamine HCl for 7 min at 5 μl/min (total volume of 35 μl). Rinse with HBS for 1–2 min to remove excess reagent.
4. Couple periodate-oxidized glycoprotein for 6 min at 5 μl/min (total volume of 30 μl). Rinse as described in *Protocol 1*.
5. **Optional**: reduce with sodium borohydride for 6 min at 5 μl/min (total volume 30 μl). Rinse with HBS and treat with 0.01–0.1 M HCl as described in *Protocol 1*.

Protocol 3. Periodate oxidation of glycoproteins

Equipment and reagents

- Protein (in this case, CEA), 1 mg/ml in 0.01 M sodium acetate, pH 5.5
- Sodium metaperiodate (freshly prepared): 10 mM in 0.01 M sodium acetate, pH 5.5 (protect from light)
- Ethylene glycol: 20% (v/v) in 0.01 M sodium acetate, pH 5.5
- Spin column (any manufacturer) loaded with Sephadex G25 or PD-10, equilibrated in 0.01 M sodium acetate, pH 4.0

Method

1. React 100 μl of CEA with 10 μl of periodate for 1 h at 4°C, protected from light.
2. Destroy excess periodate with 10 μl of ethylene glycol for 10 min at 4°C, protect from light.

Protocol 3. *Continued*

3. Remove excess reagents by gel filtration on the spin column (elute with 0.2–0.5 ml of pH 4.0, 0.1 M sodium acetate, 1000 *g.*, 10 min). Final volume should be about 0.5 ml.

3.2 Regeneration step

Successful kinetic analyses on the BIAcore also requires that the immobilized ligand surface can be regenerated after binding ligate. Since it is imperative to generate binding curves at different concentrations of ligate, the ligand surface must be regenerated to a level close to, if not exactly the same, as its original state of activity. And since there is a distinct possibility that the conditions of regeneration may irreversibly denature the ligand, careful attention to this step is a prerequisite. The most successful general approach is to use short pulses of 10–100 mM HCl. In practice, a baseline is established, the maximum concentration of ligate is bound, rinsed with buffer, and pulsed with 1–10 mM HCl. The procedure is repeated until the baseline returns to the previous baseline value. If this baseline value is not reached after 4–5 pulses, then higher concentrations (20–100 mM) of HCl are tried. If this approach fails, the BIAcore manual suggests trying a variety of other denaturants such as 1–2 M $MgCl_2$, 1–4 M urea or guanidine HCl, pH 9–10 Tris, or combinations of these. Obviously, this critical step can be time consuming and frustrating if it doesn't give a surface that can be multiply regenerated. A second approach is to bind the ligand through an indirect approach (as mentioned above) and regenerate the surface by rebinding ligand before each ligate binding experiment. This is a costly approach in terms of ligand use, and suffers from the possibility of complicating the kinetics by two dissociating phases, but is appealing in that it obviates the need to determine regeneration conditions for each new ligand. An example of this approach is given in *Protocol 4*.

4. Kinetic analysis of anti-CEA antibodies

4.1 Direct binding assays: comparison of murine and chimeric T84.12

To illustrate the determination of dissociation constants, we will use two parent murine monoclonal antibodies to CEA, namely mT84.12 and mT84.66. mT84.12 and mT84.66 recognize the same epitope on CEA, but differ completely in their CDR sequences (*Table 1*). For mT84.12, CEA was bound to the biosensor chip as described above, and mT84.12 was allowed to bind to immobilized CEA at concentrations ranging from 60 to 1000 nM (*Figure 4A*). Two 60 sec pulses of 10 mM HCl were used to regenerate the biosensor chip between injections. Analysis of the data by the integrated rate equation

Table 1. Sequence of CDRs for anti-CEA antibodies, T84.66 and T84.12, and the anti-idiotypic antibody 6G6.C4

Light chains	CDR1	CDR2	CDR3
T84.66	RAGESVDIFGV	LESGIPV	QTNEDP-YT
T84.12	FASQNV----H	RYSGVPD	QCNSYPLFT
6G6.C4	KASQNV---GN	RYSGVPD	QYDNYP-WT
Heavy chains			
T84.66	DTYMH	RIDPANGNSKYVPKFQG	FGYYVSDYAMA
T84.12	SYAMS	SISS-DGITFYVDSVKG	IDYY-GGGGFG
6G6.C4	TYWMN	RIDPYDSVTHYNQKFRD	MDY----GNHD

(including the dissociation phase) gave $k_{on} = 1.08 \times 10^5$ M^{-1} s^{-1} and $k_{off} = 1.36 \times 10^{-5}$ s^{-1} for a value of $K_d = 1.26 \times 10^{-10}$ M (*Table 2*). The value for K_d agrees well with values obtained by EIA (enzyme immunoassay). The chimeric version of the antibody, cT84.12, was expressed in Sp2/0 cells in a bioreactor, purified by a combination of Protein A affinity and ion-exchange chromatography, and was analysed on the BIAcore (*Figure 4B*). In this case, the antibody showed a significantly lower k_{on} and higher k_{off}, resulting in a K_d of only 1.3×10^{-8} M (*Table 2*). While we are unable to explain the significant loss in K_d for cT84.12, these results demonstrated that the engineered antibody was unsuitable for further animal and human studies. (There is some indication that the cysteine residue in L3 is responsible for its instability.)

4.2 Indirect binding assays: comparison of murine and chimeric T84.66

A similar analysis for mT84.66 was made difficult by its apparent irreversible binding to CEA on the biosensor chip. Previous measurements of the antibody K_d by RIA (radioimmunoassay) and EIA gave values of 5×10^{-11} M, suggesting that its extremely low K_d may be responsible for the regeneration problem. While no suitable regeneration conditions were discovered, we did note that the antibody slowly dissociated from CEA over a period of several days with buffer flow. In order to determine the binding kinetics, an indirect method was devised (*Protocol 4*). A lower affinity anti-CEA antibody recognizing a different epitope, H6C8 (12), was immobilized and a fresh solution of CEA (250 nM) was bound to the chip prior to each injection of mT84.66. The biosensor chip was regenerated with two 60 sec pulses of 10 mM HCl. The binding curves for mT84.66 to CEA are shown in *Figure 5A* (including the CEA rebinding steps). A similar analysis was performed for the engineered chimeric version, cT84.66, shown in *Figure 5B*. The kinetic values agree well between mT84.66 and cT84.66, but the K_d values are about 10-fold less than those determined by RIA or EIA. In the case of RIA determina-

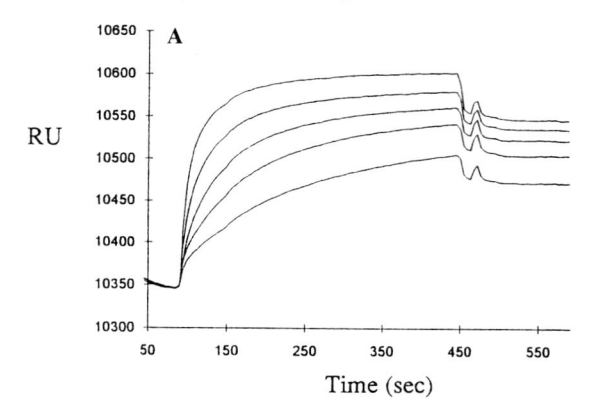

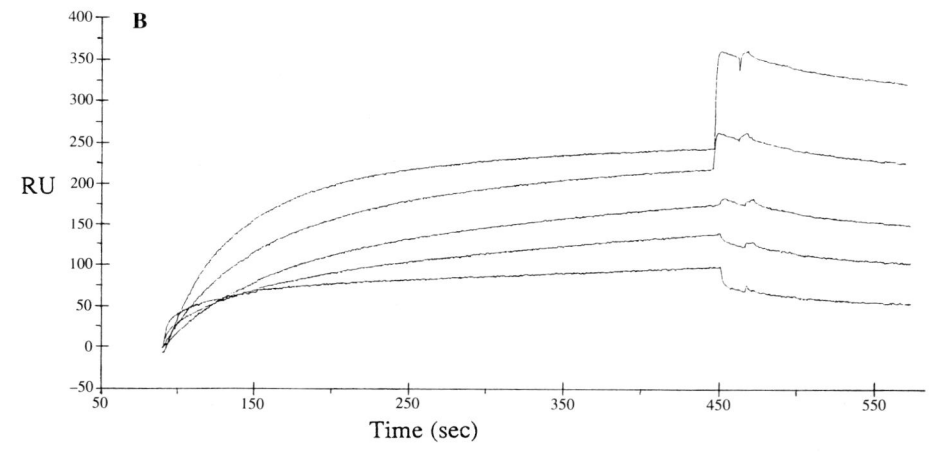

Figure 4. Binding kinetics for anti-CEA antibodies mT84.12 and cT84.12. CEA was covalently coupled to CM-dextran as shown in Figure 3. A, mT84.12: the ligate was bound to CEA (ligand) at five concentrations (bottom to top curves), 62.5 nM, 125 nM, 250 nM, 500 nM, and 1000 nM. Regeneration between binding curves was with two 60 sec pulses of 10 mM HCl. B, cT84.12: the ligate was bound to CEA (ligand) at five concentrations (bottom to top curves), 30 nM, 60 nM, 120 nM, 240 nM, and 480 nM. Regeneration between binding curves was with two 60 sec pulses of 10 mM HCl.

tions, iodination of CEA resulted in considerable loss of mT84.66 binding activity (13), and in the case of EIA the CEA antigen may be partially denatured when coated on the solid phase. Thus, the more 'native-like' conditions of the indirect assay on the BIAcore may give more accurate data, and explain why the extremely low K_d for either mT84.66 or cT84.66 prevents successful regeneration of the CEA coupled biosensor chip. In any case, the results convincingly demonstrate that chimerization of the antibody resulted in no loss of CEA binding activity.

Table 2. Kinetic and affinity constants for anti-CEA antibodies[a]

Antibody	k_{on} (M^{-1} s^{-1})	k_{off} (s^{-1})	K_d (M)
mT84.12	1.08×10^5	1.36×10^{-5}	1.26×10^{-10}
cT84.12	1.60×10^4	2.00×10^{-4}	1.30×10^{-8}
scFv-14-T84.66	1.00×10^6	2.50×10^{-4}	2.44×10^{-10}
scFv-28-T84.66	6.50×10^5	3.00×10^{-4}	4.54×10^{-10}
mini-LD-cT84.66	1.00×10^5	5.00×10^{-5}	5.00×10^{-10}
mini-Flex-cT84.66	3.00×10^5	1.10×10^{-3}	3.70×10^{-10}
mT84.66/CEA	1.78×10^6	1.42×10^{-5}	7.98×10^{-12}
cT84.66/CEA	1.64×10^6	1.41×10^{-5}	8.60×10^{-12}
mT84.66/A3B3	1.04×10^6	2.28×10^{-5}	2.19×10^{-11}
H6C8/A3B3	3.80×10^5	1.04×10^{-5}	2.74×10^{-11}
6G6.C4/mT84.66	5.65×10^5	9.32×10^{-6}	1.65×10^{-11}
6G6.C4/cT84.66	5.10×10^5	2.27×10^{-5}	4.45×10^{-11}

[a] In the first group CEA was immobilized on the biosensor chip and antibody binding was performed. In the second group H6C8 was immobilized on the biosensor chip, and CEA was bound followed by antibody binding. In the third group either mT84.66 or H6C8 was immobilized on the biosensor chip, followed by binding of A3B3. The dissociation phase calculation, k_{off}, was performed on antibody binding to CEA immobilized on the biosensor chip. In the fourth group the anti-idiotypic antibody 6G6.C4 was immobilized on the biosensor chip followed by binding of either mT84.66 or cT84.66.

Protocol 4. Indirect binding assay for the BIAcore

Equipment and reagents

- BIAcore
- Ligand 1: H6C8, 0.27 µM in 0.01 M sodium acetate, pH 4.0
- Ligand 2: CEA, 250 nM in HBS (*Protocol 1*)
- NHS (*Protocol 1*)
- Ligate: mT84.66 or cT84.66, 30–480 nM in HBS
- EDC (*Protocol 1*)
- Ethanolamine HCl (*Protocol 1*)
- HBS (*Protocol 1*)

Method

1. Activate the carboxyl groups of the hydrogel as in *Protocol 1*.
2. Couple, cap, rinse, and treat ligand 1 (H6C8) as in *Protocol 1* (steps 2–3).
3. Bind ligand 2 (CEA) for 1.3 min at 20 µl/min (total volume 26 µl). Rinse for 4 min at 20 µl/min.
4. Bind first concentration of ligate (30 nM) for 1.3 min at 20 µl/min. Rinse for 4 min at 20 µl/min. Regenerate with two 1 min pulses of 0.01 M HCl at 5 µl/min.
5. Repeat steps 3–4 using next higher concentration of ligate.

Protocol 4 can be generalized to any antibody that does not regenerate, providing that one has access to additional antibodies that bind non-overlapping epitopes. This can also be tested on the BIAcore by binding each antibody sequentially, demonstrating a roughly equivalent mass increase (increase in RUs) for each antibody. The method requires a ready supply of antigen, since the antigen (ligand 2) must be rebound to ligand 1 before the kinetic measurement is made with ligate. Finally, the kinetics are complicated by the fact that ligand 2 may dissociate from ligand 1 during the course of the kinetic measurements. The dissociation kinetics of ligand 2 should be checked before proceeding with binding of ligate. If dissociation is minimal (less than 10%) over the association phase, one may expect reasonable kinetics for binding of ligate. Problems that are encountered in the analysis are readily apparent with the BIA evaluation software. In the examples shown (*Figure 5*), dissociation of ligand was not a problem. If dissociation of ligand 2 is a problem, one may compensate by deriving a model that takes into account the dissociation phase (4).

Expression of soluble CEA domains allows a second measurement of the K_d of mT84.66. The soluble domain A3B3 was expressed in HeLa cells and purified by affinity chromatography using a low affinity anti-CEA antibody (17). Binding of A3B3 to mT84.66 was irreversible, similar to the binding of this antibody to CEA. Therefore, it was necessary to perform indirect binding as before. First we investigated the ability of A3B3 to bind H6C8. The K_d for this interaction was 2.74×10^{-11} M (*Table 2*). Then the indirect assay was performed for mT84.66. The kinetics constants are shown in *Table 2*. The calculated $K_d = 2.19 \times 10^{-11}$ M is 10-fold greater than obtained for binding of mT84.66 to CEA. The larger size of CEA, including its dimeric structure may account for the difference in K_d.

4.3 Assays for engineered antibody fragments

scFv versions of T84.66 were engineered with linker sizes of 14 (scFv-14-T84.66) and 28 (scFv-28-T84.66) amino acids (21). The secreted forms were expressed by *E. coli*, and purified from supernatants using affinity chromatography on the anti-idiotypic antibody 6G6.C4H (16). scFv-14-T84.66 migrated as a 50/50 mixture of monomers and dimers on gel permeation chromatography, while scFv-28-T84.66 was 100% monomeric (see also Filpula *et al.*, Chapter 11). The binding kinetics for both scFvs are shown in *Figure 6* and their binding constants in *Table 2*. The reversible formation of dimers for scFv-14-T84.66 on the biosensor chip may be responsible for the fluctuating values at equilibrium (*Figure 6A*). Depending on the comparison made, the K_ds of the scFvs are 10–100-fold greater than native mT84.66 (net loss in affinity). The contribution of dimers in scFv-14-T84.66 is responsible for a 2-fold decrease in K_d over scFv-28-T84.66. These data suggest that the smaller size of the scFvs, in some way, may affect the overall binding constants. Larger versions of the scFv including C_H3 domains, termed minibodies (22), were constructed and shown to be dimeric, presumably through the

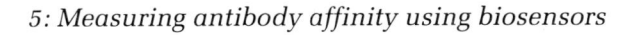

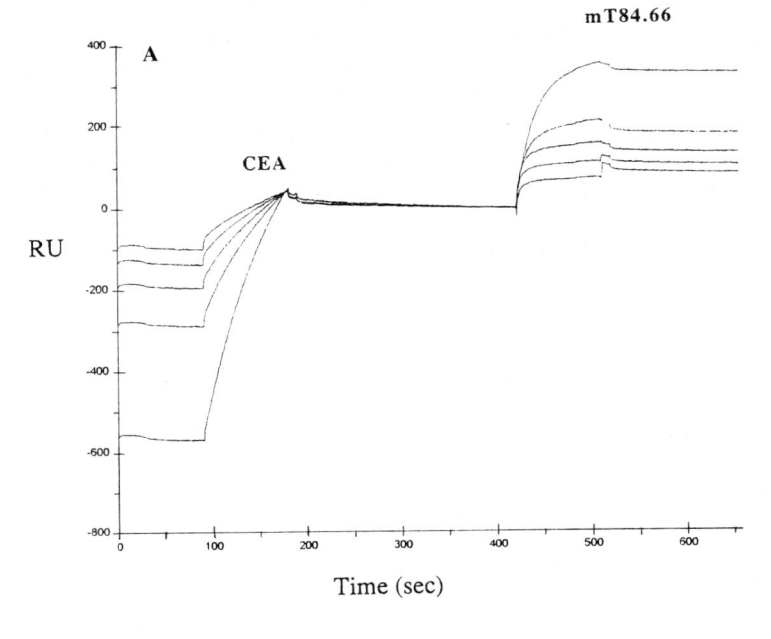

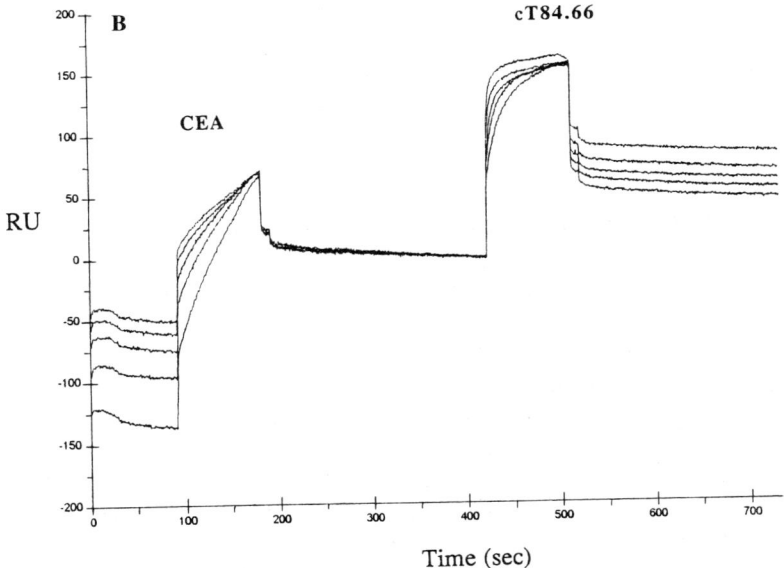

5: Measuring antibody affinity using biosensors

Figure 5. Binding kinetics for anti-CEA antibodies mT84.66 and cT84.66. In both cases the antibody H6C8 was covalently coupled to CM-dextran, followed by binding of CEA (250 nM) and antibody ligate. After ligate binding, the chip was regenerated with two 60 sec pulses of 10 mM HCl, and rebinding of CEA. The two phases of the assay are labelled. A, mT84.66: binding curves for mT84.66 at 30 nM, 60 nM, 120 nM, 240 nM, and 480 nM. B, cT84.66: binding curves for cT84.66 at 30 nM, 60 nM, 120 nM, 240 nM, and 480 nM.

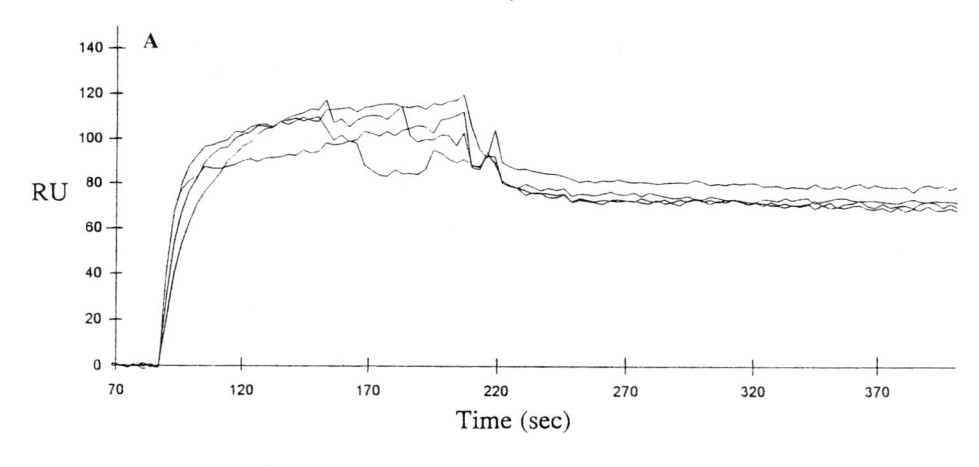

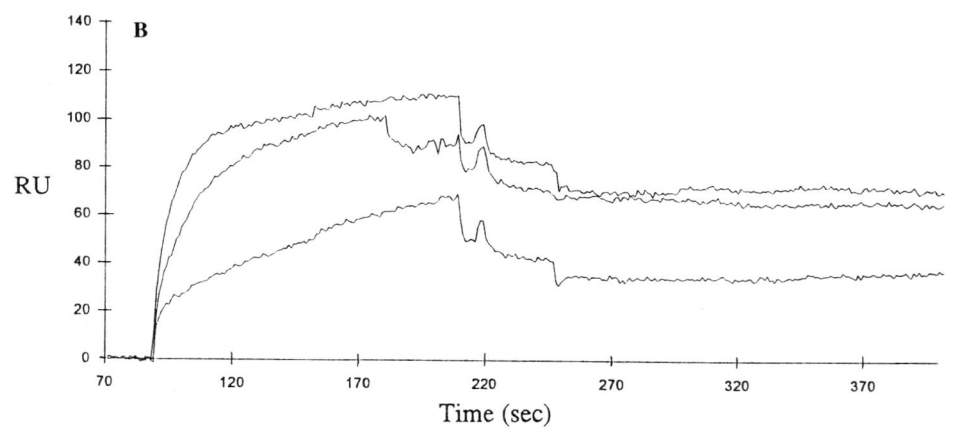

Figure 6. Binding kinetics for anti-CEA antibodies sFV-14-T84.66 and sFV-28-T84.66. CEA was covalently bound to CM-dextran as shown in Figure 3. Regeneration was performed with two 60 sec pulses of 10 mM HCl. A, sFv-14-T84.66: the ligate concentrations were 25 nM, 50 nM, 100 nM, 200 nM, and 400 nM. B, sFv-28-T84.66: the ligate concentrations were 25 nM, 50 nM, 100 nM, 200 nM, and 400 nM.

interaction of two C_H3 domains. Two versions were generated, one bearing a two amino acid linker between the V_H and C_H3 domains (mini-LD-cT84.66), and one bearing the natural human IgG1 hinge between the two domains (mini-flex-cT84.66). The binding constants for both minibodies are shown in *Table 2*. Interestingly, the K_ds are similar to the scFvs, and still higher than the parent antibody. While the loss in affinity for the divalent minibody is hard to explain, it should be noted that it is still in the very high range, and that most antibodies with low nanomolar binding constants are suitable for clinical applications. The loss in affinity for scFvs is easier to explain, since

5: *Measuring antibody affinity using biosensors*

they are either monovalent, or in our best case, a mixture of monomers and dimers. (Note that throughout this book, the term K_d has been consistently used; a gain in affinity, K_a, is a loss in K_d.)

4.4 Assays for anti-idiotypic antibody

The last antibody studied was the anti-idiotypic antibody 6G6.C4H (16) 6G6.C4H was immobilized on the biosensor chip and the binding of both mT84.66 and cT84.66 was analysed. Analysis of the binding kinetics gave K_d = 1.65 × 10^{-11} M for mT84.66 and 4.45 × 10^{-11} M for cT84.66. By this analysis, 6G6.C4H is a moderate affinity antibody, recognizing the native and engineered versions of T84.66 with nearly equal affinity.

5. Conclusions

These studies demonstrate the utility of real-time binding studies for determining the affinity constants and kinetic parameters for engineered antibodies, ranging from whole dimeric antibodies, to mono- and divalent versions of scFvs. Single chain constructs with short linkers (< 14) have a marked tendency to dimerize forming diabodies (23). In the case of scFv-14-T84.66, we observed a 50/50 mixture of monomers and dimers. We believe that the equilibrating forms were responsible for the fluctuating curves observed for the binding of this antibody to immobilized CEA (*Figure 6A*). Since the same linker was used in the minibodies, it may have affected the stability constants of these constructs. As expected, the bivalent constructs show the highest affinity constants, by virtue of their longer dissociation rates. During the course of the analysis, we encountered a major problem, namely the irreversible association of T84.66 with CEA, a problem that was overcome by setting up indirect assays. In this respect we were fortunate that we had additional anti-CEA antibodies that recognized non-overlapping epitopes on CEA, and whose affinity constants were low enough to allow regeneration of the biosensor chip. Although the indirect assay requires addition of CEA prior to each binding experiment, we were able to obtain good kinetic data. The assay also worked for the analysis of the CEA A3B3 domain which was expressed in HeLa cells.

In several cases we observed that the engineered antibodies had higher K_ds than the parent antibodies, suggesting that the engineering had affected either their antigen combining sites or their avidity for antigen. While the differences in K_ds observed cannot be explained by the kinetic data alone, the net method provides a rapid screen for the selection of optimum antibodies for use in clinical trials. We therefore recommend the routine use of kinetic and affinity screening for engineered antibodies prior to preclinical analysis.

Acknowledgements

This research was supported by NCI grant CA 43904 from the National Institutes of Health to J. E. S., and grants GA 08712 from BMFT and W56/94/Ne 2 from Deutsche Krebsshilfe to M. N.

References

1. Fagerstam, L. G., Frostell-Karlsson, A., Karlsson, R., Persson, B., and Ronnberg, I. (1992). *J. Chromatogr.*, **597**, 397–410.
2. Cush, R., Cronin, J. M., Stewart, W. J., Maule, C. H., Molloy, J., and Goddard, N. J. (1993). *Biosens. Bioelectroni*, **8**, 347.
3. Karlsson, R., Michaelsson, A., and Mattsson, L. (1991). *J. Immunol. Methods*, **145**, 229–40.
4. O'Shannessy, D. J., Brigham-Burke, M., Soneson, K. K., Hensely, P., and Brooks, I. (1993). *Anal. Biochem.*, **212**, 457–68.
5. O'Shannessy, D. J., Brigham-Burke, M., Soneson, K. K., Hensely, P., and Brooks, I. (1994). In *Methods in Enzymology* Vol. 240, pp. 323–49. Academic Press, San Diego.
6. Malmborg, A.-C., Michaelsson, A., Ohlin, M., Jansson, B., and Borrebaeck, C. A. (1992). *Scand. J. Immunol.*, **35**, 643–50.
7. Altschuh, D., Dubs, M. C., Weiss, E., Zeder-Lutz, G., and Van-Regenmortel, M. H. (1992). *Biochemistry*, **31**, 6298–304.
8. Gold, P. and Freedman, S. O. (1965). *J. Exp. Med.*, **121**, 439–62.
9. Egan, M. L., Pritchard, D. G., Todd, C. W., and Go, V. L. W. (1977). *Can. Res.*, **37**, 2638–43.
10. Fritsche, R. and Mach, J.-P. (1977). *Immunochemistry*, **14**, 119–27.
11. Shively, J. E. and Beatty, J. D. (1985). *Crit. Rev. Oncol. Hematol.*, **2**, 355–99.
12. Hass, G. M., Bolling, T. J., Kinders, R. J., Henslee, J. G., Mandecki, W., Dorwin, S. A., and Shively, J. E. (1991). *Cancer Res.*, **51**, 1876–82.
13. Wagener, C., Yang, Y. H. J., Crawford, F. G., and Shively, J. E. (1983). *J. Immunol.*, **130**, 2308–15.
14. Wagener, C., Clark, B. R., Rickard, K. J., and Shively, J. E. (1983). *J. Immunol.*, **130**, 2302–7.
15. Neumaier, M., Shively, L., Chen, F.-S., Gaida, F.-J., Ilgen, C., Paxton, R. J., Shively, J. E., and Riggs, A. D. (1990). *Cancer Res.*, **50**, 2128–34.
16. Gaida, F.-J., Pieper, D., Roder, U. W., Shively, J. E., Wagener, C., and Neumaier, M. (1993). *J. Biol. Chem.*, **268**, 14138–45.
17. Hefta, L. J. F., Chen, F.-S., Ronk, M., Sauter, S., Sarin, V., Oikawa, S., Nakazato, H., Hefta, S., and Shively, J. E. (1992). *Cancer Res.*, **52**, 5647–55.
18. Johnsson, B., Lofas, S., and Lindquist, G. (1991). *Anal. Biochem.*, **198**, 268–77.
19. Coligan, J. E., Henkart, P. A., Todd, C. W., and Terry, W. D. (1973). *Immunochemistry*, **10**, 591–9.
20. O'Shannessy, D. J., Brigham-Burke, M., and Peck, K. (1992). *Anal. Biochem.*, **205**, 132–6.

21. Wu, A. M., Chen, W., Raubitschek, A. A., Fischer, R., Williams, L. E., Neumaier, M., Hu, S. Z., Maryon, T., Wong, J. Y. C., and Shively, J. E. *Immunotechnology*, in press.
22. Hu, S.-Z., Shively, L., Raubitschek, A. A., Sherman, M., Williams, L. E., Wong, J. T. C., Shively, J. E., and Wu, A. M., submitted.
23. Holliger, P., Prospero, T., and Winter, G. (1993). *Proc. Natl Acad. Sci. USA*, **90**, 6444–8.

6

Analysis of human antibody sequences

GERALD WALTER and IAN M. TOMLINSON

1. Introduction

Diverse repertoires of antibody genes are generated during B-cell development by the combinatorial rearrangement of relatively small numbers of basic building blocks: V_H (variable), D (diversity), and J_H (joining) segments for the heavy (H) chain variable domain and V_L and J_L segments for the light (L) chain variable domain (for review see ref. 1). The antigen-binding site comprises three loops from the heavy chain variable domain (H1, H2, and H3) and three from the light chain variable domain (L1, L2, and L3), which are supported by a conserved β-sheet scaffold (2).

In humans, there are up to 51 functional V_H, about 30 functional D and six functional J_H segments which have been mapped to Chromosome 14 (3, 4). For light chains which may be either kappa (κ) or lambda (λ), up to 40 functional V_κ and five functional J_κ segments are located on chromosome 2 (5, 6) and an estimated 30 functional V_λ and four functional J_λ segments are located on chromosome 22 (7).

The rearrangement of V, D, and J segments introduces two levels of diversity into the antigen-binding site of the antibody. First, there is combinatorial diversity, that is the choice of one of several V_H, D, and J_H segments for the heavy chain variable domain and one of several V_κ or V_λ and J_κ or J_λ segments for the light chain variable domain, coupled with the association of different heavy and light chains. Secondly, a great deal of diversity can be introduced at the V–(D)–J joins due to imprecise joining, the use of different D segment reading frames and the addition of N and P nucleotides at the join (8). Rearranged V genes are further diversified by somatic mutation during an immune response (9).

2. Amplification and cloning of antibody V genes

Primers for PCR amplification of immunoglobulin genes can be based in the genes themselves, in their flanking regions or for cloned fragments, in appropriate vector or linker regions. As shown in *Figure 1* and *Table 1*, there

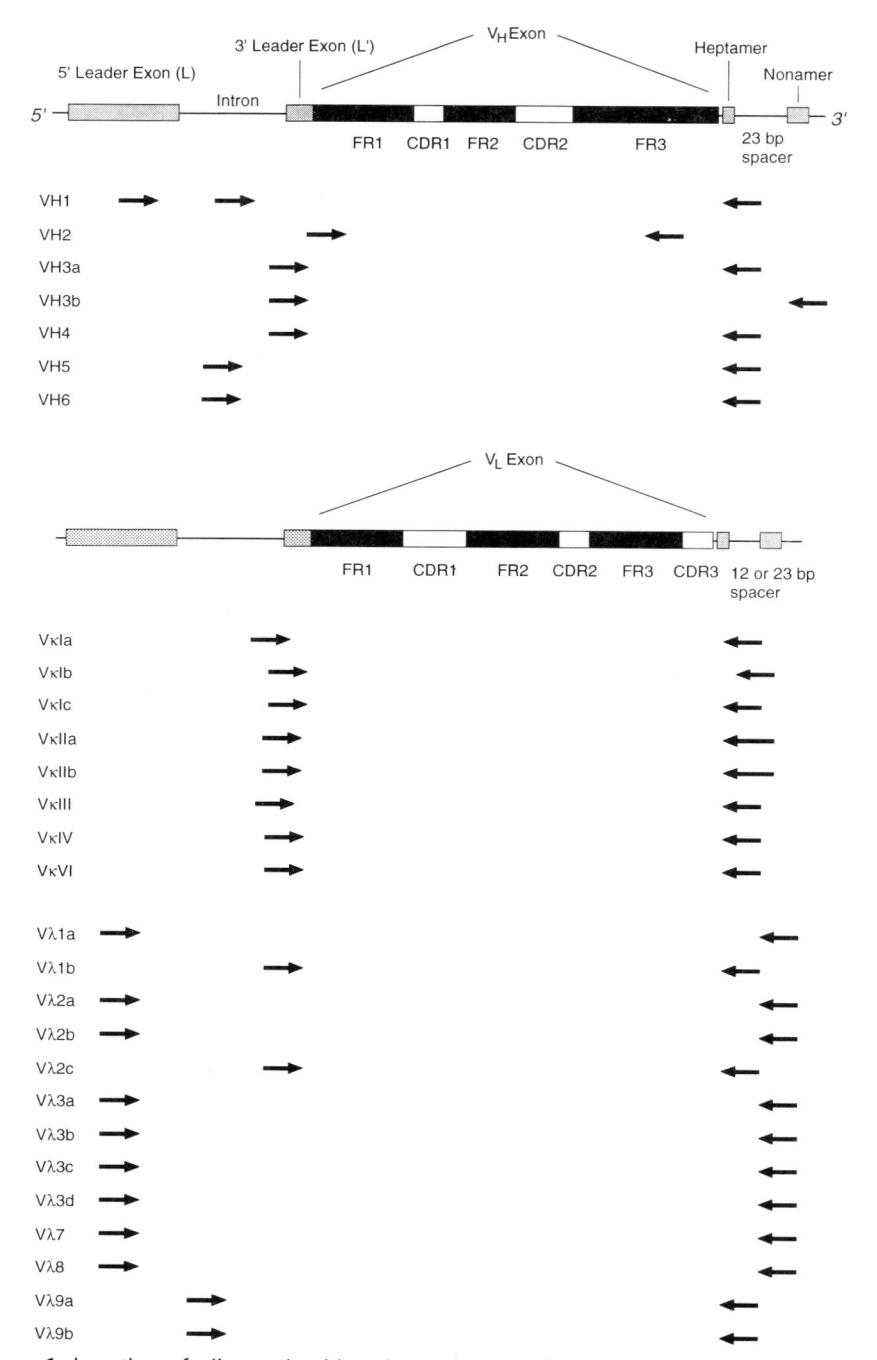

Figure 1. Location of oligonucleotide primers for amplifying human V gene segments from genomic DNA.

6: Analysis of human antibody sequences

Table 1a. Primers for amplifying V gene segments from genomic DNA

Primer set	Primers (5' or 3')	5' Sequence 3'[a]
V$_H$1 H, M[b]	VH1 LEA EX4 5'	CCA TGG ACT GGA (CT)(CT)T GGA G
	VH1 LEA IN 5'	G(AG)A (AG)G(AG) GAT T(GT)(AGT) (GT)TC CAG T
	VH1 HEPT 3'	T(CG)T GG(GT) TT(CT) TCA CAC TGT G
V$_H$2 H, M	VH2 INT B2 5'	CCT TGA (AG)GG AGT CTG GTC C
	VH2 INT FOR 3'	CCA CAG GGT CCA TGT TGG
V$_H$3a M	VH3 LEA3 5'	GT(AT) TGC A(AG)G TG(CT) CCA GTG T
	VH3 HEPT 3'	(AC)TG (AG)(CCT) TCC CCT C(AG)C T(CG)T G
V$_H$3b M	VH3 LEA3 5'	GT(AT) TGC A(AG)G TG(CT) CCA GTG T
	VH3 NON1 3'	GGT TTG TG(TC) C(TC)G GGC (GT)CA
V$_H$4	VH4 LEA 5'	CTG TTC ACA GGG GTC CTG TC
	VH4 HEPT 3'	ACT CAC CTC CCC TCA CTG TG
V$_H$5	VH5 LEA 5'	AGG TCA CAG AG(AG) AGA A(CT)G G
	VH5 HEPT 3'	GCT GGT TTC TCT CAC TGT G
V$_H$6	VH6 LEA 5'	TCA CAG CAG CAT TCA CAG A
	VH6 HEPT 3'	CTG ACT TCC CCT CAC TGT G
V$_\kappa$Ia H, M	VK1(BA1) 5'	TGT TCC TAA TAT CAG ATA
	VK1 (FA) 3'	CGG GCT TGT ATC ACA GTG
V$_\kappa$Ib H, M	VK1(BLO) 5'	AAT C(TG)C AGG T(GT)C CAG ATG
	VK1(FL) 3'	GTT (CT)(AG)G GT(GT) (GT)GT AAC ACT
V$_\kappa$Ic	VK1(BLO) 5'	AAT C(TG)C AGG T(GT)C CAG ATG
	VK1(FO) 3'	ATG (AC)CT TGT (TA)AC ACT GTG
V$_\kappa$ IIa H, M	VK2(BOA) 5'	(TA)A(TC) TTC AGG ATC CAG TG
	vk2(fa1x) 3'	GAG GTT TTC TAG A(TG)G (GA)(GT)(CT) TGT A(GC)C ACT GTG
V$_\kappa$ IIb H, M	VK2(BOA) 5'	(TA)A(TC) TTC AGG ATC CAG TG
	vk2(fa2x) 3'	GAG GTT TTC TAG AAG (GA)(GT)(CT) TGT A(GC)C ACT GTG
V$_\kappa$ III H, M	VK3(LA) 5'	TCC AAT (TC)T(CT) (AG)GA TAC CAC
	VK3(F1) 3'	T(CT)A (TA)G(CT) TGA ATC ACT GTG
V$_\kappa$ IV H, M	VK4 LEA 5'	ACT ACA GGT GCC TAC GGG
	VK4 HEPT 3'	CGA GGC TGA AGC ACT GTG
V$_\kappa$ VI H, M	VK6 LEA 5'	TTT TCA GCC TCC AGG GGT
	VK6 HEPT1 3'	GGG TTG TA(GA) CAC AGT GTG
Vλ1a	Vλ1 LEA1 5'	(AG)CC (TG)G(GC) T(CT)(CT) CCT CTC CTC
	Vλ1 NON1 3'	GGT TCT TGT CTC AGT TCC
Vλ1b	Vλ1 BACK2 5'	CTT CCA (CG)GG TCC TGG G(CT)C
	Vλ1 FOR3 3'	C(AC)(TCA) (TC)(GT)G (GC)CT GGA GCA CTG T
Vλ2a	Vλ2 LEA1 5'	ATG GCC TGG GCT CTG CTG
	Vλ2 NON1 3'	GGT TTT GGT CTC AGT TCC
Vλ2b	Vλ2 JLEAD 5'	GGG CTC TGC TGC TCC TCA C
	Vλ2 JNON 3'	GGT TTT GGT CTC AGT TCC
Vλ2c	Vλ2 BACK3 5'	(CT)TT TCC AGG (CAG)TC (CT)TG GGC
	Vλ2 FOR3 3'	A(ATG)G AAC TTG GAC C(AT)C TGT
Vλ3a	Vλ3 LEA1 5'	GCC TGG ACC CCT CTC TGG
	Vλ3 NON1 3'	GGT TTC TGT CTC ACT TCC
Vλ3b	Vλ3 JLEAD 5'	GGA CCC CTC TCT GGC TCA C
	Vλ3 JNON 3'	GGT TTC TGT CTC ACT CCT

Table 1a. *Continued*

Primer set	Primers (5' or 3')	5' Sequence 3'[a]
Vλ3c	Vλ10 BACK1 5'	AGA CCC TTA TCT TCA GAC
	Vλ3 JNON 3'	GGT TTC TGT CTC ACT CCT
Vλ3d	Vλ3 BACK4 5'	CTC GGC GTC CTT GCT TAC
	Vλ3 JNON 3'	GGT TTC TGT CTC ACT CCT
Vλ7	Vλ7 LEA1 5'	GCC TGG ACT CCT CTC TTT
	Vλ7 NON1 3'	GGT TTA TGT CTT GGT TCC
Vλ8	Vλ8 BACK1 5'	TGG ATG ATG CTT CTC CTC
	Vλ8 FOR1 3'	TAG GTT TAA ATC ACT GTG
Vλ9a	Vλ9 BACK1 5'	GCT CAT CAG CCA CCC ACC
	Vλ9 FOR1 3'	CCT CTG CCT GTG TCA CTG
Vλ9b	Vλ9 BACK2 5'	GTT TAC AGG TCT CTG TGC
	Vλ9 FOR2 3'	TTC ATC TGT GTC ACT GTG

Table 1b. Forward (3') primers for amplifying rearranged antibody V genes

Primer	Location	5' Sequence 3'[a]
HJ 1245	J_H region	T GAG GAG ACG GTG ACC AGG GT
HJ 36	J_H region	T GA(AG) GAG ACG GTG ACC (AG)T(TG) GT
IGM CH1 FOR	IgM C_H1 region	T CCG ACG GGG AAT TCT C
IGG CH1 FOR3	IgG C_H1 region	C GGT TCG GGG AAG TAG T
HU-IGM-S-F-1150	IgM secretory exon	TAC AGG GTG GGT TTA CC
HU-IGM-M-F-1210	IgM membrane exon	AGG AAG AGG ACG ATG AA
HU-IGG-S-F-1129	IgG secretory exon	GGC TGT CGC ACT CAT TT
HU-IGG-M-F-1185	IgG membrane exon	AGA TGG TGA TGG TCG TC
HU-IGG-M-F-1231	IgG membrane exon	GTG ACG GTG GCA CTG TA
HuJκ1FOR	J_κ region	TTT GAT TTC CA(CG) CTT GGT CCC
HuJκ3FOR	J_κ region	TTT GAT ATC CAC TTT GGT CCC
HuJκ5FOR	J_κ region	TTT AAT CTC CAG TCG TGT CCC
CKF 19	C_κ region	GGC GGG AAG ATG AAG ACA G
CKF 67	C_κ region	AGC AGG CAC ACA ACA GAG G
HuJλ2-3FOR	J_λ region	TAG GAC GGT CAG CTT GGT CCC
HuJλXFOR	J_λ region	TAG GGC GGT CAG CAT GGT CCC
CLFOR1	C_λ region	AGT GTG GCC TTG TTG GC
CLFOR2	C_λ region	AGC TCC TCA GAG GAG GG

[a] Restriction sites have been omitted and are to be added according to vectors used for cloning.
[b] Primer sets marked *H* or *M* can be used to amplify human V gene segments from hamster or mouse backgrounds, respectively.

is a vast range of primers which have been used successfully for amplifying human variable region genes. Although the majority are degenerate and family-specific, some segment-specific primers have been designed for special applications (10). Combinations of primers can be used in 'nested' approaches. This is particularly appropriate if the amount of template is very small as in 'haploid genome equivalent' (10) or 'single chromosome' PCR

6: Analysis of human antibody sequences

(4). A choice of forward (3') primers is available for amplifying germline or rearranged V genes.

Due to the introduction of PCR artefacts during amplification (single base substitutions or crossovers between related segments), whenever possible sequences should always be confirmed in clones from at least two independent amplifications (11).

If antibody V genes are to be cloned for sequencing only and subsequent expression of the insert is not required, they can be cloned into M13 phage vectors (12).

2.1 Germline V segments

For amplifying germline V segments from genomic DNA (*Protocol 1*), we designed family-specific forward (3') primers based on the heptamer or nonamer recombination signals and parts of the recombination spacer at the 3' end of the V exons (*Figure 1, Table 1a*). Priming from the heptamer has been used before (13, 14) and has the advantage that since the heptamer and nonamer are lost during recombination, rearranged V genes are not amplified. Back (5') primers can be based in the leader peptide or leader intron.

Protocol 1. Amplification and cloning of germline V gene segments from genomic DNA

Equipment and reagents

- Thermal cycler (0.5 ml tube or 96-well plate model; Biometra or Perkin Elmer)
- 0.5 ml microcentrifuge tubes (Treff Lab) or 96-well thermostable microtitre plate (Costar)
- Millipore-filtered water
- 10 × Taq polymerase buffer: 500 mM KCl, 100 mM Tris–HCl, pH 8.8, 15 mM $MgCl_2$, 1% Triton X-100
- 5 mM (each) dNTP mix (Pharmacia)
- 10 μM (10 pmol/μl) oligonucleotide primers
- Taq polymerase (5 U/μl)
- Mineral oil (Sigma)
- Agarose (electrophoresis grade)
- 10 × TBE (Tris–borate–EDTA) buffer: 108 g Tris-base, 55 g boric acid, 8.3 g EDTA
- Molecular weight markers (123 bp ladder; Life Technologies)
- Wizard PCR Preps, DNA Clean-up and Minipreps kits (Promega)
- 2 ml disposable syringes (Plastipack, Becton Dickinson)
- 80% propan-2-ol (isopropanol)
- Restriction enzymes (New England Biolabs)
- Phenol (saturated with 0.1 M Tris–HCl, pH 8.0)
- Chloroform/3-methyl-1-butanol 98% (isoamyl alcohol) (24 : 1 ratio)
- Ethanol/Na-acetate mix: 96 ml ethanol, 4 ml 3 M Na-acetate, pH 4.8
- 70% ethanol
- QIAGEN Blood & Cell Culture DNA Mini kit (QIAGEN)
- 10 ml screw-capped tubes
- M13mp19 (Boehringer Mannheim)
- Gene Pulser (Bio-Rad)
- Electro-competent *E. coli* BMH 71–18 (15) or TG1 (16) bacteria
- 1 M IPTG: 2.38 g isopropyl β-D-thio-galactopyranoside in 10 ml Millipore filtered-water, store at –20°C
- 5% X-Gal: 0.5 g 5-bromo-4-chloro-3-indolyl-β-D-galactopyranoside in 10 ml dimethylformamide, wrap tube in aluminium foil and store at –20°C
- H-top: 8 g Bacto-agar, 10 g Bacto-tryptone, 8 g NaCl, in 1 litre deionized water, pH 7.0. To 50 ml H-top, add 267 μl 5% X-Gal, 56 μl 1 M IPTG
- TYE plates: 15 g Bacto-agar, 10 g Bacto-tryptone, 8 g NaCl, 5 g yeast extract, in 1 litre deionized water, pH 7.0
- Wooden toothpicks

Protocol 1. *Continued*

Method

1. Prepare genomic DNA with QIAGEN Blood & Cell Culture DNA Mini kit.

2. Set up 50 μl PCR reactions in 0.5 ml microcentrifuge tubes or a 96-well thermostable microtitre plate containing:

    - 10 × Taq polymerase buffer   5.0 μl
    - dNTP mix (5 mM each dNTP)   2.0 μl
    - back primer (10 μM)   2.5 μl
    - forward primer (10 μM)   2.5 μl
    - Taq polymerase (5 U/μl)   0.5 μl

    Add genomic DNA (100 ng–10 μg, or equivalent quantity of phage lambda, cosmid or YAC DNA) and Millipore-filtered water to 50 μl. Allow a negative control (no DNA added) for each primer combination.

3. Overlay the reaction mixture with mineral oil and spin briefly in a microcentrifuge.

4. Transfer to thermal cycler preheated to 94°C. Hold at 94°C for 5 min, then perform 25–35 cycles of amplification (depending on amount of template): 94°C for 1 min (denaturation), 55°C for 1 min (annealing), and 72°C for 1 min (extension); final extension at 72°C for 5 min. Hold at 4°C.

5. After cycling, run 5 μl of the PCR mixture on a 1.5% agarose gel in 1 × TBE buffer alongside molecular weight markers (123 bp ladder).

6. Purify remaining 45 μl using Wizard PCR Preps kit (see *Protocol 4*, step 5).

7. Cut overnight using appropriate restriction enzymes in recommended buffer and isolate DNA from digestion mixtures using Wizard DNA Clean-Up System (same protocol as for Wizard PCR preps (*Protocol 4*, step 5), except that no Direct Purification Buffer is required).

8. Ligate overnight into precut and purified M13mp19.

9. Clean ligated product using Wizard Minipreps kit (same protocol as for Wizard PCR Preps (*Protocol 4*, step 5), except that no Direct Purification Buffer is required and buffered wash solution is provided). Alternatively, extract twice with phenol/chloroform and precipitate with ethanol (see *Protocol 3*, steps 8–12).

10. Transform electro-competent *E. coli* (BMH 71-18 or TG1) bacteria using the Bio-Rad Gene Pulser system.

11. Add transformed bacteria to 3 ml of molten H-top media containing X-Gal and IPTG and pour on to warmed TYE plates. Grow overnight and toothpick colourless plaques for sequencing.

2.2 Rearranged V genes

Forward (3') primers for amplification of rearranged V genes are usually based either in J or C regions (*Table 1b*). If the starting material is polyA$^+$ mRNA (*Protocol 2*), back (5') primers must be based in the leader peptide or V exon sequences and not in the leader intron. If rearranged V genes are to be cloned for expression, back (5') primers for PCR should be based at the start of FR1 (see Chapter 1).

Protocol 2. Amplification and cloning of rearranged V genes from mRNA

Equipment and reagents

- As in *Protocol 1*
- RNAzol B (Cinna/Biotecx)
- First-strand cDNA synthesis kit (Pharmacia)

Method

1. Prepare mRNA as in Chapter 1 (Section 3.1). Alternatively, RNA can be prepared by the RNAzol B method (Cinna/Biotecx). Briefly, cells are homogenized in RNAzol B, RNA is extracted with chloroform, precipitated with propan-2-ol and washed with 70% ethanol.

2. Synthesize cDNA with First-strand Synthesis kit using oligo(dT) (Pharmacia) random hexamer (Chapter 1, *Protocol 2*) or specific immunoglobulin constant region primers (*Table 1b*). Briefly, 5 μg of RNA are mixed with 1 μl primer (25 μM), 1 μl DTT, and 5 μl First Strand Mix of the kit and incubated at 37°C for 1 h. RNA:cDNA duplexes are denatured at 90°C for 5 min and kept on ice.

3. Set up PCR reactions and amplify DNA as in *Protocol 1*, steps 2–4.

4. Proceed for cloning as in *Protocol 1*, steps 5–11.

---

3. Sequencing of immunoglobulin genes

The first antibody sequence was obtained by protein sequencing using the method of Edman degradation (17). Since then, recombinant DNA technology has taken over and now most antibodies are sequenced at the DNA level. To date, the chain termination method of DNA sequencing (Sanger technique, ref. 18) is by far the most widely used. This technique is based on the polymerization of a single strand of DNA by extension of an oligonucleotide primer annealed to the template of interest. With reagent kits and instrumentation readily available, the most crucial components are primer and template.

3.1 Sequencing primers

Sequencing primers have to fulfil certain criteria which are of particular importance when using real-time detection systems (like automated sequencers) where signal strength is critical. Such primers should **not**:

- contain long runs of a single base (i.e. more than three or four, especially of Gs or Cs);
- form secondary structures or hybridize to form dimers;
- be too short (less than 16 bases long) to ensure good and specific annealing;
- have melting temperatures below 45°C.

Despite these guidelines, it is still almost impossible to predict the performance of a certain primer, and the ultimate test is to try it for a specific sequencing application. For this reason, it is preferable to choose sequencing primers based in vectors, linkers, or other conserved regions where they can be optimized and used for any project irrespective of the insert. Such primers are often identical to PCR primers used in another context, and we have therefore included their sequences in Appendix 2.

3.2 Template preparation

Having selected the best primer, obtaining high-quality template is the next hurdle in the sequencing protocol. Almost all sequencing kits can be used with single- or double-stranded DNA. The only crucial factor is its purity.

3.2.1 Single-stranded DNA

Single-stranded DNA gives longer reads with fewer ambiguities than double-stranded DNA. Before PCR cycle sequencing (see below) became available, single-stranded DNA was also more convenient than double-stranded DNA, which required an additional denaturation step before sequencing. For sequencing of antibody V genes, single-stranded M13 phage DNA is prepared as described in *Protocol 3*.

Protocol 3. Preparation of single-stranded DNA templates for sequencing

Equipment and reagents

- As in *Protocol 1*
- 10 ml glass or polyethylene tubes
- 20% PEG 6000/2.5 M NaCl (Sigma)
- 2 × TY medium: 16 g Bacto-tryptone, 10 g Bacto-yeast extract, 5 g NaCl in 1 litre deionized water, pH 7.0

Method

1. Grow *E. coli* bacteria overnight at 37°C in 2 × TY medium.

6: Analysis of human antibody sequences

2. Dilute 1 ml of this culture into 100 ml 2 × TY medium and place 1.5 ml aliquots in 10 ml glass or polyethylene tubes.
3. Toothpick single phage plaques from an H-top agar plate (*Protocol 1*) into the 1.5 ml aliquots.
4. Grow at 37 °C for 5 h with vigorous shaking.
5. Transfer cultures to 1.5 ml microcentrifuge tubes and spin for 5 min in a microcentrifuge at top speed (around 13 000 *g*).
6. Transfer 1 ml of the supernatant to a new 1.5 ml microcentrifuge tube, add 200 µl of 20% PEG/2.5 M NaCl and incubate for 30 min at room temperature.
7. Spin for 5 min, discard the supernatant, respin briefly, remove last traces of supernatant and resuspend the phage pellets in 200 µl of Millipore-filtered water.
8. Add 100 µl phenol and 100 µl chloroform, vortex hard for at least 6 sec, and spin for 10 min.
9. Transfer 190 µl of the aqueous phase to a new tube and repeat phenol/chloroform extraction.
10. Transfer 180 µl of the aqueous phase to a new tube, add 500 µl ethanol/Na-acetate mix, vortex, and keep at −20 °C or −70 °C for at least 15 min.
11. Spin in a microcentrifuge for 20 min at high speed, wash pellet with 500 µl 70% ethanol and respin for 5 min.
12. Remove bulk of ethanol, let the rest evaporate, dissolve the pellet in 25 µl of Millipore-filtered water and store at −20 °C.

3.2.2 Double-stranded DNA
Although double-stranded DNA tends to give shorter sequencing reads than single-stranded DNA, it is normally more than adequate for sequencing antibody V genes which are between 300 and 400 nucleotides in length. Direct sequencing of plasmids or phagemids requires high-purity preparations of double-stranded vector DNA, free from host DNA. This can be achieved using commercially available miniprep kits, such as Wizard Minipreps (Promega), according to the recommended protocol.

3.2.3 PCR products
PCR products are the most popular choice of sequencing templates today. Their preparation is very straightforward, does not involve any overnight growth steps, and can be automated. Subject to the above criteria (see Section 3.1), the same primers used for template amplification can be used for sequencing. Furthermore, the template can be sequenced from either end.

The general procedure for the preparation of PCR templates is outlined in

Protocol 4. The starting material can be cloned double- or single-stranded DNA, colonies or plaques picked from a plate, or, alternatively, a glycerol stock.

However, direct amplification from genomic DNA (see *Protocol 1*) is also possible. The main drawback of this approach is the possibility of mispriming elsewhere in the genome. Even if a band of the correct length is excised from a gel, misprimed amplification products the same size as the target sequence may cause mixed sequencing reads which are impossible to resolve.

Protocol 4. Preparation of PCR templates

Equipment and reagents
- As in *Protocol 1*
- U-form 96-well microtitre plate (Greiner)
- 0.22 µm 96-well filtration plate from Millipore Multiscreen Assay System
- Plate sealer (Greiner)
- 8-well multichannel pipette

Method

1. Set up 50 µl PCR reactions as in *Protocol 1*, step 2.

2. Toothpick bacterial colonies or phage plaques into the reaction mixtures and overlay with a drop of mineral oil.

3. Immediately transfer microcentrifuge tubes or microtitre plate into the preheated thermal cycler, hold at 94°C for 2 min and amplify for 20 cycles of 94°C for 1 min, 55°C for 1 min and 72°C for 1 min, with a final extension at 72°C for 10 min. Hold at 4°C.

4. After cycling, run 5 µl of the PCR mixture on a 1.5% agarose gel in 1 × TBE buffer alongside molecular weight markers (123 bp ladder).

5. *Either* Purify with Wizard PCR Preps as described in the manufacturer's instructions. Briefly, remove remaining 45 µl from under the oil, transfer into a 1.5 ml microcentrifuge tube containing 100 µl of Direct Purification Buffer and vortex to mix. Add 1 ml of Wizard PCR Preps resin. Vortex briefly three times over a 1 min period. Apply the resin/DNA mix to the minicolumn/syringe barrel assembly and push or suck through the minicolumn. Wash with 2 ml 80% propan-2-ol, dry by centrifugation, elute in 50 µl Millipore-filtered water, and discard the minicolumn. Store the purified PCR templates in the microcentrifuge tube at −20°C.

Or alternatively

Aliquot 250 µl of Wizard PCR Preps resin/Direct Purification premix (mixed in a 10:1 ratio) to each well of a 96-well filtration plate. Trans-

6: Analysis of human antibody sequences

fer remaining 45 μl of PCR mixture to filtration plate using an 8-well multichannel pipette and mix by pipetting up and down. Stack on top of a used microtitre plate to collect waste and centrifuge at 1500 *g* for 3 min. Add 250 μl of 80% propan-2-ol to each well. Centrifuge at 1500 *g* for 3 min. Add a further 250 μl of 80% propan-2-ol to each well. Centrifuge at 1500 *g* for 3 min. Add 50 μl Millipore-filtered water to each well. Wait 1 min. Stack on top of a new U-form 96-well microtitre plate for sample collection and centrifuge at 1500 *g* for 3 min. Apply plate sealer and store the purified PCR templates in the microtitre plate at −20 °C.

3.3 Sequencing techniques

The basic principle of the Sanger technique of DNA sequencing has not changed since its introduction. It involves the primer-initiated and dideoxynucleotide-terminated polymerization of a single DNA strand on a template DNA strand, followed by size fractionation of the synthesized chains on a polyacrylamide gel. Each chain length is characterized by one of the four terminators (ddA, ddC, ddG, ddT), which corresponds to the complementary nucleotide at that position in the template DNA strand. When run in a high-resolution gel, the reactions form a ladder in steps of one nucleotide (each 'rung' on the ladder corresponds to A, C, G, or T). The DNA ladder can be visualized using fluorescent or radioactive labels which are either attached to the primer, the terminators, or incorporated into the DNA during extension. Reading the ladder from bottom to top gives the reverse-complement sequence of the template. Technical refinements of this basic principle were focused in three areas:

- different polymerases used for chain extension
- different labels
- reading of sequencing gels performed manually or by automated methods.

3.3.1 Manual sequencing

Theoretically, there is no difference between manual and automated sequencing, apart from reading the gel. The most popular polymerases, Sequenase (USB) and Taq polymerase, can be used for both techniques. Therefore, the main differences are in the labels used, with radioactive ^{35}S being the favoured isotope for manual sequencing. Although DNA labelled with ^{35}S requires longer exposure to photographic film than ^{32}P, the resulting sequence ladder is much sharper. Protocols using the Sequenase kit for sequencing single-stranded DNA and the fmol kit (Promega) for double-stranded plasmid DNA are described in *Protocols 5* and *7*, respectively.

Protocol 5. Sequencing of single-stranded DNA with Sequenase and ^{35}S

Equipment and reagents
- As in *Protocol 1*
- Sequenase kit (USB)
- [α-^{35}S]dATP (10 μCi/μl, Amersham)

Method
1. Place 7 μl (1 μg) of single-stranded DNA template in a 0.5 ml microcentrifuge tube.
2. Add 3 μl priming mix (1 μl primer (0.5 μM) and 2 μl Reaction Buffer), mix by vortexing, and spin briefly in microcentrifuge.
3. Heat-denature at 65°C for 2 min in thermal cycler (tube model), and let cool to room temperature over 30 min. Place on ice.
4. Make up labelling solution (amounts for 24 templates): 26 μl DTT, 10.4 μl Labelling Mix from kit (7.5 μM dGTP, 7.5 μM dCTP, 7.5 μM dTTP or 15 μM dITP, 7.5 μM dCTP, 7.5 μM dTTP), 13 μl [α-^{35}S] dATP, 41.6 μl Millipore-filtered water.
5. Add 3.5 μl labelling solution to each primer–template tube and place on ice.
6. Aliquot 2.5 μl of each termination mix (T, C, G, A for each template) into a 96-well thermostable microtitre plate.
7. For 24 samples, dilute 9 μl Sequenase with 51 μl Enzyme Dilution Buffer.
8. Add 2 μl diluted Sequenase to the first primer–template–label tube and place into thermal cycler (tube model) set to 15°C. Proceed with the remaining tubes until only two tubes remain to be labelled.
9. Place 96-well thermostable microtitre plate with termination mixes into thermal cycler (plate model) set to 42°C. After Sequenase has been added to the last tube (or after 2 min from adding Sequenase to the first tube), begin terminating the reactions by adding 3.5 μl reaction mix to each of the four termination mixes in the microtitre plate, pipetting up and down once.
10. When 5 min has elapsed, add 4 μl Stop Solution.

3.3.2 Automated sequencing

The introduction of automated sequencing machines during the 1980s has revolutionized DNA sequencing and is the basis of all current large-scale sequencing projects. The major advantages of automated sequencing are the computerized analysis of the gel and electronic data storage.

6: Analysis of human antibody sequences

Protocol 6. Preparation, running, and processing of gradient sequencing gels for manual sequencing

Equipment and reagents

- Electrophoresis apparatus Base Runner (IBI Kodak) with glass plates (one mirrored and one 'notched' plate), gel spacers, bulldog clamps, and sharkstooth comb
- Chromatography paper (Whatman)
- Gel dryer
- Film cassette
- X-ray film (Biomax, IBI Kodak)
- Fuji X-Ray Film Processor RGII
- 40% acrylamide/bisacrylamide (19/1; Severn)
- Urea (enzyme grade)
- Bromophenol Blue
- 25% ammoniumpersulfate
- *N,N,N',N'*-Tetramethylethylenediamine (TEMED)
- Dimethyldichlorosilane
- 10% methanol/10% acetic acid fixation solution
- 2.5 cm electrical tape (Scotlab)
- 10 × TBE: 0.9 M Tris base, 0.9 M boric acid, 22 mM EDTA (108 g Tris, 55 g boric acid, 8.3 g EDTA, water to 1 litre, filtered through 0.45 µm filter, pH should be 8.0–8.5)

Method

1. Make up for gradient sequencing gel: *5 × TBE mix* (8.4 g urea, 3 ml 40% acrylamide/bisacrylamide, 10 ml 10 × TBE, Millipore-filtered water to 20 ml, 1.5 mg Bromophenol Blue; add 28 µl each of 25% ammoniumpersulfate and TEMED immediately before *pouring)* and *0.5 × TBE mix* (21 g urea, 7.5 ml 40% acrylamide/bisacrylamide, 2.5 ml 10 × TBE, Millipore-filtered water to 50 ml; add 90 µl each of 25% ammoniumpersulfate and TEMED immediately before pouring).

2. Wash glass plates and spacers in detergent (e.g. Alconox), rinse with deionized water and dry. Wipe and polish with ethanol. Treat one plate with dimethyldichlorosilane, wipe and polish with ethanol.

3. Assemble glass plates and spacers and seal gel chamber all round with electrical tape.

4. Take up 12 ml 5 × TBE mix (blue) and 10 ml 0.5 × TBE mix (colourless) in the same pipette (allow air to pass through pipette to mix the two solutions) and pour down the centre of the gel chamber. Take up the rest of the 0.5 × TBE mix with a 50 ml disposable syringe and fill the gel chamber up to the top. Avoid air bubbles! If there are bubbles, hold gel chamber vertically and squeeze plates together just below the bubble, forcing it upwards.

5. Insert sharkstooth comb up-side-down across the top to form a slot. Clamp the gel chamber with two bulldog clamps across the top and three on each side and allow to polymerize for at least 1 hour.

6. Remove clamps, tape, and comb from the gel chamber. Flush the slot with deionized water and mount the gel chamber on to the Base Runner electrophoresis apparatus. Insert the sharkstooth comb with the teeth just penetrating the gel surface. Assemble the buffer tanks and fill up with approximately 1 litre 1 × TBE buffer.

Protocol 6. *Continued*

7. Flush the sample slots with 1 × TBE buffer and load a few test slots with Stop Solution to test for leaks. Prerun the gel at 60 W for 10 min (or alternatively at 42 W for 30 min) to prewarm the gel chamber.
8. When ready to load the samples (*Protocols 5* and *7*) on to the gel, set the thermal cycler (plate model) to 75 °C, denature the samples for 5 min, and immediately load on to the gel in the order of T–C–G–A.
9. Run the gel at 47 W until the Bromophenol Blue (first) dye front is just about to run off the gel.
10. Disassemble the gel chamber. Take off the silanized plate, place the gel (sticking to the other plate) into approximately 2 litres of 10% methanol/10% acetic acid fixation mixture and fix for 15–20 min.
11. Carefully, remove the gel on the glass plate from the fixation solution. Gently lay chromatography paper on to the gel. Run your hand from the centre of the paper outwards to the edges of the plate to exclude bubbles or wrinkles from the gel–paper sandwich. Lift gel–paper sandwich off the glass plate. Cover with Saran wrap, trim edges of paper and Saran wrap and cut to fit X-ray cassette (discard the top of gel). Dry using gel dryer at 80 °C for 45 min.
12. Remove Saran wrap and expose to X-ray film in cassette (without intensifying screen) at room temperature for 12–36 h. Develop in X-Ray Film Processor.
13. Read the film using a light box and enter sequence ladder (TCGA) into a sequence editor (e.g. MacVector).

Protocol 7. Sequencing of double-stranded plasmid DNA with the fmol DNA Sequencing System and ^{35}S

Equipment and reagents

- As in *Protocols 1* and *5*
- fmol DNA Sequencing System (Promega)

Method

1. Aliquot 2 μl of each termination mix (T, C, G, A, for each template) into a 96-well thermostable microtitre plate and store at 4 °C.
2. For each template, mix the following reagents in a microcentrifuge tube: 9.7 μl template DNA (approx. 1 μg), 0.3 μl primer (10 μM), 5 μl 5 × fmol Sequencing Buffer, 1 μl [α-^{35}S]dATP (10 μCi/μl), and 1 μl of Taq polymerase (5 U/μl), and mix briefly by pipetting up and down.
3. Add 4 μl of the primer–template–enzyme mix to each of the termination mix aliquots in the microtitre plate, spin briefly, and add one drop of mineral oil to each well.

6: Analysis of human antibody sequences

4. Place the thermostable microtitre plate into the thermal cycler pre-heated to 95°C and start the following cycling programme: 2 min at 95°C followed by 60 cycles of 95°C for 30 sec, 42°C for 30 sec, and 70°C for 1 min. Hold at 4°C.
5. Prepare gradient sequencing gel as in *Protocol 6*.
6. After the cycling programme has been completed, add 3 μl of fmol Sequencing Stop Solution to each well and spin briefly to terminate the reactions.
7. Denature, load and run samples on a gradient sequencing gel as in *Protocol 6*.

In our laboratory, we use the Perkin Elmer Applied Biosystems (ABI) 373A Automated Sequencer. Since termination at each of the four nucleotides is characterized by a different coloured dye, all four reactions can be run in a single lane, rather than in four separate lanes. Using dye-labelled terminators (DyeDeoxyTerminators, ABI), all four reactions can be performed simultaneously in the same tube (*Protocol 8*). Alternatively, if dye-labelled primers are used, four separate sequencing reactions are performed, each with one of the four terminators and a particular dye-labelled primer (*Protocol 9*). Although Sequenase works well in both techniques, we favour Taq polymerase using the cycle sequencing protocol.

The choice of enzyme (Taq or Sequenase), label (dye terminators or dye primers) and template (single- or double-stranded DNA) for automated sequencing is a trade-off between the length of reliable read, the time taken to produce template, the time taken to set up the sequencing reactions, and any post-reaction purification which is required. In general, longer and more accurate reads are produced using Sequenase, dye primers and single-stranded DNA, whereas cycle sequencing with dye terminators and double-stranded PCR templates is the easiest way of sequencing short stretches of DNA such as antibody V genes.

Protocol 8. Sequencing with DyeDeoxyTerminators for the ABI 373A Automated Sequencer

Equipment and reagents
- As in *Protocol 1*
- PRISM Ready Reaction Premix (ABI) or: DyeDeoxyTerminators (ABI), 5 × TACS buffer (Terminator Ammonium Cycle Sequencing buffer: 400 mM Tris–HCl, pH 9.0, 10 mM MgCl$_2$, 100 mM (NH$_4$)$_2$SO$_4$), dNTP mix (750 μM dITP, 150 μM dATP, 150 μM dCTP, 150 μM dTTP), Taq polymerase (5 U/μl)
- Phenol/water/chloroform mix (ABI)

Protocol 8. *Continued*

Method

1. Make up reaction premix in 1.5 ml microcentrifuge tubes. For each template, use *either* 4 µl 5 × TACS buffer, 1 µl dNTP mix, 1 µl of each DyeDeoxyTerminator (ddA, ddC, ddG, ddT), and 0.8 µl Taq polymerase (5 U/µl), *or* 9.5 µl PRISM Ready Reaction Premix and 0.3 µl Millipore-filtered water.
2. Aliquot 9.8 µl reaction premix into 0.5 ml microcentrifuge tubes or a 96-well thermostable microtitre plate.
3. Add 0.2 µl primer (16 µM) and up to 10 µl template (usually 2 µl of purified PCR product from *Protocol 4*). Make up to 20 µl with Millipore-filtered water.
4. Mix and spin briefly. Add one drop of mineral oil to each tube/well.
5. Place the tubes/plate into the thermal cycler preheated to 96 °C and start the following programme: 25 cycles of 96 °C for 30 sec, 50 °C for 15 sec, and 60 °C for 4 min. Hold at 4 °C.
6. Aliquot 75 µl Millipore-filtered water into new 0.5 ml microcentrifuge tubes and take the 20 µl reaction mixes from under the oil. Add 100 µl phenol/water/chloroform mix, vortex hard for at least 6 sec, and spin in a microcentrifuge at high speed (approximately 13000 *g*) for 5 min.
7. Transfer 90 µl of the aqueous layer to fresh tubes and extract with phenol/water/chloroform as before.
8. After centrifugation, transfer 75 µl of the aqueous layer to fresh tubes, add 250 µl ethanol/Na-acetate mix and precipitate on ice for at least 15 min (or leave at −20 °C for longer).
9. Spin in a microcentrifuge at high speed for 15 min, wash pellets with 250 µl 70% ethanol, centrifuge for 5 min, and evaporate remaining ethanol in the thermal cycler at 60 °C. If samples are to be stored, store them as dried pellets at −20 °C.

Protocol 9. Preparation and running of sequencing gels on the ABI 373A Automated Sequencer

Equipment and reagents

- As in *Protocol 6*
- 10 × TBE: 108 g Tris base, 55 g boric acid, 8.3 g EDTA, Millipore-filtered water to 1 litre, filter through 0.45 µm filter, pH should be 8.0–8.5
- Deionized formamide/EDTA: 100 ml formamide, 10 g Amberlite MB-1 resin, stir gently for 30 min, filter through a 0.45 µm filter, take 200 µl deionized formamide and add 40 µl 50 mM EDTA, pH 8.0
- Gelmix: 250 g urea, 28.5 g acrylamide, 1.5 g bis-acrylamide, add Millipore-filtered water to approximately 450 ml, stir at 50 °C until dissolved, adjust to 450 ml, add 55 g Amberlite MB-1 resin, stir gently for 30 min, filter through a 0.45 µm filter and degas for 5 min, add 50 ml 10 × TBE, mix, and aliquot into 50 ml Falcon tubes, store at 4 °C for no longer than a month

6: Analysis of human antibody sequences

- One 'notched' and one 'plain' glass sequencing plate, pair of gel spacers, gel former, and sharkstooth comb
- 373A Automated Sequencer (ABI) with Apple Macintosh computer

Method

1. Wash glass plates and spacers in deionized water and wipe dry with window wiper or Kimwipes. Do **not** use organic solvents!
2. Assemble glass plates and spacers. For vertical pouring, seal plates with electrical tape. For horizontal pouring, leave plates untaped.
3. To a 50 ml aliquot of premade gelmix add 100 µl 25% ammonium persulfate and 22.5 µl TEMED. Invert four times and either pour (vertically) directly from the tube down a corner of the plate assembly or take the gelmix up into a 50 ml disposable syringe and inject (horizontally) from the centre of the plate assembly. Tap the plates whilst pouring horizontally to avoid air bubbles.
4. Insert gel former and clamp the plate assembly with two bulldog clamps across the top (plus two on each side when poured horizontally). Allow the gel to polymerize for at least 1.5 h. Gels can be kept overnight at 4°C by wrapping gel former and top of plate assembly in a Kimwipe moistened with 1 × TBE and covered with Saran wrap.
5. Remove tape, rinse and remove gel former, flush slot and wash the outside of the gel chamber with deionized water. Wipe dry as before, put in the lower buffer chamber and mount the gel chamber on to the ABI 373A.
6. Switch on machine and computer and select '*Plate Check*' option. Reclean plates if necessary (blue, red, yellow, green lines should all be straight). Alternatively, blow off dust particles with a rubber bulb or lift up plates by placing pipette tips under the bottom left and/or right hand corners. Set the PMT voltage so that the blue line is at about 900.
7. Insert the appropriate sharkstooth comb with the teeth just penetrating the gel surface.
8. Assemble the upper buffer chamber and fill chambers with a total of 1.5 litres of 1 × TBE buffer. Fill upper chamber to approx. 1 cm from top, with remaining buffer in lower chamber.
9. Dissolve samples (*Protocols 8* and *10*) in 4 µl of deionized formamide/EDTA, denature at 90°C for 2.5 min and place on ice.
10. Flush sample slots with 1 × TBE buffer and load every other sample.
11. Select '*Set up run*' (30 watts, 40°C, 8–10 h for reading approx. 400 bases) and '*Prerun*' for 5 min. Restart computer, check the '*Settings*' and fill in a new '*Sample sheet*'.

Protocol 9. *Continued*

12. *'Abort'* prerun. Flush sample slots as before and load remaining samples.
13. Select *'Start run'* on machine and *'Collect data'* on computer.
14. After the run, check the gel image for correct lane tracking, retrack and reanalyse data if necessary.

Protocol 10. Sequencing with Dye Primers for the ABI 373A Automated Sequencer

Equipment and reagents

- As in *Protocol 8*
- 5 × Cycle Sequencing Buffer: 400 mM Tris–HCl, pH 8.9, 25 mM MgCl$_2$, 100 mM (NH$_4$)$_2$SO$_4$
- Dye Primers (JOE for A, FAM for C, TAMRA for G, ROX for T)
- d/ddNTP-mixes:
  — d/ddA mix: 1.5 mM ddATP, 62.5 μM dATP, 250 μM dCTP, 375 μM c^{7}dGTP, 250 μM dTTP

— d/ddC mix: 750 μM ddCTP, 250 μM dATP, 62.5 μM dCTP, 375 μM c^{7}dGTP, 250 μM dTTP
— d/ddG mix: 125 μM ddGTP, 250 μM dATP, 250 μM dCTP, 94 μM c^{7}dGTP, 250 μM dTTP
— d/ddT mix: 1.25 mM ddTTP, 250 μM dATP, 250 μM dCTP, 375 μM c^{7}dGTP, 62.5 μM dTTP

Method

1. For each template, dilute 0.8 μl Taq polymerase (5 U/μl) with 1 μl 5 × Cycle Sequencing Buffer and 5.2 μl water.
2. For each template, aliquot 5 × Cycle Sequencing Buffer, d/ddNTP mixes, Dye Primers (0.4 μM), diluted Taq polymerase, and DNA template into four wells (A, C, G, T) of a 96-well thermostable microtitre plate: 1 μl each for A and C mixes and 2 μl each for G and T mixes. Overlay with a drop of mineral oil and spin briefly.
3. Place the microtitre plate into the thermal cycler preheated to 95°C and start the following programme: 15 cycles of 95°C for 30 sec, 55°C for 30 sec, and 70°C for 1 min, followed by 15 cycles of 95°C for 30 sec and 70°C for 1 min. Hold at 4°C.
4. Prepare sequencing gel as in *Protocol 9*.
5. For each template, combine the four reaction mixes (A, C, G, T) in a microcentrifuge tube and add 80 μl ethanol/Na-acetate solution. Mix thoroughly and precipitate on ice for at least 15 min (or leave at −20°C for longer).
6. Spin, wash, dry, dissolve, denature, and run the samples on the ABI 373A Automated Sequencer as in *Protocols 8* and *9*.

4. Analysis of antibody sequences

4.1 Software packages for sequence analysis

There are several software packages on the market for the analysis of DNA sequences. If using an automated sequencer, the analysis software should be compatible with its output, so that sequences can be directly imported. The ABI 373A automated sequencer software runs on an Apple Macintosh. The ABI sequence editor SeqEd (the updated version is called SequenceNavigator) and the MacVector package (IBI Kodak), which run on Apple Macintosh, are both used in our laboratory for sequence analysis.

4.2 Editing, translating, and comparing sequences

For each lane of the sequencing gel the ABI 373A sequencer analysis software produces a large chromatogram file of about 100 K (depending on the length of the run) and a smaller ASCII sequence file of about 2 K. Although the smaller sequence file can be directly imported into MacVector, the larger chromatogram file can only be analysed using the SeqEd package. SeqEd allows multiple chromatograms from the same or from different runs to be displayed simultaneously (*Figure 2*). The sequences can be compared and the

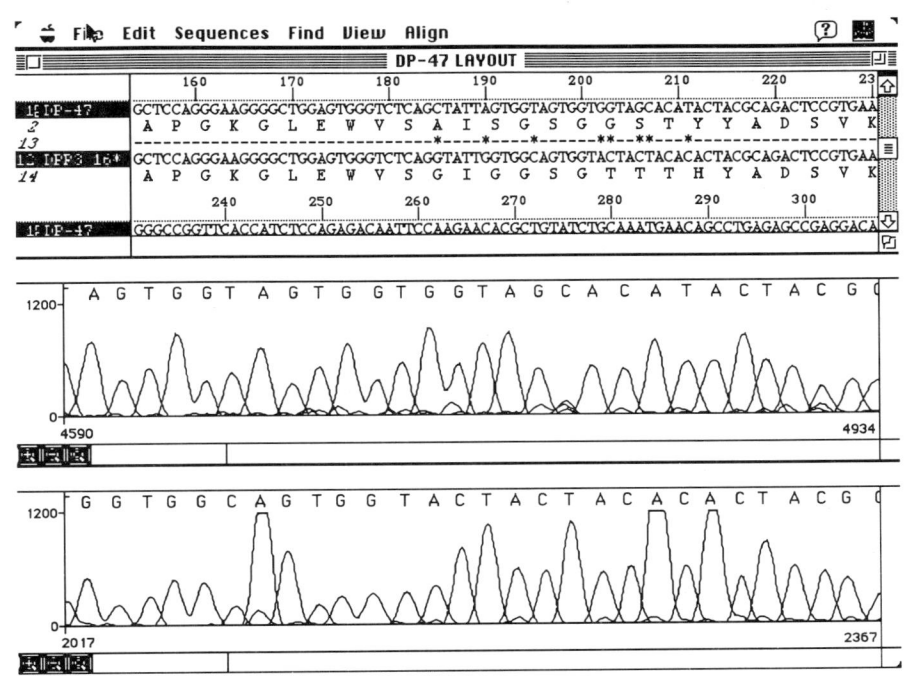

Figure 2. SeqEd (ABI) can be used to display multiple chromatograms alongside their sequences and protein translations. Differences between the two sequences are denoted by stars.

differences between them highlighted. This is particularly useful for resolving ambiguities, miscalled bases, or PCR errors. In addition, SeqEd can reverse-complement sequences and produce amino acid translations. Having edited the sequence, the updated chromatogram can be saved, or the sequence itself can be exported as an ASCII file, or in a range of other formats (e.g. Staden), or simply copied and pasted into other applications.

4.3 Multiple alignments

Alignment of an immunoglobulin V gene sequence to a small database (fewer than a thousand sequences) is most conveniently performed using MacVector's *'align to folder'* option. This is particularly useful for determining the germline counterparts of rearranged V genes as well as for highlighting sequence differences between different germline alleles. Both DNA and protein alignments can be produced, and sequence homologies can be clearly illustrated (*Figure 3*). The degree of homology for each sequence in the align-

```
Alignment List

Search Analysis for Sequence: Translation of DP-47/V3-23...Matrix: pam250 matrix
Search from 1 to 98 where origin = 1      Score Region from 1 to 98
Date: September 10,1994                   Maximum possible score: 480
Time: 13:52:21

Database: UserFolder: DP-47 REAs

                        10        20        30        40        50        60        70        80        90
                         *         *         *         *         *         *         *         *         *
          Translatio   EVQLLESGGGLVQPGGSLRLSCAASGFTFSSYAMSWVRQAPGKGLEWVSAISGSGGSTYYADSVKGRFTISRDNSKNTLYLQMNSLRAEDTAVYYCAK

   1. 0515             10        40        70       100       130       160       190       220       250
   [ 435 ]            _____...................................................................................>

   2. 0320             10        40        70       100       130       160       190       220       250
   [ 414 ]            _____.............T..KT.............P................................V....L.F...>

   3. 0334             10        40        70       100       130       160       190       220       250
   [ 413 ]            _____....V....R.............................T..N.................R.............D......>

   4. 0223             10        40        70       100       130       160       190       220       250
   [ 404 ]            _____...............G...I....G...............G.RT..D.....H......T.....R...F..............T.>

   5. 0208             10        40        70       100       130       160       190       220       250
   [ 402 ]            _____..............D...S.GG...................S......NI.................R......T...G.G........>

   6. 0314             10        40        70       100       130       160       190       220       250
   [ 398 ]            _____...................N.G.NL..........I.S......D*...........V.................V....I.....>

   7. 0316             10        40        70       100       130       160       190       220       250
   [ 379 ]            _____.........V...V...RI.....H..............G.G...TT.H..........S....E..............L....Q>

   8. 0220             10        40        70       100       130       160       190       220       250
   [ 365 ]            _____................R.D.TKSQ.T........E......T.GD..DW.R.T.........H..F.E.S..............>

   9. 0329             10        40        70       100       130       160       190       220       250
   [ 358 ]            _____....E...E...V.......N...N...R..........G.NVD.QRIR.....Q.........ANGI...G.S.....AG......>

  10. 0008                                 10        40        70       100       130       160       190
  [ 345 ]            _____...................GS...........................................T.>

  11. 0123                                 10        40        70       100       130       160       190
  [ 315 ]            _____...T...I.................A..E................S.N...S...........L.F...>
```

Figure 3. MacVector (IBI Kodak) can be used to align rearranged V gene sequences to their germline precursor. Rearranged sequences are ranked according to a homology score which is shown in square brackets.

6: Analysis of human antibody sequences

ment is given as a score, and multiple sequences are ranked accordingly. Because homology is summed across the entire length of each sequence, rather than being expressed as a percentage, shorter sequences are always ranked lower in the alignment. Another drawback of the homology algorithm is that mismatches at the very 5' or 3' ends are not displayed (because they would reduce the score), and the sequence appears to terminate prematurely. These problems can be solved using dedicated software.

4.4 Databases

Although there appear to be a finite number of germline V segment alleles, rearranged V region sequences have an almost unlimited potential for diversity. Consequently, the number of rearranged sequences in the public domain is rapidly increasing. In order to make the best use of all the available information, several sequence databases are available.

4.4.1 GenBank, EMBL, and Entrez

The largest sequence databases are the GenBank (NCBI, Bethesda, USA) and EMBL (EBI, Hinxton Hall, Cambridge, UK) data libraries. Their contents are virtually identical and they exchange entries on a daily basis. Release 39 of EMBL (June 1994) contained nearly 200 million bases from 183 000 entries (19). The entries are distributed in a flat-file format to a number of sites around the world, where they can be accessed via the Internet (World Wide Web address for GenBank searches, http://www2.ncbi.nlm.nih.gov/genbank/query-form.html and for EMBL searches, http://www.ebi.ac.uk/queries/queries.html). A typical entry contains the sequence, a brief description for cataloguing purposes, the taxonomic description of the source organism, bibliographic information, and the features table, with locations of coding regions and other biologically significant sites. Submission of nucleotide sequences to EMBL can also be performed over the Internet (http://www.ebi.ac.uk/ebi_docs/embl-db/ebi/databasehome.html).

The Entrez database (NCBI, Bethesda, USA) integrates the GenBank and EMBL sequence data libraries with the MedLine citation index (National Library of Medicine, Bethesda, USA). The '*Entrez*: Sequences' and the '*Entrez*: References' volumes are updated every two months and are available either on CD-ROM or via the Internet (for information contact: info @ ncbi.nlm.nih.gov).

4.4.2 Databases of immunoglobulin molecules

First published in 1973, Kabat's *Sequences of proteins of immunological interest* was for many years the only database dedicated to molecules of the immunoglobulin superfamily. The most recent edition was published in 1991 and contains several thousand antibody sequences from a number of different species (20). Although some laboratories have now set up their own electronic systems using dedicated software (e.g. SAW, H. Schroeder, University of Alabama, USA), there is also an international collaboration aimed at establishing an advanced database of immunoglobulin super-family sequences and structures (IMGT, Marie-Paule Lefranc, Montpellier, France).

4.4.3 V BASE: a directory of human immunoglobulin V gene sequences

We have compiled a comprehensive directory of human germline V_H, D, J_H, V_κ, J_κ, V_λ, and J_λ sequences (V BASE). Each sub-directory contains an extensive reference list of published sequences and a folder containing the sequences as text files (those with identical nucleotide sequences are grouped). This can be used for determining the germline counterparts of rearranged genes and the extent and position of somatic mutation in these genes. By identifying the position of the V–(D)–J joins and any N- and P-nucleotide addition during recombination, it should also facilitate investigation of the mechanisms involved in recombination. The V BASE sequence directory can be accessed via the Internet (World Wide Web address http://www.mrc-cpe.cam.ac.uk/imt-doc/vbase-home-page.html) or can be provided on a floppy disk. For further information contact V BASE Sequence Directory, c/o Ian Tomlinson, MRC Centre for Protein Engineering, Hills Road, Cambridge, CB2 2QH (fax: +44 1223 402140).

4.4.4 Homology searches with large databases

Although the MacVector package can be used to align an immunoglobulin sequence to the entire Entrez database, in practice it would take several hours, maybe even days, to complete the search. Faster algorithms are available which can complete the search in a much shorter time.

We have used the BLAST algorithm (Basic Local Alignment Search Tool, ref. 21) on a UNIX workstation (Silicon Graphics Ltd) to assign each rearranged human V gene sequence in the EMBL data library to its closest germline counterpart in our V BASE sequence directory. The output is displayed in a multiple sequence alignment viewer called BLIXEM (22), which also enables the annotation for each rearranged sequence to be displayed. These alignments have been used for the various statistical analyses described below.

4.5 Statistical analyses

By assigning germline counterparts to somatically mutated rearranged genes, the germline and somatic components of total antibody diversity can be ascertained. Macintosh-based software has been developed (P. Dear, unpublished) which can score somatic mutations per gene or across several genes in an alignment. Alternatively, the number of mutations occurring at each position can be calculated across the alignment. Analyses can be performed on

6: Analysis of human antibody sequences

both the amino acid and nucleotide level. This software can also be used to calculate germline or somatic diversity using a number of different formulae.

4.5.1 Germline variability

In order to quantify germline diversity, the variability at each amino acid position can be calculated across the aligned repertoire of germline sequences. The most frequently used formula calculates variability at position *i* as:

$$\text{Var (i)} = \frac{\text{No. of different amino acids}}{\text{Frequency of the most common amino acid}} \quad \text{(ref. 23)}.$$

There are, however, a number of drawbacks with this formula, particularly when calculating variability across a large number of sequences. It does not account for the numbers of times each amino acid is seen, only whether or not it occurs. As the dataset gets very large, almost all amino acids are seen at each position, albeit some very rarely. In this case, the formula simply becomes the inverse of the frequency of the most common amino acid, which does not account for the variability of the other amino acids at this position. These problems can be avoided by calculating variability using the *information-theoretical entropy* $S = -\Sigma p_i \log_2 p_i$ (24) with $p_i = n_i/N$, where n_i is the number of occurrence of an amino acid *i* at a position in the alignment and N is the total number of amino acids at this position. However, none of these variability measures take into account the different characteristics of different amino acids—that leucine and isoleucine are more similar to one another in structural terms than leucine and tryptophan, for example. More sophisticated formulae using protein evolution matrices can be used to correct for these differences.

4.5.2 Somatic mutation

Somatic mutation is confined to the variable domain of the antibody. The extent of somatic mutation in an antibody gene is largely associated with the isotype of its heavy chain constant domains. Molecules of the IgM isotype tend to be rarely mutated or unmutated. The switch to the IgG isotype triggers somatic mutation in both the heavy and light chain genes.

We have performed an extensive comparison of rearranged V_H and V_κ segments to their germline counterparts and scored somatic mutation at each amino acid position (*Figure 4*).

One of the most striking features of somatic mutation is that certain nucleotide changes are more frequently observed than others. For example, transitions are favoured over transversions, so that changes from T to C are more frequent than changes from T to A or G, changes from C to T are more frequent than changes from C to A or G, changes from A to G are more frequent than changes from A to T or C, and changes from G to A are more

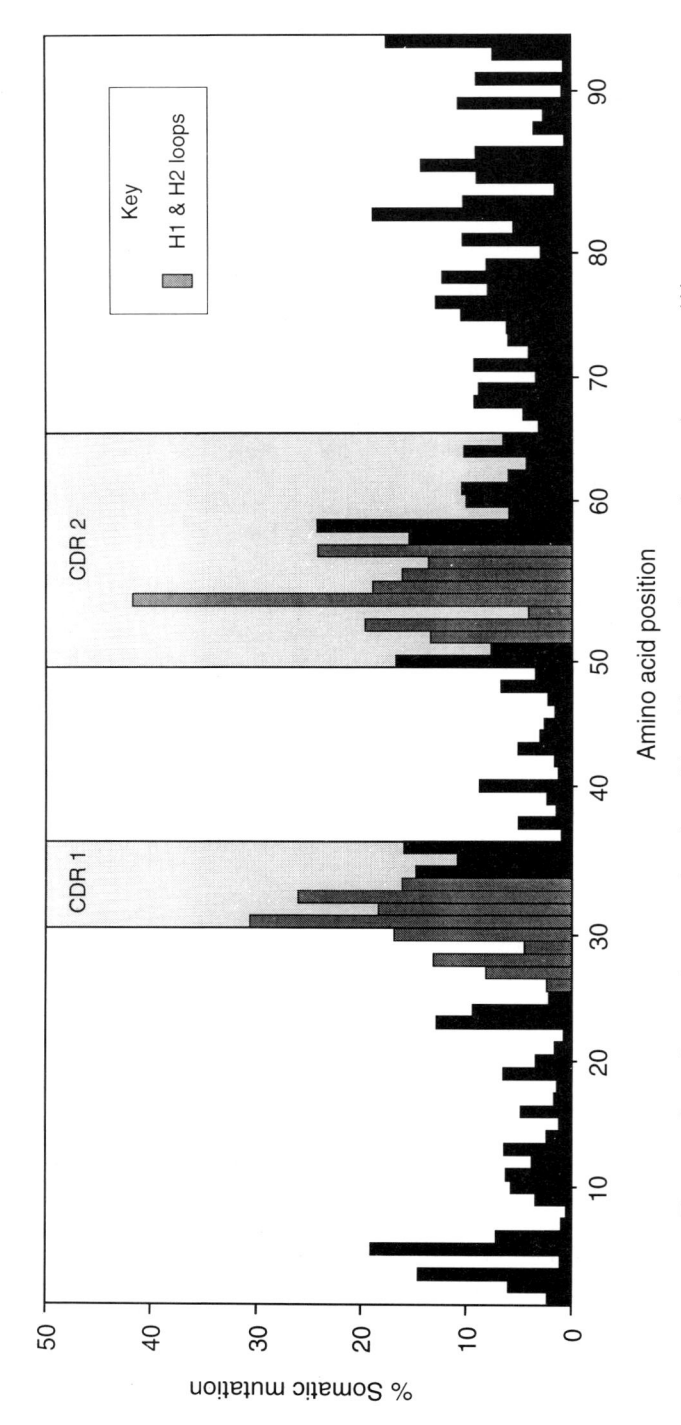

Figure 4. Somatic mutation at each amino acid position scored across a thousand rearranged V_H genes.

6: Analysis of human antibody sequences

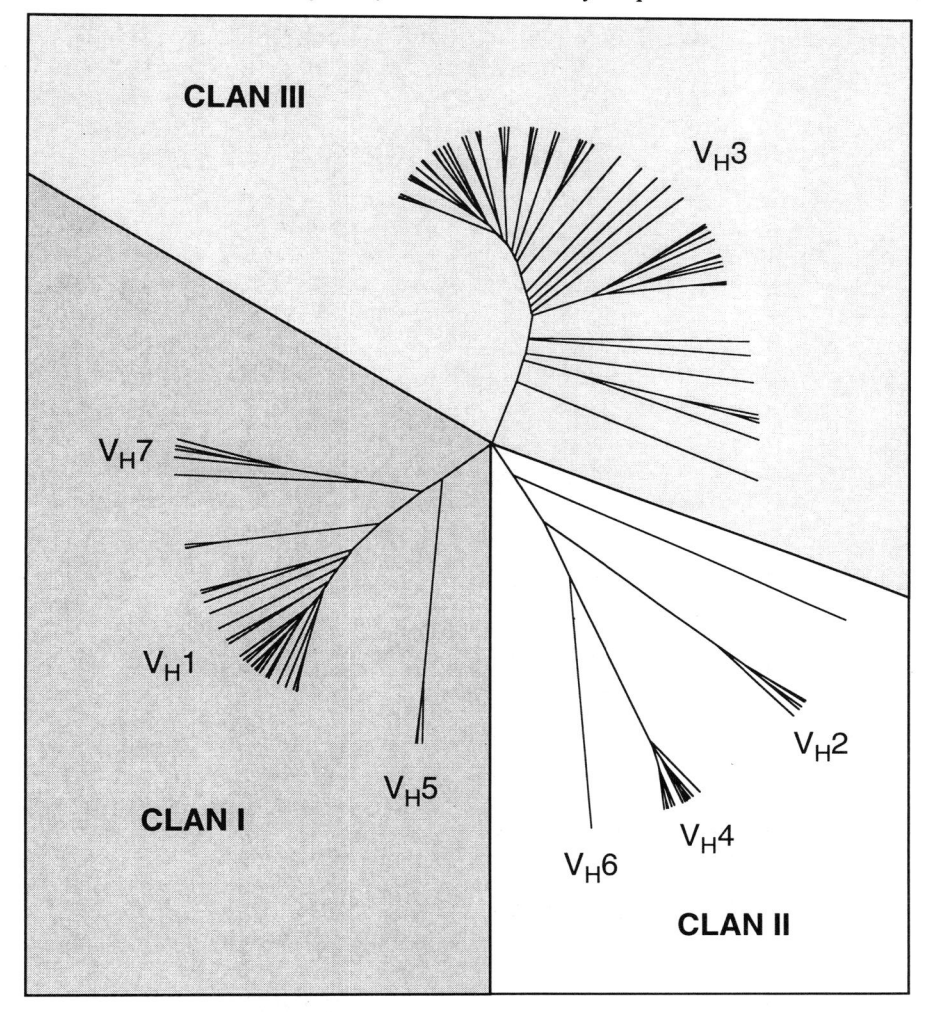

Figure 5. Phylogenetic tree for human germline V_H segments.

frequent than changes from G to T or C. However, there is also a marked strand polarity. In particular, there is a strong bias against mutations occurring in T. This strand polarity is indicative of an error-prone repair mechanism which acts on one DNA strand only.

Another interesting feature of somatic mutation in antibody V genes is the ratio of replacement to silent mutations (R/S ratio). The R/S ratio of somatic mutation in the CDRs is higher than it is in the FRs. When making such comparisons, it is important to calculate the predicted R/S ratio given random mutation of the germline sequence irrespective of positive or negative selection. For human V_H segments the predicted R/S ratios of the CDRs is higher

143

than those of the FRs (25). This means that the CDRs would appear to be set up for replacement mutations. However, mutation is not random and certain nucleotide substitutions are more likely than others (see above). The observed number of replacement and silent substitutions should therefore be interpreted with extreme caution and always in light of the codon composition of the original germline sequence.

4.5.3 Phylogenetic trees

The relatedness of nucleotide or amino acid sequences can be measured in terms of percentage homology. Comparisons between the nucleotide sequences of different members of the same multigene family allow the degree of evolutionary conservation within the family to be determined. This can be illustrated in the form of a phylogenetic tree. Using the PHYLIP package (26) a phylogenetic tree was drawn for the human V_H segments (*Figure 5*). They are clearly divided into seven families. Members of some families are very highly related (e.g. V_H4s), whilst others are more diverged (e.g. V_H3s). Alternatively, human V_H segments are divided into three 'subgroups' (23), also known as 'clans' (27), which can be used to classify V_H segments from other species.

References

1. Tonegawa, S. (1983). *Nature*, **302**, 575.
2. Chothia, C., Lesk, A. M., Tramontano, A., Levitt, M., Smith-Gill, S. J., Air, G., Sheriff, S., Padlan, E. A., Davies, D., Tulip, W. R., Colman, P. M., Spinelli, S., Alzari, P. M., and Poljak, R. J. (1989). *Nature*, **342**, 877.
3. Matsuda, F., Kyun Shin, E., Nagaoka, H., Matsumura, R., Haino, M., Fukita, Y., Taka-ishi, S., Imai, T., Riley, J. H., Anand, R., Soeda, E., and Honjo, T. (1993). *Nature Genet.*, **3**, 88.
4. Cook, G. P., Tomlinson, I. M., Walter, G., Riethman, H., Carter, N. P., Buluwela, L., Winter, G., and Rabbitts, T. H. (1994). *Nature Genet.*, **7**, 162.
5. Zachau, H. G. (1993). *Gene*, **135**, 167.
6. Cox, J. P. L., Tomlinson, I. M., and Winter, G. (1994). *Eur. J. Immunol.*, **24**, 827.
7. Williams, S. C. and Winter, G. (1993). *Eur. J. Immunol.*, **23**, 1456.
8. Alt, F. W. and Baltimore, D. (1982). *Proc. Natl Acad. Sci. USA*, **79**, 4118.
9. Berek, C. and Milstein, C. (1988). *Immunol. Rev.*, **105**, 5.
10. Walter, G., Tomlinson, I. M., Cook, G. P., Winter, G., Rabbitts, T. H., and Dear, P. H. (1993). *Nucl. Acids Res.*, **21**, 4524.
11. Tomlinson, I. M., Walter, G., Marks, J. D., Llewelyn, M. B., and Winter, G. (1992). *J. Mol. Biol.*, **227**, 776.
12. Messing, J., Gronenborn, B., Muller, H. B., and Hans, H. P. (1977). *Proc. Natl Acad. Sci. USA*, **74**, 3642.
13. Sanz, I., Kelly, P., Williams, C., Scholl, S., Tucker, P., and Capra, J. D. (1989). *EMBO J.*, **8**, 3741.
14. Borghesi-Nicoletti, C. and Schulze, D. H. (1991). *Anal. Biochem.*, **192**, 449.

15. Gronenborn, B. (1976). *Mol. Gen. Genet.*, **148**, 243.
16. Gibson, T. J. (1984). PhD thesis, University of Cambridge.
17. Niall, H. and Edman, P. (1967). *Nature*, **216**, 262.
18. Sanger, F., Nicklen, S., and Coulson, A. R. (1977). *Proc. Natl Acad. Sci. USA*, **74**, 5463.
19. Emmert, D. B., Stoehr, P. J., Stoesser, G., and Cameron, G. N. (1994). *Nucl. Acids Res.*, **22**, 3445.
20. Kabat, E. A., Wu, T. T., Perry, H. M., Gottesman, K. S., and Foeller, C. (1991). *Sequences of proteins of immunological interest* (5th edn). US Department of Health and Human Services, Bethesda, Maryland.
21. Altschul, S. F., Gish, W., Miller, W., Myers, E. W., and Lipman, D. J. (1990). *J. Mol. Biol.*, **215**, 403.
22. Sonnhammer, E. L. L. and Durbin, R. (1994). *Comput. Applic. Biosci.*, **10**, 300.
23. Wu, T. T. and Kabat, E. A. (1970). *J. Exp. Med.*, **132**, 211.
24. Shenkin, P. S., Erman, B., and Mastrandrea, L. D. (1991). *Proteins: Structure, Function, and Genetics*, **11**, 297.
25. Chang, B. and Casali, P. (1994). *Immunol. Today*, **15**, 367.
26. Felsenstein, J. (1993). PHYLIP (Phylogeny Inference Package), 3.5c. Department of Genetics, University of Washington, Seattle.
27. Kirkham, P. M., Mortari, F., Newton, J. A., and Schroeder, H. J. (1992). *EMBO J.*, **11**, 603.

7

Rodent to human antibodies by CDR grafting

MARY M. BENDIG and S. TARRAN JONES

1. Introduction

The concept that a functional human-like antibody could be created by grafting the antigen binding complementarity determining regions (CDRs) from variable domains of rodent antibodies on to human variable domains was first demonstrated by Dr Greg Winter and his colleagues at the MRC Laboratory for Molecular Biology in 1986 (1). Since then several different research groups have developed and improved the methods for designing and constructing humanized antibodies via CDR grafting (2). There are reports of at least 80 antibodies being humanized by CDR grafting and at least eight of these antibodies being evaluated in clinical trials (3, 4). This chapter describes the procedures currently used at the MRC Collaborative Centre to create a 'reshaped human' antibody from a mouse antibody via CDR grafting.

2. Cloning and sequencing mouse variable regions

The first step is to clone and sequence the cDNAs coding for the variable domains of the mouse antibody to be humanized. The variable domain genes (V genes) are cloned following polymerase chain reaction (PCR) using specially designed primers that hybridize to the 5' ends of the mouse leader sequences and to the 5' ends of the mouse constant regions (*Protocols 1, 2, and 3*; *Tables 1, 2,* and *3*). By using primers that hybridize to sequences external to the DNA sequences coding for the variable domains, the full, accurate sequences of the mouse V genes, are obtained. In addition, the cloned mouse leader and variable region sequences are useful in the construction of chimeric mouse–human light and heavy chains and sometimes also in the construction of the reshaped human light and heavy chains.

Table 1. PCR primers for cloning mouse kappa light chain variable regions

Name	Sequence[c]
MKV1[a] (30mer)	ATGAAGTTGCCTGTTAGGCTGTTGGTGCTG
MKV2 (30mer)	ATGGAGWCAGACACACTCCTGYTATGGGTG
MKV3 (30mer)	ATGAGTGTGCTCACTCAGGTCCTGGSGTTG
MKV4 (33mer)	ATGAGGRCCCCTGCTCAGWTTYTTGGMWTCTTG
MKV5 (30mer)	ATGGATTTWCAGGTGCAGATTWTCAGCTTC
MKV6 (27mer)	ATGAGGTKCYYTGYTSAGYTYCTGRGG
MKV7 (31mer)	ATGGGCWTCAAGATGGAGTCACAKWYYCWGG
MKV8 (31mer)	ATGTGGGGAYCTKTTTYCMMTTTTTCAATTG
MKV9 (25mer)	ATGGTRTCCWCASCTCAGTTCCTTG
MKV10 (27mer)	ATGTATATATGTTTGTTGTCTATTTCT
MKV11 (28mer)	ATGGAAGCCCCAGCTCAGCTTCTCTTCC
MKC[b] (20mer)	ACTGGATGGTGGGAAGATGG

[a] MKV indicates primers that hybridize to leader sequences of mouse kappa light chain variable region genes.
[b] MKC indicates the primer that hybridizes to the mouse kappa constant region gene.
[c] Ambiguity codes: M = A or C; R = A or G; W = A or T; S = C or G; Y = C or T; K = G or T.

Table 2. PCR primers for cloning mouse heavy chain variable regions

Name	Sequence[c]
MHV1[a] (27mer)	ATGAAATGCAGCTGGGGCATSTTCTTC
MHV2 (26mer)	ATGGGATGGAGCTRTATCATSYTCTT
MHV3 (27mer)	ATGAAGWTGTGGTTAAACTGGGTTTTT
MHV4 (25mer)	ATGRACTTTGGGYTCAGCTTGRTTT
MHV5 (30mer)	ATGGACTCCAGGCTCAATTTAGTTTTCCTT
MHV6 (27mer)	ATGGCTGTCYTRGSGCTRCTCTTCTGC
MHV7 (26mer)	ATGGRATGGAGCKGGRTCTTTMTCTT
MHV8 (23mer)	ATGAGAGTGCTGATTCTTTTGTG
MHV9 (30mer)	ATGGMTTGGGTGTGGAMCTTGCTATTCCTG
MHV10 (27mer)	ATGGGCAGACTTACATTCTCATTCCTG
MHV11 (28mer)	ATGGATTTTGGGCTGATTTTTTTTATTG
MHV12 (27mer)	ATGATGGTGTTAAGTCTTCTGTACCTG
MHCG1[b] (21mer)	CAGTGGATAGACAGATGGGGG
MHCG2a (21mer)	CAGTGGATAGACCGATGGGGC
MHCG2b (21mer)	CAGTGGATAGACTGATGGGGG
MHCG3 (21mer)	CAAGGGATAGACAGATGGGGC

[a] MHV indicates primers that hybridize to leader sequences of mouse heavy chain variable region genes.
[b] MHCG indicates primers that hybridize to mouse constant region genes.
[c] Ambiguity codes are as used in Table 1.

7: Rodent to human antibodies by CDR grafting

Table 3. Primers for PCR screening

Name	Sequence
pCR™II Forward Primer (18mer)	C T A G A T G C A T G C T C G A G C
pCR™II Reverse Primer (21mer)	T A C C G A G C T C G G A T C C A C T A G

Protocol 1. Recommendations on the preparation and use of a PCR room

Equipment and reagents

- A designated PCR room containing a dedicated refrigerator/freezer and a class II microbiological safety cabinet fitted with a UV-lamp (Walker Safety Cabinets Ltd)
- Three sets of Gilson Pipetman (Anachem, cat. no. H44801 [P2], cat. no. H44802 [P10], cat. no. H23600 [P20], cat. no. H23615 [P100], cat, no, H23601 [P200], and cat no. H23602 [P1000])
- Aerosol-resistant pipette tips to fit Gilson Pipetman (Stratagene, cat. no. 410136 [P2/P10], cat. no. 410130 [P20], cat. no. 410138 [P100], cat. no. 410132 [P200], and cat. no. 410136 [P1000])
- 0.25 M HCl
- PCR reagents (see *Protocol 2*)

Recommendations

1. Wear a laboratory coat (designated for use in the PCR room only) and gloves at all times. Change gloves frequently.
2. Wipe down the inside of the cabinet with 0.25 M HCl both before and after use to depurinate any contaminating DNA. Make sure that the UV lamp is routinely on *overnight* to damage any remaining DNA. **NB**: UV light is dangerous, particularly to the eyes.
3. Use aerosol-resistant pipette tips when preparing a PCR reaction.
4. Store the components of the PCR reaction in a refrigerator/freezer located inside the PCR room. If this is not possible, store them in a designated box in a refrigerator/freezer away from any potential sources of contaminating DNA.
5. Bring the template DNA into the PCR room in a sealed container (i.e. microcentrifuge tube) and open it only inside the cabinet.
6. Prepare the PCR reactions in two stages. The first stage may be done on the bench in the PCR room. The second stage must be carried out in the cabinet.
7. In the first stage, combine (in this order) the sterile water, 10 × PCR buffer II, $MgCl_2$, and dNTPs using the first set of Gilson Pipetman.
8. In the second stage, using the second set of Pipetman, add the PCR primers. Finally, with the third set of Pipetman, add the template DNA (or RNA:cDNA duplex), thermophilic DNA polymerase, and mineral oil.

Protocol 1. *Continued*

9. Always prepare a negative control, omitting the template DNA, to confirm the absence of contaminating DNA in the PCR reactions.

10. Be aware of the risk of contamination of the PCR product once outside the PCR room. Contamination during electrophoresis or during the extraction of DNA from gel slices is possible. Treat electrophoresis equipment, such as gel boxes and well-formers, with 0.25 M HCl overnight. Place gels on clean transparent film when using a UV transilluminator to cut bands from a gel.

Protocol 2. PCR cloning of the mouse variable regions

Equipment and reagents

- DNA thermal cycler (e.g. Perkin Elmer, cat. no. N801-0177)
- GeneAmp™ PCR reaction tubes (Perkin Elmer, cat. no. N801-0180)
- Benchtop centrifuge (Fisons, cat. no. CEK-126-010N)
- Glasstic® disposable cell-counting slide (Bio-stat Diagnostic, cat. no. 887144)
- Hybridoma cell line
- Sterile water. Treat deionized, distilled water with 1 ml/litre of DEPC (Sigma, cat. no. D-5758) for 12 h at 37°C. Autoclave for 20 min at 115°C and 15 p.s.i.
- Phosphate buffered saline (PBS): 8.0 g/litre NaCl, 0.2 g/litre KCl, 1.15 g/litre Na$_2$HPO$_4$, 0.2 g/litre KH$_2$PO$_4$, pH 7.2
- RNA isolation kit (Stratagene, cat. no. 200345)
- First-strand cDNA synthesis kit (Pharmacia, cat. no. 27-9261-01)
- PCR cloning primers (see *Tables 1* and *2*). Prepare separate 10 µM stock solutions of MHV 1-12, MKV 1-11, MHC, and MKC primers in sterile water
- AmpliTaq® DNA polymerase (5 U/µl; Perkin Elmer, cat. no. N801-0060)
- 10 × PCR buffer II: 500 mM KCl, 100 mM Tris–HCl, pH 8.3; and 25 mM MgCl$_2$ (Perkin Elmer, cat. no. N808-0010)
- Trypan Blue (Sigma, cat. no. T-8154)
- LB medium: 10 g Bacto-tryptone, 5 g Bacto-yeast extract, 10 g NaCl in 1 litre, pH adjusted to 7.0 and autoclaved.
- GeneAmp® dNTPs: separate 10 mM stock solutions of dATP, dCTP, dGTP, and dTTP in distilled water, titrated to pH 7.0 with NaOH (Perkin Elmer, cat. no. N808-0007)
- Mineral oil (Sigma, cat. no. M-5904)
- 10 × TBE buffer: 1.35 M Tris base, 0.45 M boric acid, 26 mM EDTA, pH 8.8 (this has been modified to prevent precipitation upon storage) by increasing Tris base and reducing boric acid
- Agarose (UltraPure™) (Life Technologies, cat. no. 15510-019)
- 10 mg/ml ethidium bromide (Sigma, cat. no. E-1510)
- TA Cloning® kit (Invitrogen, cat. no. K2000-01)
- LB agar plates: LB medium with 15 g/L Bacto-agar (Difco).
- Ampicillin (Sigma, cat. no. A-2804). Prepare a 100 mg/ml stock solution in water and sterilize by filtration through a 0.22 µm filter. Store at -20°C
- X-Gal (Sigma, cat. no. B-9146). Prepare a 40 mg/ml stock solution with dimethylformamide in a glass or polypropylene tube. Wrap the tube in aluminium foil and store at -20°C

A. *Counting cells*

1. Grow the mouse hybridoma cell line in an appropriate culture medium to provide a total viable cell count of at least 10^8 cells.

2. Pellet the cells in a benchtop centrifuge (250 *g*, 5 min). Gently resuspend the cells in 20 ml of PBS.

3. Add 100 µl of cells to 200 µl of PBS and 200 µl of Trypan Blue and mix gently. Pipette 10 µl of this mixture into a disposable cell-counting slide. Count the cells per square and determine the number of cells/ml according to the manufacturer's instructions.
4. Pellet approximately 10^8 cells (250 g, 5 min).

B. *Preparation of RNA and cDNA*

1. Use the RNA isolation kit as described by the manufacturer to purify total RNA from the cells. The kit uses a guanidinium thiocyanate phenol–chloroform single-step extraction procedure (5).
2. Determine the quantity and quality of the total RNA by measuring the OD_{260} and OD_{280} and by testing 1–5 µg aliquots on a non-denaturing 1% (w/v) agarose gel in 1 × TBE buffer containing 0.5 µg/ml ethidium bromide. The concentration of RNA = OD_{260} × 40 µg/ml. The quality is satisfactory if $OD_{260}:OD_{280} > 1.9$ and distinct bands are seen on the agarose gel representing 28S and 18S RNA, the 28S band being more intensely stained.
3. Following the manufacturer's instructions, use the first-strand cDNA synthesis kit to produce a single-stranded DNA copy of the hybridoma mRNA using the *Not*I-(dT)$_{18}$ primer. Use 5 µg of total RNA in a 33 µl final reaction volume.
4. Following the reaction, heat at 90°C for 5 min to denature the RNA:cDNA duplex and to inactivate the reverse transcriptase. Chill on ice.

C. *Preparation and cloning of PCR product*

1. Label eleven GeneAmp™ PCR reaction tubes MKV1 to MKV11. In each tube, prepare a 100 µl reaction containing:

   - sterile water   69.5 µl
   - 10 × PCR buffer II   10.0 µl
   - 25 mM MgCl$_2$ 6.0 µl
   - 10 mM stock solutions of dNTPs   2.0 µl each
   - 10 µM MKC primer   2.5 µl
   - one of the 10 µM MKV primers   2.5 µl
   - RNA:cDNA template mix   1.0 µl

   Finally, add 0.5 µl of AmpliTaq® DNA polymerase and overlay the reaction mix with 50 µl of mineral oil.
2. Prepare a similar series of reaction mixes to PCR-clone the mouse heavy chain variable region gene using the 12 MHV primers and the appropriate MHC primer.
3. Place the reaction tubes into a DNA thermal cycler and cycle (after an initial melt at 94°C for 1.5 min) at 94°C for 1 min, 50°C for 1 min, and

Protocol 2. *Continued*

72 °C for 1 min over 25 cycles. Follow the last cycle with a final extension step at 72 °C for 10 min before cooling to 4 °C. Use a ramp time of 2.5 min between the annealing (50 °C) and extension (72 °C) steps and a 30 sec ramp time between all other steps of the cycle.

4. Run a 10 µl aliquot from each PCR reaction on a 1% (w/v) agarose/1 × TBE buffer gel, containing 0.5 µg/ml ethidium bromide, to determine which of the leader primers produces a PCR product. Positive PCR products will be about 420–500 bp in size.

5. For those PCR reactions that appear to produce full-length PCR products, repeat the procedure (part B, step 3 to part C, step 3) to obtain at least two independent PCR reactions that give full-length PCR products.

6. Directly clone a 1 µl aliquot of any potential PCR product into the pCR™II vector provided by the TA Cloning® kit as described in the manufacturer's instructions.[a] Pipette out 10.0% (v/v), 1.0% (v/v), and 0.1% (v/v) aliquots of the transformed *E. coli* cells on to individual 90 mm diameter LB agar plates containing 50 µg/ml ampicillin and overlaid with 25 µl of the X-Gal stock solution. Incubate overnight at 37 °C.

7. Identify positive colonies by PCR screening (see *Protocol 3*).

[a] This kit allows the direct cloning of PCR products without prior purification and takes advantage of the preference of AmpliTaq® DNA polymerase to insert a 3′-overhanging thymidine (T) at each end of the PCR product.

Protocol 3. PCR screening of bacterial colonies for positive clones

Equipment and reagents

- Techne PHC-3 DNA thermal cycler with Techne Hi-temp 96® multiwell plate (Techne, cat. no. FPHC3MD)
- Techne Hi-Temp 96® microplate (Techne, cat. no. FMW11)
- Nunc inoculating needles (Life Technologies, cat. no. 254399)
- 30 ml universal container (Sterilin, cat. no. 128C)
- PCR screening primers that bracket the site of insertion. When screening for inserts within pCR™II, use the forward and reverse primers described in *Table 3*. Stock primer solutions should be 10 µM
- Gilson P20 Pipetman (see *Protocol 1*)
- Innova® benchtop incubator shaker (New Brunswick Scientific, cat. no. 4000)
- Sterile water (see *Protocol 2*)
- AmpliTaq® DNA polymerase (see *Protocol 2*)
- 10 × PCR buffer II and 25 mM MgCl₂ (see *Protocol 2*)
- GeneAmp® dNTPs (see *Protocol 2*)
- Mineral oil (see *Protocol 2*)
- LB medium (see *Protocol 2*).
- Ampicillin (see *Protocol 2*)
- 10 × TBE buffer (see *Protocol 2*)
- Agarose (see *Protocol 2*)
- Ethidium bromide (see *Protocol 2*)
- Aerosol resistant tips (see *Protocol 1*)

7: Rodent to human antibodies by CDR grafting

Method

1. Prepare a bulk solution of the PCR reaction mix (sufficient for 20 samples) by combining the following:
   - sterile water   280 µl
   - 10 × PCR buffer II   40 µl
   - 25 mM MgCl$_2$ 24 µl
   - 10 mM stock solutions of dNTPs   8 µl each
   - 10 mM stock solutions of the pCR™II forward and reverse primers (*Table 3*)   10 µl each

   Finally, add 4 µl of AmpliTaq® DNA polymerase to the bulk PCR reaction mix.

2. To avoid bacterial contamination in the PCR room, all subsequent steps must be done on the bench outside the PCR room (see *Protocol 1*).

3. Dispense the above 'mastermix' in 20 µl aliquots into the 96-well microplate using a designated P20 pipetman used solely for this purpose.

4. Label twenty 30 ml universal containers and add to each 3 ml of LB medium containing 50 µg/ml ampicillin.

5. Using an inoculating needle, gently 'stab' an individual colony from a putative transformation mix grown overnight on selective agar plate (see *Protocol 2*). Then, stab the needle into one of the 20 µl aliquots of mastermix, making sure that the base of the microplate well is touched gently with the needle.[a]

6. With the same needle, immediately inoculate a universal container (step 4) and incubate this culture overnight at 37 °C and 300 r.p.m. in the shaking incubator.

7. Prepare a negative control (*Protocol 1*) and a positive control where possible.[b]

8. Overlay each of the inoculated PCR reactions with 2 drops of mineral oil per well.

9. Load the microplate into the Techne PHC-3 and cycle (after an initial melt at 94 °C for 5 min) at 94 °C for 1 min, 50 °C for 1 min, and 72 °C for 1 min over 25 cycles. Complete the PCR reaction with a final extension step at 72 °C for 10 min before cooling to 4 °C. Use a ramp time of 30 sec between each step.

10. Run a 10 µl aliquot from each PCR reaction on a 1% (w/v) agarose/1 × TBE buffer gel containing 0.5 µg/ml ethidium bromide, and estimate the size of any PCR products.[c]

11. Sequence the DNA from at least two independently isolated positive clones of each variable region to identify possible errors introduced by PCR.

> **Protocol 3.** *Continued*
> [a] Do not swirl the needle in the solution or rub it along the inside of the microplate well as excess template DNA can have a negative effect on the efficiency of a PCR reaction.
> [b] As a positive control, add 1 μl of template DNA (5 ng/μl) to 20 μl of mastermix.
> [c] Bands of 520–600 bp will be seen using the pCR™II forward and reverse primers when the pCR™II vector contains a variable region gene. A negative result (i.e. no insert in the pCR™II vector) will produce an approximately 100 bp band. Treat any bands smaller than 500 bp with caution as they may be pseudogenes.

3. Construction of a chimeric antibody

In most cases, a chimeric antibody is constructed and tested for its ability to bind to antigen prior to constructing a reshaped human antibody. There are two reasons for this. One is to confirm, in a functional assay, that the correct mouse variable regions have been cloned and sequenced. The second is to create a valuable positive control for evaluating the reshaped human antibody. In a chimeric antibody, no alterations have been made to the protein domains that constitute the antigen-binding site. A bivalent chimeric antibody is expected, therefore, to bind to antigen as well as the parent bivalent mouse antibody. Since the chimeric antibody is usually constructed with the same human constant regions that will be used in the reshaped human antibody, it is possible to compare directly the chimeric and reshaped human antibodies in antigen-binding assays that employ anti-human constant region antibody–enzyme conjugates for detection.

As a first step in the construction of chimeric light and heavy chains, the cloned mouse leader-variable regions are modified at the 5'- and 3'-ends using PCR primers to create restriction enzyme sites for convenient insertion into the expression vectors, Kozak sequences for efficient eukaryotic translation (6), and splice-donor sites for RNA splicing of the variable and constant regions (*Table 4*). The adapted mouse light and heavy chain leader-variable

Table 4. Sequences important for the efficient expression of immunoglobulin genes in mammalian cells

Name	Reference	Consensus DNA Sequence[a, b]
Kozak translation initiation site	6	+1 GCCGCC**R**CC**AUGG**
Kappa light chain splice-donor site		AC::**GT**RAGT
Heavy chain splice-donor site		MAG::**GT**RAGT
Immunoglobulin splice-acceptor site		YYYYYYYYYYNC**AG**::G

[a] Bases shown in bold are considered to be invariant within each consensus sequence.
[b] Ambiguity codes are as used in Table 1.

regions are then inserted into vectors designed to express chimeric or reshaped human light and heavy chains in mammalian cells (7). These vectors contain the human cytomegalovirus (HCMV) enhancer and promoter for transcription, a human light or heavy chain constant region, a gene such as *neo* for selection of transformed cells, and the SV40 origin of replication for DNA replication in *cos* cells (8). For detailed protocols on the transformation and expression of recombinant antibodies in eukaryotic cells see Bebbington, Chapter 12.

4. Design and construction of a reshaped human antibody

4.1 Analysis of the mouse variable regions

Prior to actually beginning to design the reshaped human variable regions, it is important to carefully analyse the amino acid sequences of the mouse variable regions to identify the residues that are most critical in forming the antigen-binding site (*Protocol 4*, part A).

In addition to studying and comparing the primary amino acid sequences of the mouse variable regions, a structural model of the mouse variable regions is built based on homology to known protein structures, in particular, to the structures of other antibody variable regions. Molecular modelling is carried out using a Silicon Graphics IRIS 4D workstation, the molecular modelling package QUANTA (Polygen Corp.), and the Brookhaven database of protein structures with a few additional, as yet unpublished, immunoglobulin structures. As a first step in the modelling exercise, the framework regions (FRs) of the new variable regions are modelled on FRs from similar, structurally-solved immunoglobulin variable regions. Most of the CDRs of the new variable regions are modelled based on the canonical structures for CDRs (9, 10, 11, 12). Those CDRs which do not appear to belong to any known group of canonical structures, for example CDR3 of the heavy chain variable region, are modelled based on similar loop structures present in any structurally-solved protein. In order to relieve unfavourable atomic contacts and to optimize Van der Waals and electrostatic interactions, the model is subjected to energy minimization using the CHARMM potential (13) as implemented in QUANTA.

4.2 Design of the reshaped human antibody

The first step in the design process is to select the human light and heavy chain variable regions that will serve as templates for the design of the reshaped human variable regions (*Protocol 4*, part B. See also Walter and Tomlinson, Chapter 6). In most cases, the selected human light and heavy chains come from two different human antibodies. By not restricting the selection of human variable regions to variable regions that are paired in the same antibody, it is possible to obtain much better homologies between the mouse variable regions to be humanized and the human variable regions selected to serve as templates. In practice, the use of human variable regions from different antibodies as the basis of the design of a reshaped human antibody has not been a problem. This is probably because the packing of light and heavy chain variable regions is highly conserved.

The next step in the design process is to join the mouse CDRs to the FRs from the selected human variable regions. The preliminary amino acid sequences are then carefully analysed to judge whether or not they will re-create an antigen-binding site that mimics that present in the original mouse antibody. At this stage, the model of the mouse variable regions is particularly useful in evaluating the relative importance of each amino acid in the formation of the antigen-binding site. Within the FRs, the amino acid differences between the mouse and the human sequences should be examined. In

Table 5. Important residues for the maintenance of CDR loop conformation[a]

CDR loop	Canonical structure (loop size[b])	Residues important for loop conformation[c, d] (most common amino acids)
L1	1(10)[e]	2(I), 25(A), 30(V), 33(M, L), and **71**(Y)
	2(11)	2(I), 25(A), 29(V, I), 33(L), and **71**(F,Y)
	2(12)[f]	2(I, N), 25(A), 28(V, I), 33(L), and **71**(F, Y)
	3(17)	2(I), 25(S), 27b(V, L), 33(L), and **71**(F)
	4(15)[f]	2(I), 25(A), 27b(V), 33(M), and **71**(F)
	4(16)	2(V, I), 25(S), 27b(I, L), 33(L), and **71**(F)
L2	1(7)	**48**(I), **51**(A, T,), **52**(S, T), and **64**(G)
L3	1(9)	90(Q, N, H) and 95(P)
	2(9)[e]	90(Q) and 94(P)
	3(8)	90(Q) and 95(P)
H1	1(5)	**24**(A, V, G), **26**(G), **27**(F, Y), 29(F), 34(M, W, I), and **94** (R, K)
	2(6)	**24**(V, F), **26**(G), **27**(F, Y, G), 29(I, L), 35(W, C), and **94** (R, H)
	3(7)	**24**(G, F), **26**(G), **27**(G, F, D), 29(L, I, V), 35a(W, V), and **94**(R, H)
H2	1(16)	55(G, D) and **71**(V, K, R)
	2(17)	52(P, T, A), 55(G, S), and **71**(A, T, L)
	3(17)	54(G, S, N) and **71**(R)
	4(19)	54(S), 55(Y) and **71**(R)
	5(18)	52(Y), 54(K), 55(W) and **71**(P)

[a] This Table summarizes information presented in Chothia and Lesk (9), Chothia et al. (10), Tramontano et al. (11), and Chothia et al. (12).
[b] Loop size is the number of residues in the CDR loop as defined by Kabat et al. (14).
[c] Numbering is according to Kabat et al. (14). Note that in Chothia et al. (10) L1 and H1 are numbered differently.
[d] The residue numbers printed in bold are located within the FRs of the variable region. Residues 26–30 of the heavy chain variable region are defined as FR residues by Kabat et al. (14); however, structurally they are part of the H1 loop (10).
[e] These canonical structures have been observed only in mouse antibodies and not in human antibodies.
[f] Approximately 25% of human and 20% of mouse sequences have 13 residues in canonical structure 2 or 14 residues in canonical structure 4. These minor variations in loop size result in changes at the tip of the L1 loop but do not significantly alter loop conformation (10).

7: Rodent to human antibodies by CDR grafting

Table 6. Conserved residues found at the V_L/V_H interface[a]

Variable region	Residue position[b]	Number of sequences analysed	Number of different amino acids observed	Principal amino acids at this position (number of occurrences[c])
V_L	34	1365	16	A(326), H(306), N(280)
	36	1324	7	Y(1057), F(143)
	38	1312	11	Q(1158)
	44[d]	1244	14	P(1060)
	46	1252	17	L(827)
	87	1222	8	Y(874), F(319)
	89	1238	16	Q(654)
	91	1234	17	W(175), Y(216), G(209), S(169)
	96[d]	1034	20	L(220), Y(203), W(196), R(121)
	98[d]	1066	6	F(1058)
V_H	35	1459	19	H(378), N(356), S(287)
	37	1398	10	V(1212), I(151)
	39	1397	13	Q(1315)
	45[d]	1397	10	L(1362)
	47	1357	14	W(1252)
	91	1689	9	Y(1332), F(340)
	93	1683	16	A(1426)
	95	1451	20	D(285), G(212), S(187)
	100-100K[d,e]	1211	19	F(707), M(224)
	103[d]	1276	10	W(1251)

[a] The positions of interdomain residues are as defined by Chothia et al. (19). The immunoglobulin sequences analysed are from the database of Kabat et al. (14).
[b] Numbering is according to Kabat et al. (14). The residue numbers printed in bold are located within the FRs of the variable region.
[c] Only those residues that displayed a frequency of occurrence of > 10% are shown.
[d] One of six residues that constitute the core of the V_L/V_H interface as defined by Chothia et al. (19).
[e] The residue that is immediately N-terminal to residue 101 in CDR3 is the amino acid that is part of the core of the V_L/V_H interface. The numbering of this residue varies.

addition, any unusual amino acid sequences in the FRs of either the mouse or human sequences should be studied. Finally, any potential glycosylation sites in the FRs of either the mouse or human sequences should be identified and their possible influence on antigen binding considered (*Protocol 4*, part C). It is important to make the minimum number of changes in the human FRs. The goal is to achieve good binding to antigen while retaining human FRs that closely match the sequences from natural human antibodies.

4.3 Construction of the reshaped human antibody
4.3.1 Construction of the reshaped human variable regions
Once the amino acid sequences of the reshaped human variable regions have been designed, it is necessary to decide how DNA sequences coding for these amino acid sequences will be constructed. There are two fundamental

approaches. One is to take an existing DNA sequence coding for a variable region that is very similar to the newly designed reshaped human variable region and to modify the existing DNA sequence so that it will code for the newly designed reshaped human variable region. Modification to existing sequences are usually carried out using PCR and specially designed synthetic oligonucleotide primers (20). The second approach is to make a DNA sequence synthetically that will code for the newly designed reshaped human variable region (*Protocol 5*).

Protocol 4. Design of a reshaped human antibody

Equipment
- Genetics Computer Group (GCG) sequence analysis software package (University of Wisconsin Biotechnology Center, Madison, WI, USA)
- EMBL Data Library including the Kabat database (European Molecular Biology Laboratory, Heidelberg, Germany)
- Leeds database of protein sequences (Department of Biochemistry and Molecular Biology, University of Leeds, Leeds, UK)
- SPARC® station 1 (Sun Microsystems Inc.)
- Molecular model of the mouse variable regions

Method

A. *Analysis of the amino acid sequences of the mouse variable regions*

1. Use the '*SeqEd*' program in the GCG package to create a series of files containing the consensus amino acid sequences of the subgroups of mouse and human light and heavy chain variable regions as defined by Kabat *et al.* (14).

2. With the same program, create two files containing the amino acid sequences of the light chain and heavy chain variable regions of the mouse antibody.

3. Compare the amino acid sequences of the mouse variable regions to the mouse consensus sequences using the '*Gap*' program in the GCG package and identify the mouse subgroups to which the mouse variable regions belong.

4. Analyse the amino acid sequences of the mouse variable regions and locate the following features within them:
   - CDRs and FRs (14)
   - residues that are part of the canonical sequences for loop structure (see *Table 5*)
   - residues located at the V_L/V_H interface (see *Table 6*)
   - residues in the FRs that are unusual or unique for that position when compared to the consensus sequence for that mouse subgroup (14)
   - potential glycosylation sites.[a]

7: Rodent to human antibodies by CDR grafting

B. *Selection of the human variable regions to serve as templates for reshaping*

1. Compare the amino acid sequences of the mouse variable regions to the human consensus sequences and identify the most similar human subgroup for each mouse variable region (see part A, steps 1 and 3).
2. Compare the amino acid sequences of the mouse variable regions to all human variable region sequences in the database and, for each mouse variable region, identify the 10 most similar human sequences. Use the *'FastA'* program in the GCG package.
3. Analyse the selected human variable regions for the following characteristics:
   - per cent similarity with the mouse variable region;
   - per cent identity with the mouse variable region noting the location of regions of non-identity;
   - length of the CDRs in comparison to the mouse CDRs;
   - identity to the mouse sequence in the residues in the FRs that are part of the canonical sequences for loop structure (see *Table 5*);
   - identity to the mouse sequence in the residues located at the V_L/V_H interface (see *Table 6*);
   - residues in the FRs that are unusual or unique for a particular position when compared to the consensus sequence for that human subgroup (Kabat *et al.*, ref. 14);
   - potential glycosylation sites.[a]
4. Make a subjective decision as to the most appropriate human sequences, one human light chain variable region sequence and one human heavy chain variable region sequence, to serve as templates for the design of a reshaped human antibody.

C. *Design of the first versions of reshaped human variable regions*

1. Write out the sequences of the proposed reshaped human variable regions with the CDRs from the mouse variable regions joined to the FRs from the selected human variable regions.
2. Highlight the amino acids in the human FRs that are different from those that were present in the mouse FRs. Use the structural model of the mouse variable regions to help evaluate the significance of the proposed amino acid changes. Consider conserving the following mouse residues:
   - Residues that belong to canonical sequences for loop structure (see part A, step 4 and *Table 5*).
   - Residues that the model suggests have a role in supporting a CDR loop. Carefully examine buried residues and residues in the 'Vernier' zone[b] (15). Carefully examine the H3 loop where there are no defined canonical structures to use for guidance.

Protocol 4. *Continued*

- Residues that the model suggests are on the surface near the antigen-binding site.
- Residues located at the V_L/V_H interface (see part A, step 4 and *Table 6*).

3. Examine the revised sequences and consider the following points:
   - Removing any potential *N*-glycosylation sites within the human FRs and conserving, or removing, any potential *N*-glycosylation sites that are present in the mouse FRs. Use the model to predict whether potential glycosylation sites are located at positions that are on the surface and accessible and, therefore, likely to be used.
   - Role of mouse residues that are atypical when compared to the consensus sequence for that subgroup of mouse variable regions (see part A, step 4). It is possible that atypical amino acid residues have been selected for, at certain positions, to improve binding to antigen.
   - Location of human residues that are atypical when compared to the consensus sequence for that subgroup of human variable regions (see part B, step 3). It is possible that the potential immunogenicity of the reshaped human antibody will be increased if the human FRs contain unusual human sequences.

D. *Design of the additional versions of reshaped human variable regions*

1. When preliminary assays indicate that the first reshaped human antibody has a binding affinity equal to or better than the mouse or chimeric antibody, make additional versions to further reduce the number of substitutions of mouse residues into the human FRs.

2. When preliminary assays indicate that the first reshaped human antibody has a poor binding affinity, determine whether one or both reshaped human variable regions is the cause. Using *Protocol 6*, express the reshaped human light and heavy chains in all combinations with the chimeric light and chains and determine the relative binding affinities of antibodies expressed.

3. Re-analyse the model and ask if any additional substitutions of mouse residues into the human FRs are required. Be particularly cautious about any amino acid differences between the mouse and human FRs that occur in buried residues.

4. Reconsider the removal or inclusion of any potential glycosylation sites.

[a] The consensus sequence for *N*-glycosylation is Asn–Xaa–(Ser/Thr)–Yaa. When Yaa is a proline, the probability of glycosylation is reduced by approximately 50%. When Xaa is a proline, the probability is reduced by approximately 90%. When both Xaa and Yaa are proline, the

7: Rodent to human antibodies by CDR grafting

probability of glycosylation is even further reduced (16). There is no single consensus sequence for O-glycosylation. The Thr/Ser acceptor sites tend to reside in helical segments containing other serine or threonine residues and adjacent prolines. Four motifs that predict O-glycosylation (17, 18) are:
- Xaa–Pro–Xaa–Xaa where at least one Xaa is Thr
- Thr–Xaa–Xaa–Xaa where at least one Xaa is Thr
- Xaa–Xaa–Thr–Xaa where at least one Xaa is Arg or Lys
- Ser–Xaa–Xaa–Xaa where at least one Xaa is Ser.

[b] The Vernier zone (15) describes a series of residues which do not necessarily contact the antigen but affect the conformation of the CDR. These are:
- heavy chain (2, 27–30, 47–49, 67, 69, 71, 73, 78, 93–94, 103)
- light chain (2, 4, 35–36, 46–49, 64, 66, 68–69, 71, 98) (numbering according to Kabat).

Protocol 5. Synthesis of a DNA sequence coding for the reshaped human variable region

Equipment and reagents
- DNA thermal cycler (see *Protocol 2*)
- GeneAmp™ PCR reaction tubes (see *Protocol 2*)
- Sterile water (see *Protocol 2*)
- AmpliTaq® DNA polymerase (see *Protocol 2*)
- 10 × PCR buffer II and 25 mM MgCl$_2$ (see *Protocol 2*)
- GeneAmp® dNTPs (see *Protocol 2*)
- PCR primers. Prepare separate 10 µM stock solutions of the primers in sterile water
- Mineral oil (see *Protocol 2*)
- Wizard™ PCR-Preps DNA Purification System (Promega, cat. no. A7170)
- TA Cloning® kit (see *Protocol 2*)
- LB agar plates (see *Protocol 2*)
- Ampicillin (see *Protocol 2*)
- X-Gal (see *Protocol 2*)
- 10 × TBE buffer (see *Protocol 2*)
- Agarose (see *Protocol 2*)
- Ethidium bromide (see *Protocol 2*)

Method

A. *Design of the oligonucleotides and primers*

1. Using the '*SeqEd*' program in the GCG package, create a file of the DNA sequence of the human leader-variable region that has been selected as the template for reshaping[a] (see *Protocol 4*, part B).

2. Substitute the DNA sequences coding for the mouse CDRs for the DNA sequences coding for the human CDRs. Make minor modifications in the DNA sequences coding for the human FRs as required to make any amino acid changes that were specified in the design of the reshaped human variable region (see *Protocol 4*, part C). Base the codon usage on the mouse variable region or refer to the database of immunoglobulin genes (14) and try to avoid rare codon usage.

3. In order to function in the expression vectors previously described (8), the reshaped human leader-variable region will require the following DNA sequences:
    - Kozak translation initiation sequence (see *Table 4*)
    - splice-donor sequence (see *Table 4*)
    - restriction sites for cloning into the expression vector.

161

Protocol 5. *Continued*

4. Check the DNA sequence for the presence of unintentional splice-donor sites that might interfere with RNA processing (see *Table 4*). Check that there are no internal restriction sites that will interfere with cloning into the expression vector. Remove undesirable DNA sequences by altering the codon usage (see part A, step 2).

5. Divide the DNA sequence (approximately 410–470 bp in size) into eight oligonucleotides[b] approximately 90–100 bp in size with overlaps of 24 bp (see oligonucleotides 1–8 in *Figure 1*). Using the '*Map*' program in the GCG package, identify unique restriction sites already present in the DNA sequence and sites where unique restriction sites could be inserted without altering the amino acid sequence. Where possible, engineer unique restriction sites into each overlapping region (see part A, step 2).

6. Using the '*Stemloop*' program in the GCG package, identify potential stemloops within the DNA sequence. Remove any potential stemloops with a melting temperature of over 40°C by modifying the codon usage (see part A, step 2).

7. Design four nested 24 bp primers[b] (primers A–D in *Figure 1*) and two 24 bp nested primers[b] (primers I and II in *Figure 1*) to assist in the assembly of the oligonucleotides.

8. Synthesize and, if necessary, purify the oligonucleotides and primers.

B. *PCR assembly of the oligonucleotides*

1. Prepare two PCR reactions containing:

    - sterile water   65.5 μl
    - 10 × PCR buffer II   10.0 μl
    - 25 mM MgCl$_2$ 6.0 μl
    - 10 mM stock solutions of dNTPs   2.0 μl each
    - *either* primers 1, 2, 3, and 4 *or*
      primers 5, 6, 7, and 8 (*see Figure 1*)   2.5 μl each (at 10 μM)

    Overlay each reaction with 50 μl of mineral oil and place in a DNA thermal cycler.

2. 'Hot-start' the reactions by incubating for 5 min at 94°C, cooling to approximately 80°C, and adding 0.5 μl of AmpliTaq® DNA polymerase[c] to each reaction. Cycle at 94°C for 2 min and 72°C for 5 min for eight cycles using a 30 sec ramp time between each step. Follow the last cycle with a final extension step at 72°C for 10 min before cooling to 4°C.

3. Prepare two PCR mixes containing:

    - sterile water   60.5 μl
    - 10 × PCR buffer II   10.0 μl

7: Rodent to human antibodies by CDR grafting

- 25 mM MgCl$_2$ 6.0 µl
- 10 mM stock solutions of dNTPs   2.0 µl each
- *Either* 10 µl of PCR product '1/4' plus 2.5 µl each of the nested primers A and B (at 10 µM) *or* 10 µl of PCR product '5/8' plus 2.5 µl each of the nested primers C and D (at 10 µM) (see *Figure 1*)

Overlay each reaction with 50 µl of mineral oil and place in a DNA thermal cycler. 'Hot start' adding enzyme as described in part B, step 2. Cycle at 94 °C for 1.5 min and 72 °C for 2.5 min for 20 cycles. Extend and cool as described in part B, step 2.

4. Purify the PCR products from part B, step 3[d] using the Wizard™ PCR-Preps DNA Purification kit according to the manufacturer's instructions. Evaluate the quality and quantity of the PCR products by measuring the OD_{260} and by testing aliquots on a 1.5% (w/v) agarose gel in 1 × TBE buffer containing 0.5 µg/ml ethidium bromide.

5. Assemble the two PCR products from part B, step 3 by preparing a PCR reaction containing:

   - PCR products A/B and C/D (at 10 µM)   2.5 µl each
   - sterile water   70.5 µl
   - 10 × PCR buffer II   10.0 µl
   - 25 mM MgCl$_2$ 6.0 µl
   - 10 mM stock solutions of dNTPs   2.0 µl each

   Overlay with oil, 'hot start', and cycle as described in part B, step 2.

6. Amplify the PCR product 'A/D' from the previous step by preparing a PCR reaction containing:

   - sterile water   60.5 µl
   - 10 × PCR buffer II   10.0 µl
   - 25 mM MgCl$_2$ 6.0 µl
   - 10 mM stock solutions of dNTPs   2 µl each
   - 10 µl of PCR product 'A/D', 2.5 µl each of the nested primers I and II (at 10 µM) (see *Figure 1*)

   Overlay with oil, 'hot start', and cycle as described in step 3.

7. Clone the PCR product (see *Protocol 2*, part C, step 6).

8. PCR screen the transformants using *Protocol 3*.

9. Sequence the DNA from several clones that have inserts of the correct size. If errors are found in the DNA sequences, make use of the unique restriction sites within the reshaped human variable region to assemble one correct DNA sequence from a few sequences with errors in different sections of the sequence.

[a] If the DNA sequence of the selected human variable region is not available, use the DNA sequence from a closely related human variable region gene modifying the DNA sequence as necessary to obtain the required coding sequence.

Protocol 5. *Continued*

[b] AmpliTaq® DNA polymerase can add an extra thymidine (T) to the 3'-end of the PCR product. To avoid possible PCR errors due to this effect, design the oligonucleotides that will serve as PCR primers so that the 5'-end of each primer hybridizes to the complementary strand of template DNA one base downstream of an adenine (A).
[c] AmpliTaq® DNA polymerase is used rather than Pfu or Vent. AmpliTaq® permits the use of the TA Cloning® kit.
[d] As a precaution, consider TA-cloning the PCR products from part B, steps 3 and 4 above as described in *Protocol 2*, part C, step 5.

4.3.2 Linking the reshaped human variable regions to human constant regions

The reshaped human variable regions together with their leader sequences are cloned into mammalian cell vectors that already contain human constant regions. Each reshaped human variable region is linked via an intron to the desired human constant region. The expression vectors are identical or similar to the vectors that were described for the construction of chimeric light and heavy chains (7, 8).

5. Preliminary expression and analysis of the reshaped human antibodies

The two mammalian cell-expression vectors, one coding for the reshaped human light chain and one coding for the reshaped human heavy chain, are co-transfected into *cos* cells by electroporation (see Chapter 12, Bebbington). The vectors will replicate in the *cos* cells and transiently express and secrete reshaped human antibody. The medium is collected three days after transfection and analysed by ELISA to determine the approximate amount of antibody present (*Protocol 6*).

In most cases, *cos* cells will also have been transfected with the vectors that express the chimeric antibody. The chimeric and reshaped human antibodies as produced in the *cos* cells can be tested and compared for their relative abilities to bind to antigen. If purified antigen is available, the simplest approach is to use an ELISA format where the antigen is coated on the immunoplate and bound chimeric or reshaped human antibody is detected using a goat anti-human antibody–enzyme conjugate.

Protocol 6. ELISA method for measuring antibody concentration[a]

Equipment and reagents
- Nunc-Immuno Plate MaxiSorb™ (Life Technologies, cat. no. 43945A)
- Titertek® plate sealer (ICN Biomedicals, cat. no. 76–401–05)

- Titertek® Plus 8 Channel *Digital* Pipette (ICN Biomedicals, cat. no. 79-713-00)
- Benchtop centrifuge (see *Protocol 2*)
- Microplate Manager® data analysis software package (Bio-Rad, cat. no. 170-6618)
- Model 3550 microplate reader (Bio-Rad, cat. no. 170-6602)
- Model 1550 microplate washer (Bio-Rad, cat. no. 170-6541)
- Cell culture medium from transfected *cos* cells
- Goat anti-human IgG antibody, Fc$_\gamma$ fragment-specific (Jackson ImmunoResearch Laboratories Inc. via Stratech Scientific, cat. no. 109-005-098)
- Human IgG1/kappa antibody (Sigma, cat. no. I-3889)
- Goat anti-human kappa light chain alkaline phosphatase conjugate (Sigma, cat. no. A-3813)
- 15 mg *p*-nitrophenyl phosphate (*p*NPP) tablets (Sigma, cat. no. N-2640)
- PBS (see *Protocol 2*)
- Coating buffer: 0.05 M carbonate–bicarbonate buffer, pH 9.6. Add together 16 ml of 0.2 M anhydrous bicarbonate, 34 ml of 0.2 M sodium hydrogen carbonate, and 150 ml deionized, distilled water
- Blocking buffer: PBS with 2% (w/v) BSA (Sigma, cat. no. A-2135)
- Washing buffer: PBS with 0.05% (v/v) Tween-20 (Sigma, cat. no. P-1379)
- Sample–enzyme conjugate buffer: 0.1 M Tris–HCl, pH 7.0, 0.1 M NaCl, 0.02% (v/v) Tween-20 (Sigma, cat. no. P-1379), and 0.2% (w/v) BSA (Sigma, cat. no. A-2153)
- 10% diethanolamine buffer. Add together 10 ml of a solution containing 1 mg/ml MgCl$_2$ and 2 mg/ml sodium azide, 9.7 ml of diethanolamine, 6.6 ml of 2 M HCl, pH 9.8, and 73.7 ml of deionized, distilled water. Wrap in aluminium foil and store at 4°C
- 1 M NaOH

Method

1. Coat each well of a 96-well immunoplate with 100 µl aliquots of 0.4 µg/ml goat antihuman IgG antibody, diluted in coating buffer, using the multichannel pipette. Seal with a plate sealer and incubate overnight at 4°C.[b]

2. Remove the excess coating solution and block with 200 µl/well of blocking buffer for 2 h at room temperature. Wash the plate six times with 200 µl/well of washing buffer. Use the microplate washer as described by the manufacturer.

3. Dispense 100 µl of sample–enzyme conjugate buffer into all wells except the wells in column 2, rows B–G.

4. Prepare a 1 µg/ml solution of the human IgG1/kappa antibody in sample–enzyme conjugate buffer to serve as a standard. Pipette 200 µl/well into the wells in column 2, rows B and C.

5. Centrifuge the medium from transfected *cos* cells (250 *g*, 5 min) and save the supernatant.

6. Pipette 200 µl of the supernatant from the 'no DNA' control (where *cos* cells were transfected in the absence of DNA) into the well in column 2, row D.

7. Pipette 200 µl/well of experimental supernatants into the wells in column 2, rows E, F, and G.

8. Mix the 200 µl aliquots in the wells of column 2, rows B–G, with a pipette tip attached to the multichannel pipette and then transfer 100 µl to the neighbouring wells in column 3. Continue to column 11 with a series of 2-fold dilutions of the standard, control, and experimental samples.[c]

Protocol 5. *Continued*

9. Incubate at 37°C for 1 h. Rinse all the wells six times with 200 μl aliquots of washing buffer using the microplate washer.
10. Dilute the goat anti-human kappa light chain alkaline phosphatase conjugate 5000-fold[d] in sample–enzyme conjugate buffer and add 100 μl to each well. Repeat the incubation and washing steps (step 9).
11. Dissolve a 15 mg tablet of the *p*NPP substrate in 15 ml of 10% diethanolamine buffer. Add 100 μl to each well and incubate for 30 min at 37°C.
12. Stop the reaction by adding 100 μl of 1 M NaOH to each well. Read the optical density at 405 nm using the microplate reader in conjunction with Microplate Manager®.

[a] This assay is designed to detect human IgG1/kappa antibodies. It can be adapted to detect other isotypes of human, chimeric, or reshaped human antibodies. To assay for human IgG4/kappa antibodies, use human IgG4/kappa antibody (Sigma, cat. no. I-4639) as the standard and dilute the antibody–enzyme conjugate 1000-fold.[d] To assay for human IgG1/lambda antibodies, use human IgG1/lambda antibody (Sigma, cat. no. I-4014) as the standard and goat anti-human lambda light chain alkaline phosphatase (Sigma, cat. no. A-2904), diluted 2500-fold[d] in sample–enzyme conjugate buffer, as the antibody–enzyme conjugate. To assay for human IgG4/lambda antibodies, use human IgG4/lambda antibody (Sigma, cat. no. I-4764) as the standard and the anti-human lambda antibody–enzyme conjugate described above diluted 1000-fold.[d]
[b] The immunoplates may be stored at this stage for up to 1 month at 4°C.
[c] To avoid possible aberrant results caused by 'edge effects', the wells on the outside edges of the immunoplate are not used.
[d] The optimal dilution of the antibody–enzyme conjugate should be determined for each lot.

Acknowledgements

The authors wish to acknowledge the contributions of all those working in the Antibody Engineering Group at the MRC Collaborative Centre in developing and testing the methods outlined in this chapter, especially MRC Collaborative Centre scientists Dr C. A. Kettleborough, Dr J. Saldanha, and Dr O. J. Léger and visiting scientists Dr H. Maeda of Chemo-Sero-Therapeutics Research Institute, Dr F. Kolbinger of Ciba-Geigy, and Dr M. Tsuchiya and Dr K. Sato of Chugai Pharmaceuticals.

7: *Rodent to human antibodies by CDR grafting*

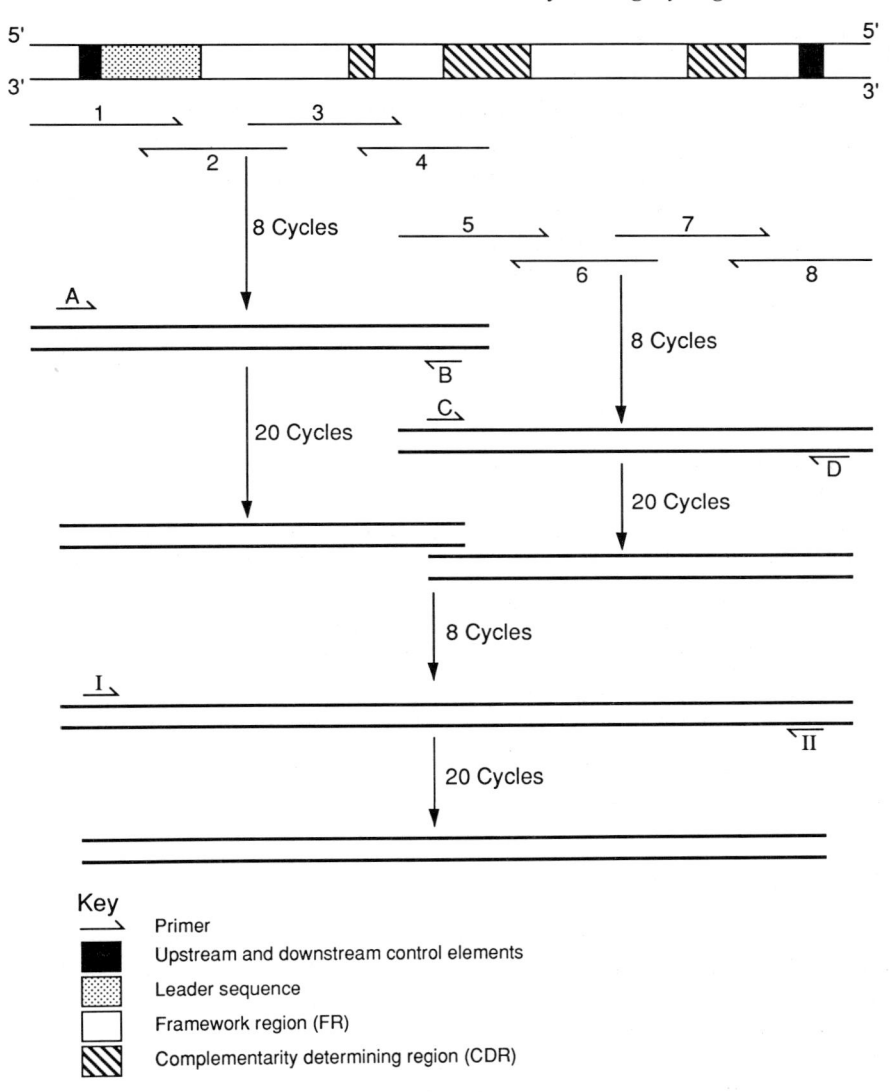

Figure 1. Diagram of the construction of a synthetic DNA sequence coding for a reshaped human variable region. A full explanation of the design of the oligonucleotide sequences and of their assembly into a full-length leader-variable region is presented in *Protocol 5*.

References

1. Jones, P. T., Dear, P. H., Foote, J., Neuberger, M. S., and Winter, G. (1986). *Nature*, **321**, 522.
2. Mountain, A. and Adair, J. R. (1992). *Biotech. and Gen. Eng. Rev.*, **10**, 1.
3. Bendig, M. M. (1994). *ImmunoMethods*, **8**, 83.
4. Emery, S. C. and Adair, J. R. (1994). *Exp. Opin. Invest. Drugs*, **3**, 241.
5. Chomczynski, P. and Sacchi, N. (1987). *Anal. Biochem.*, **162**, 156.
6. Kozak, M. (1987). *J. Mol. Biol.*, **196**, 947.
7. Kettleborough, C. A., Saldanha, J., Heath, V. J., Morrison, C. J., and Bendig, M. M. (1991). *Protein Eng.*, **4**, 773.
8. Maeda, H., Matsushita, S., Eda, Y., Kimachi, K., Tokiyoshi, S., and Bendig, M. M. (1991). *Hum. Antibod. Hybridomas*, **2**, 124.
9. Chothia, C. and Lesk, A. M. (1987). *J. Mol. Biol.*, **196**, 901.
10. Chothia, C., Lesk, A. M., Tramontano, A., Levitt, M., Smith-Gill, S. J., Air, G., Sheriff, S., Padlan, E. A., Davies, A., Tulip, W. R., Colman, P. M., Spinelli, S., Alzari, P. M., and Poljak, R. J. (1989). *Nature*, **34**, 877.
11. Tramontano, A., Chothia, C., and Lesk, A. M. (1990). *J. Mol. Biol.*, **215**, 175.
12. Chothia, C., Lesk, A. M., Gherardi, E., Tomlinson, I. M., Walter, G., Marks, J. D., Llewelyn, M. B., and Winter, G. (1992). *J. Mol. Biol.*, **227**, 799.
13. Brooks, B. R., Bruccoleri, R. E., Olafson, B. D., States, D. J., Swaminathan, S., and Karplus, M. (1983). *J. Comp. Chem.*, **4**, 187.
14. Kabat, E. A., Wu, T. T., Perry, H. M., Gottesman, K. S., and Foeller, C. (1991). *Sequences of proteins of immunological interest* (5th edn). US Department of Health and Human Services, US Government Printing Office.
15. Foote, J. and Winter, G. (1992). *J. Mol. Biol.*, **224**, 487.
16. Gavel, Y. and von Heinje, G. (1990). *Protein Eng.*, **3**, 43.
17. Gooley, A. A., Classon, B. J., Marschalek, R., and William, K. L. (1991). *Biochem. Biophys. Res. Commun.*, **178**, 1194.
18. Pisano, A., Redmond, J. W., Williams, K. L., and Gooley, A. A. (1993). *Glycobiology*, **3**, 429.
19. Chothia, C., Novotny, J., Bruccoleri, R., and Karplus, M. (1985). *J. Mol. Biol.*, **186**, 651.
20. Sato, K., Tsuchiya, M., Saldanha, J., Koishihara, Y., Ohsugi, Y., Kishimoto, T., and Bendig, M. M. (1993). *Cancer Res.*, **53**, 851.

8

Converting rodent into human antibodies by guided selection

HENNIE R. HOOGENBOOM, DEBORAH J. ALLEN, and
ANDREW J. ROBERTS

1. Introduction

Hybridoma technology, developed 18 years ago, has given rise to many rodent antibodies with exquisite specificity. Unfortunately, the therapeutic potential of rodent antibodies in humans has not yet been fully realized owing, to a great extent, to the immune responses they elicit, which can hamper or abolish their activity. Obtaining human antibodies directly by hybridoma technology has been relatively difficult, particularly in the case of those antibodies with specificity for 'self' antigens. Efforts to convert rodent antibodies into less immunogenic, more human structures have led to the construction of humanized antibodies (1), where the regions of the murine antibody responsible for antigen binding are transferred on to a human backbone (or framework) (Chapter 7). As a consequence, a substantial portion of the antibody (the six complementarity-determining regions or CDRs) remains non-human, which, in some cases can still lead to an anti-idiotype response, compromising the therapeutic benefit of the antibody treatment.

Recombinant DNA technology combined with the power of selection on antigen has, for the first time, allowed access to completely human antibodies, with the lowest immunogenicity achievable. Human antibodies with predefined specificities can be selected from repertoires displayed on filamentous phage by fusion to a gene encoding a minor coat protein (2, 3). Representative phage antibody libraries are created by direct cloning of randomly combined V-gene pairs (4, 5); very large repertoires are constructed by combinatorial infection (6). Antibodies that bind as strongly as those isolated using hybridoma technology have been isolated against a range of antigens, including highly conserved self-antigens, intracellular antigens, viruses, and cell-surface molecules.

There remains, however, an enormous pool of rodent antibodies against antigens of therapeutic interest, and methods to directly transfer some of these specificities to completely human antibodies would be desirable. We

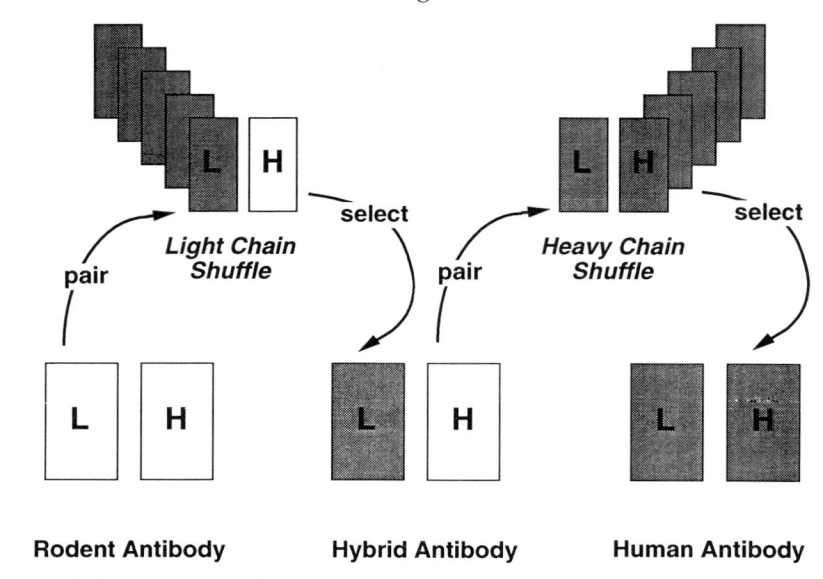

Figure 1. Guided selection for the conversion of rodent into human antibodies. The heavy chain of a rodent antibody (white) is shuffled with a repertoire of human light chains (grey) and selected to give a hybrid antibody. The selected human light chain(s) are then paired with a repertoire of human heavy chains which, following selection, gives the full human antibody. L, light chain; H, heavy chain.

have developed a methodology, based on chain shuffling of V genes and selection on antigen, to convert rodent antibodies into completely human antibodies with similar binding characteristics (7). The principle of the method, termed 'guided selection', is depicted in *Figure 1*. In the first step, one of the V domains of the starting rodent antibody (heavy or light chain) is paired with a repertoire of naturally occurring human partner domains, derived, for example, from human donor peripheral blood lymphocyte (PBL) mRNA. This repertoire is selected on the original antigen; the rodent domain will guide the selection of a partner human chain(s) that will allow the binding of antigen in the same way as the original rodent antibody. In a second step, the 'half-human' repertoire is used as a source of selected human V genes, which are then allowed to pair with a repertoire of human partner domains and are again selected on antigen. The endpoint of this sequential chain-shuffling procedure is a set of completely human antibodies, binding to the same epitope as the original rodent antibody.

Even though most of the structural diversity of antibodies resides in the heavy chain, there is no rule of thumb as to whether the V_H or the V_L of the rodent antibody should be shuffled first with partner human chains. To be able to choose one route in preference to the other would require some prior knowledge of the relative contribution of heavy and light chains to the antigen-combining site. Although this is sometimes the case for anti-hapten antibodies or antibodies whose three-dimensional structure has been determined, in most instances one should assume equivalent importance of V_H and V_L in the rodent antibody, making both domains equally suitable for the first shuffle. Alternatively, both routes are followed at the same time, and two hybrid repertoires are made (murine V_H–human V_L repertoire and murine V_L–human V_H repertoire). As well as increasing the potential for diversity, this approach offers the opportunity of 'parallel shuffling', to combine the respective properties of selected human V_Hs and V_Ls into a single selected human repertoire. It may be that most of the clones selected from the parallel chain-shuffled library are the same as those selected by sequential chain shuffling. However, in our experience the parallel chain-shuffling approach has been a great benefit, yielding clones with a higher affinity than those from sequential chain shuffling alone.

Two published examples demonstrate the power of the system. In a first model experiment, the light chain from murine anti-hapten phenyl-oxazolone antibody NQ10.12.5, of known 3-D structure, was shuffled with a repertoire of human heavy chain domains (8). The selected heavy chain partners retained many critical antigen-binding features found in the original murine heavy chain, indicating the conservation of binding interactions with the hapten. Entirely human antibodies were generated after a second shuffle of the selected human heavy chains with a repertoire of human light chains. A similar approach, starting with the murine heavy chain, was used to convert a murine anti-TNF (tumour necrosis factor) antibody into a human version, yielding five human antibodies (7); one of the full human antibodies has a similar affinity and competes with the murine antibody for binding to TNF. It also shows the same unique *in vivo* properties as the murine antibody (enhancement of the anti-tumour effect of TNF).

Thus, guided selection by rodent antibody V genes provides a powerful tool for interspecies conversion of antibodies. The combination of shuffling of V genes and selection on antigen provides a means to direct the isolation of a human V-gene pair with largely similar binding characteristics as the starting antibody. The process favours the isolation of antibodies binding to the original epitope, and, unlike CDR grafting, can give rise to antibodies with a range of different affinities and fine specificities.

Guided selection can be carried out in a number of different formats (*Figure 2*): either as single chain Fvs (scFv) or as Fab fragments, using a one or two replicon system. Two methods for guided selection in Fab format have recently been described (7, 8). This chapter will concentrate on guided selection in scFv format, which is faster, and more compatible with existing cloned human V-gene repertoires (9) than shuffling with Fab repertoires.

The full procedure is depicted in *Figure 3*. The first step in the procedure is to clone the variable domain genes of the rodent antibody. Since one of the cornerstones of the guided selection technique is the selection of phage

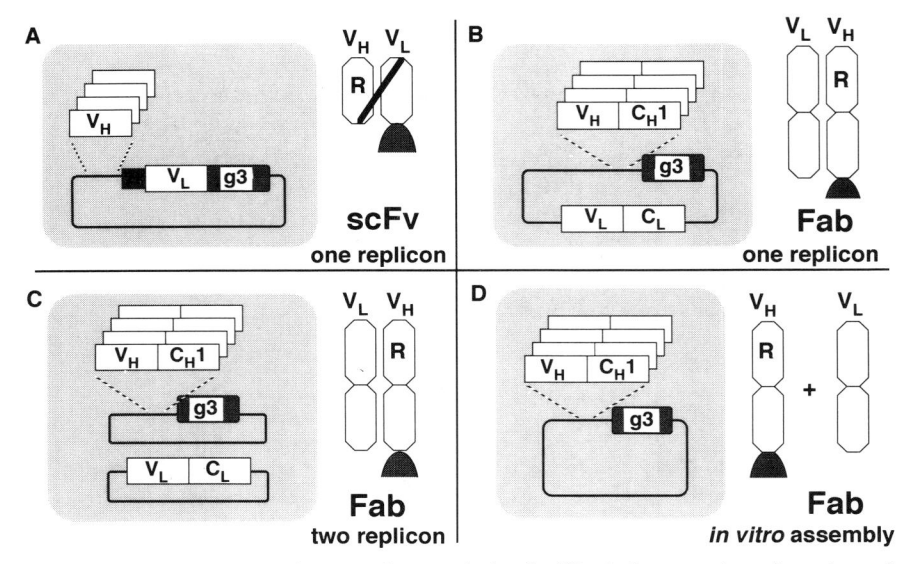

Figure 2. Guided selection formats (heavy chain shuffling). Formats based on the scFv (A) and Fab (B, C, D) configuration and using one or two replicons have been developed. In each panel the DNA is depicted on the left and the antibody format as encoded proteins on the right. In A and B heavy and light chains are on the same replicon, in C heavy and light chains are on separate replicons, and in D the light chain is provided *in trans* as protein. Illustrated here is a heavy chain shuffle, the repertoire denoted by R.

antibody fragments on antigen, it is advisable to clone the rodent antibody V genes in the format which will be used for the guided selection and test the binding activity and specificity of the antibody. It is also useful to check that the rodent antibody expressed as a phage antibody will work in the selection procedure of choice.

Methods and reagents used to amplify mouse and human antibody V genes are described in Chapter 1 (*Protocols 2* and *3*). Mouse V genes are amplified using consensus 5′ primers together with 3′ primers based on the J segment (10). This provides a means of amplification using a minimum number of primers, and in most instances works very well. A variety of different primer sequences have been described (see Chapter 1); some are also commercially available (Pharmacia, catalogue number 27–1581–01; see Table 1).

Figure 3 can be used to follow the various steps and protocols described in this chapter. First, the murine antibody genes are cloned as scFv fragments and checked for binding activity. For the guided selection the murine V_H is shuffled with a repertoire of human kappa and lambda light chains. This phage antibody repertoire is subjected to rounds of selection on antigen. At each round of selection, 24–48 clones are assayed for antigen binding either by soluble or phage ELISA (Chapter 1, *Protocols 9, 11,* and *13*). After achieving some 50–70% of positives for antigen binding, the murine V_H of

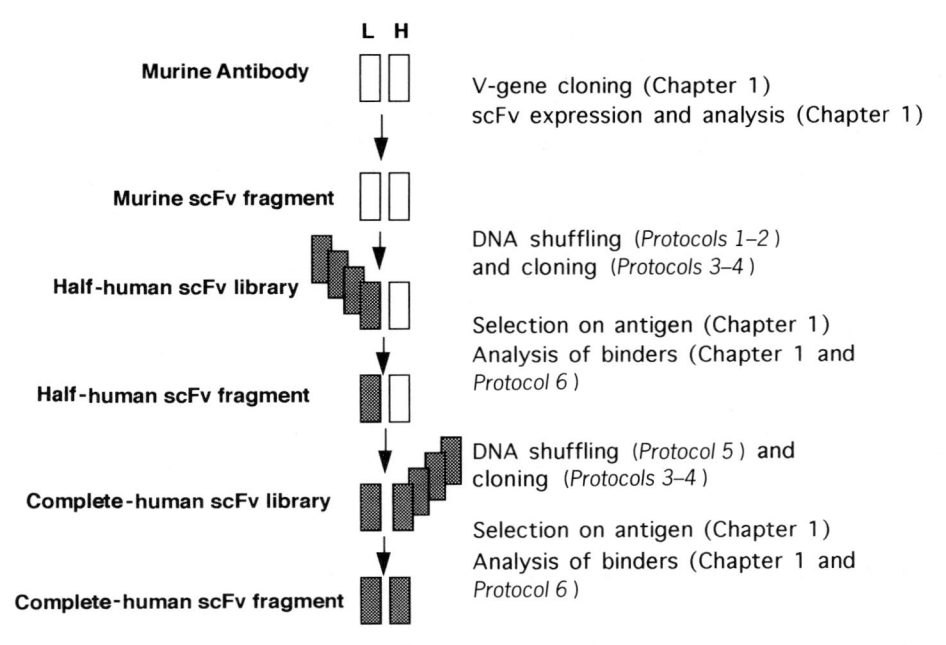

Figure 3. The sequential chain-shuffling strategy. Protocol numbers are indicated for each step of the procedure of humanization.

these antibodies is exchanged by a human V_H repertoire and the same shuffling and selection procedure followed to obtain fully human antibodies.

2. Cloning, expression, and characterization of rodent scFv fragments

Assembled rodent scFv are cloned into an expression vector based on vector pUC19, such as pUC19SNmyc (8), or phagemid pHEN1 (3), or pCANTAB5myc plasmids (see Chapter 1). These vectors allow for expression as a soluble scFv with a *myc* tag incorporated at the carboxyl terminus of the light chain, allowing immunodetection using the anti-c-*myc* monoclonal antibody 9E10 (3, 4). Expression of the rodent scFv should be confirmed by analysing a sufficient number (20–50) of colonies from the cloning plate. These can be grown as individual cultures and induced from the *lac* promoter as soluble scFv using isopropyl-β-D-thio-galactopyranoside (IPTG). The initial analysis of binding of the rodent scFv is an important step to ensure that the original antigen-combining site has been maintained and that antigen can be detected in the assay of choice, be it ELISA, cell-surface binding, immunoprecipitation, etc. Having isolated a number of clones scoring positive for antigen binding by ELISA, they should be sequenced and aligned to find

Table 1. Oligonucleotide list

PUCREVERSE	5'-AGC GGA TAA CAA TTT CAC ACA GG-3'
VH1FOR2	5'-TGA GGA GAC GGT GAC CGT GGT CCC TTG GCC CC-3'
REVJHNOXHO	5'-TGG GGC CAA GGT ACC CTG GTC ACC GTC TC-3'

HUJHFOR is an equimolar mixture of the following 4:

HUJH1–2FOR	5'-TGA GGA GAC GGT GAC CAG GGT GCC-3'
HUJH3FOR	5'-TGA AGA GAC GGT GAC CAT TGT CCC-3'
HUH4–5FOR	5'-TGA GGA GAC GGT GAC CAG GGT TCC-3'
HUJH6FOR	5'-TGA GGA GAC GGT GAC CGT GGT CCC-3'
FDTSEQ24	5'-TTT GTC GTC TTT CCA GAC GTT AGT-3'
PUCREVERSETAG	5'-GAC ACC TCG ATC AGC GGA TAA CAA TTT CAC ACA GG-3'
FDTSEQ24TAG	5'-ATT CGT CCT ATA CCG TTC TTT GTC GTC TTT CCA GAC GTT AGT-3'
REVERSETAG	5'-GAC ACC TCG ATC AGC G-3'
FDTAG	5'-ATT CGT CCT ATA CCG TTC-3'

Ambiguity codes: M = A or C; R = A or G; S = G or C; W = A or T

a consensus sequence. The clone with the sequence closest to the consensus should be used as the building block for subsequent chain shuffling steps.

3. Construction of large chain-shuffled repertoires in guided selection

One of the critical steps in evaluating the success of the guided selection procedure is the construction of large chain-shuffled repertoires with a high number of functional inserts. There are several examples in the literature where the ease with which a high-affinity antibody can be isolated by phage display correlates with the size of the starting repertoire (4, 6, 9). Since the aim of guided selection is to confer all the properties of binding specificity and affinity from the rodent antibody on to a human equivalent, maximizing library sizes during chain shuffling ensures that as many permutations as possible of heavy–light chain pairings are available for antigen selection.

Protocols for the construction of large 'primary' phage antibody repertoires are described extensively in Chapter 1. In this chapter, shuffling of rodent V genes with such cloned human V_H and V_L gene pools is described; these human V gene pools are easily accessible by PCR from the 'primary' scFv repertoire (4), or, alternatively, with appropriate oligonucleotides directly from mRNA of B cells. Chain-shuffled repertoires can be generated using rodent and human V_H and V_L genes combined together in a number of different ways (*Figure 2*). Irrespective of the route taken, the main criteria, which should be optimized in all stages of repertoire construction, are:

- the quality of restriction enzyme digested PCR product used for sub-cloning

- the efficiency of ligation and recovery of ligated product
- the transformation efficiency of the *E. coli* host

Often the conditions used for cloning the initial cDNA library of the rodent hybridoma are sufficiently stringent for construction of subsequent half-human or fully human chain-shuffled repertoires. The main distinction is the scale on which the procedure is carried out; relatively large quantities of V-gene input material are essential in chain shuffling to ensure that the maximum library diversity is achieved. The cloned rodent V_H (or V_L) gene will have been obtained as a PCR fragment; this may be cloned either by assembly of a PCR product whose termini overlap with the respective human V-gene repertoire (as described in *Protocols 1* and *2*), or directly by restriction fragment cloning. The latter can be employed if suitable restriction sites have been incorporated into the regions flanking the V genes. Particular care has to be taken to choose sites that occur infrequently in V genes. However, somatic mutation within V genes may generate restriction sites, in which case it would be better to use PCR assembly as a means of chain shuffling. It may also be possible to build individual repertoires of heavy or light chains, such that rodent V_H can be cloned into vectors containing repertoires of human V_L genes, or vice versa. We will describe only one system, where the murine V_H is combined with a human V_L repertoire by PCR assembly.

3.1 The first DNA shuffle: combining murine V_H with a human V_L repertoire

Both the rodent V_H and the human V_L repertoire fragments can be reamplified from scFv fragments cloned into their respective expression vectors. Assuming a pUC-based vector has been used for the murine scFv (such as pUC19SNmyc or pHEN1), its V_H can be amplified using a vector-based primer, PUCREVERSE, in combination with VH1FOR2. The human V_L repertoire is amplified (together with the linker segment) using REVJH-NOXHO and FDTSEQ24. The high degree of homology between the last few residues of V_H framework 4 in mouse and human allow for their assembly; primers VH1FOR2 and REVJHNOXHO are complementary to one another, which allows the murine V_H to be assembled to the human V_L repertoire/linker fragment by overlapping PCR. This assembled product can then be 'pulled through' with PUCREVERSE and FDTSEQ24, digested with *Sfi*I/*Not*I, and ligated into phage expression vector pCANTAB5myc. Due to the choice of primers, the restriction digestion afterwards will be very efficient: for both *Sfi*I and *Not*I digestions, a large overhang of DNA is created (150–200 bp), which ensures efficient digestion and gel purification.

The assembly reaction is performed in two stages. The first stage assembles the V_H and V_L + linker fragments together in the absence of the flanking primers. The product of this reaction is added to a second reaction with the 'pullthrough' primers to amplify the assembled V_H–linker-V_L product.

Protocol 1. Large-scale PCR of rodent V-gene to make half-human antibodies

Equipment and reagents

For this protocol you will require the following basic equipment and reagents:

- Microcentrifuge
- Sterile H_2O
- Sterile microcentrifuge tubes
- Sterile pipette tips
- Agarose
- 5 mg/ml ethidium bromide
- Agarose gel electrophoresis tank (Pharmacia, model GNA 100)
- 50 × stock TAE buffer: dissolve 242 g Tris base and 57.1 ml of glacial acetic acid in a final volume of 900 ml and then add 100 ml of 0.5 M EDTA (pH 8.0)

- 10 × Taq polymerase buffer as provided by the supplier of the Taq polymerase, or: 100 mM Tris–HCl pH 8.3, 500 mM KCl, 15 mM $MgCl_2$, and 0.01% (w/v) gelatin
- 5 mM dNTPs is an equimolar mixture of dATP, dCTP, dGTP, and dTTP with a total concentration of 5 mM nucleotide (i.e. 1.25 mM of each dNTP)
- DNA thermal cycler for PCR
- Mineral oil (paraffin oil) (Sigma, cat. no. M-3516)
- Geneclean (BIO 101) or Wizard PCR prep DNA purification system (Promega)
- UV transilluminator

In addition to basic equipment and reagents, the following will also be required for this protocol:

- Taq DNA polymerase (Boehringer, under licence from Perkin Elmer/Cetus)

Method

1. Set up the following PCR reaction mix:

    - H_2O 37.0 µl
    - 10 × Taq buffer   5.0 µl
    - 5 mM dNTP   1.5 µl
    - BACK primer[a]   2.5 µl
    - FOR primer[b]   2.5 µl
    - DNA[c]   1.0 µl
    - Taq DNA polymerase (5 U/µl)   0.5 µl

    Overlay with 2 drops mineral oil. Amplify by PCR using 25 cycles of 94°C for 1 min, 55°C for 1 min, 72°C for 2 min, followed by incubation at 72°C for 10 min.

2. Check each PCR amplification with a 5 µl sample on a 2% agarose gel.

3. Gel-purify fragments on a 2% lmp (low melting point) agarose/TAE gel. Carefully excise the V_H (350 bp) and V_L + linker (420 bp) bands (using a fresh sterile scalpel or razor blade for each) and transfer each band to a separate sterile microcentrifuge tube.

4. Purify the fragments from the agarose using, for example, Geneclean (BIO 101), Elutip (Schleicher and Schuell), or Wizard PCR prep DNA purification system (Promega). Recover the DNA in 50 µl H_2O.

8: Human antibodies by guided selection

[a,b] Primer pairs (at 10 µM):
Mouse V_H: PUCREVERSE; VH1FOR2
Mouse V_L + linker: REVJHNOXHO; FDTSEQ24
Human V_H: PUCREVERSE; HUJHFOR
Human V_L + linker: REVJHNOXHO; FDTSEQ24
[c] Template DNA is 10–25 ng of the murine scFv DNA, or of the human scFv IgM-derived repertoire described in ref. 4.

Protocol 2. Assembly of V_H and V_L/linker fragments

Equipment and reagents
As for *Protocol 1*.

Method

1. Estimate the quantities of V_H and V_L/linker DNA prepared by the primary PCR reactions on an agarose gel. Adjust the volumes of V_H and V_L/linker to give roughly equal masses of DNA fragments added to the assembly reaction (approximately 50 ng).

2. Set up the following assembly PCR reaction:

    - V_H.DNA (~50 ng)   x µl
    - V_L + linker DNA   y µl
    - 10 × Taq buffer   2.0 µl
    - 5 mM dNTP   1.0 µl
    - H_2O to 20 µl
    - Taq DNA polymerase (5 U/µl)   0.2 µl

    Overlay with 2 drops mineral oil. Amplify by PCR using 20 cycles of 94°C for 1.5 min, and 65°C for 3 min.

3. From the above reaction take 2 µl and add to fresh reaction mixture containing:

    - 10 × Taq buffer   5.0 µl
    - 5 mM dNTP   2.5 µl
    - PUCREVERSE (10 pmol/µl)   2.5 µl
    - FDTSEQ24 (10 pmol/µl)   2.5 µl
    - H_2O 37.0 µl
    - Taq DNA polymerase (5 U/µl)   0.5 µl

4. Overlay with 2 drops mineral oil. Amplify by PCR using 25 cycles of 94°C for 1 min, 55°C for 1 min, 72°C for 2 min, followed by incubation at 72°C for 10 min.

5. Gel-purify the assembled fragment (750 bp) on a 1% lmp (low melting point) agarose/TAE gel as in *Protocol 1*.

At this point the DNA is further treated as in *Protocol 7* of Chapter 1 (digestion, purification), and cloned into pCANTAB5myc as described below in *Protocols 3* and *4*.

For the ligation, there is no way to guarantee high-ligation efficiencies under all circumstances; however, we achieve much greater success using the 30 minute ligation kit commercially available from Amersham (DNA ligation system), which we find to be superior to standard ligation reagents. In order to optimize conditions for a particular insert, it is best to keep the amount of vector constant and to use varying quantities of insert in the ligations. A vector-only only control (ligation I below) should always be done so that the background of re-ligated vector can be assessed.

Protocol 3. Ligation protocol (with kit from Amersham)

Equipment and reagents

- Microcentrifuge
- Sterile H$_2$O
- Sterile microcentrifuge tubes
- Sterile pipette tips
- DNA Ligation system (Amersham, cat. no. RPN 1507)
- 10 × Tris–Mg buffer: 1 M Tris–HCl pH 7.6, 50 mM MgCl$_2$
- Vector DNA digested with *Sfi*I and *Not*I
- DNA insert fragment digested with *Sfi*I and *Not*I
- Phenol
- Chloroform
- Ethanol at –20 °C
- 70% ethanol at –20 °C
- Sodium acetate, 3 M, pH 5.2

Method

1. Set up the following two ligation reactions:

	I	II
vector DNA (50 ng/ml)	2 μl	2 μl
digested fragment (20–50 ng/ml)	–	1 μl
10 × Tris–Mg buffer	1 μl	1 μl
H$_2$O	7 μl	6 μl
buffer (Solution A from kit)	40 μl	40 μl

   - Mix and spin briefly in a microcentrifuge, then add:

enzyme (Solution B from kit)	10 μl	10 μl

2. Incubate at 16 °C, for 30–60 min.

3. Following ligation clean up the reaction using Geneclean (BIO 101), Wizard Clean Up (Promega) or other appropriate clean-up procedure.

4. Resuspend the ligated product in 10 μl H$_2$O and either store at –20 °C or use directly for electroporation.

8: Human antibodies by guided selection

Protocol 4. Preparation and electroporation of competent TG1

Preparation of competent *E. coli* using this method routinely gives $1-3 \times 10^{10}$ colonies/µg pBR322 DNA. Maximum efficiencies are obtained when the cells are prepared and electroporated on the same day; however, they may also be snap-frozen and stored at −70°C, resulting in approximately 10-fold lower efficiencies a week later.

Equipment and reagents

For part A:

- Sterile H₂O (ice-cold)
- Sterile 10% glycerol in water (ice-cold)
- Sterile pipette tips
- *E. coli* TG1
- 2 litre shake flask
- Shaking incubator at 37°C
- High-speed centrifuge (Sorvall or equivalent) and rotor
- Benchtop centrifuge
- Sterile centrifuge pots
- Sterile microcentrifuge tubes
- Sterile 50 ml Falcon tubes
- 2 × TY broth: 16 g tryptone, 10 g yeast extract, 5 g NaCl per litre. Sterilize by autoclaving
- Minimal agar: (per litre) 12.8 g Na₂HPO₄.7H₂O, 3 g KH₂PO₄, 0.5 g NaCl, 1 g NH₄Cl, 0.2 g MgCl₂, 5 mg thiamine hydrochloride, 4 g glucose, and 15 g agar

For part B:

- Bio-Rad Gene Pulser Plus
- 0.2 cm cuvettes, prechilled
- pBR322 DNA at 10 ng
- Cell recovery medium: 2 × TY + 2% glucose
- 2 × TYAG culture broth: 2 × TY, with 100 µg/ml ampicillin (A) and 2% (w/v) filter-sterilized glucose added (G)
- Sterile glycerol
- 2 × TYAG plates: 2 × TYAG containing 15 g agar per litre. Both large (243 × 243 mm, Nunc) and small (90 mm) plates are required
- Sterile pipette tips
- 15 ml polypropylene culture tubes
- Polypropylene freezing tubes
- Vortex mixer

A. *Preparation of electrocompetent cells*

1. Inoculate an overnight culture of *E. coli* TG1 (from a single colony on minimal agar) in 2 × TY broth. Grow at 37°C.

2. Next day, prewarm 500 ml 2 × TY broth in a sterile 2 litre flask to 37°C, then add 2.5 ml of the fresh overnight culture.

3. Grow with shaking (minimum 250 r.p.m.) for 90 min, or until the OD_{600} is approximately 0.5.

4. Chill flasks on ice for at least 30 min. Transfer cultures to sterile centrifugation bottles, for use in the Sorvall GSA rotor.

5. Spin at 6500 *g*, 4°C for 15 min. Decant the supernatant and resuspend the pellet (gently) in the starting volume of ice-cold sterile H₂O.

6. Stand on ice for 15 min.

7. Spin at 6500 *g*, 4°C for 15 min. Decant the supernatant and resuspend in 0.5 × starting volume ice-cold sterile H₂O.

Protocol 4. *Continued*

8. Stand on ice for 15 min.

9. Spin at 6500 g, 4 °C for 15 min. Decant the supernatant and resuspend in 20 ml ice-cold 10% glycerol in H_2O. Transfer to 50 ml Falcon tube and stand on ice for 15 min.

10. Spin at 2500 g, 4 °C for 10 min in a benchtop centrifuge. Decant the supernatant and resuspend the final bacterial pellet in 3 ml 10% glycerol.

11. Divide into 100 µl aliquots in sterile microcentrifuge tubes on ice; either use immediately or freeze as described above.

B. *Electroporation of ligation reactions using the Bio-Rad Gene Pulser Plus*

1. If frozen TG1 are to be used, thaw the appropriate number of aliquots on ice (allow 100 µl cells per 10 µl ligation reaction). If using fresh cells, add 100 µl directly to the resuspended ligation reaction (prepared as in *Protocol 3* above) on ice.

2. Transfer to a prechilled 0.2 cm cuvette (of a type compatible with the Bio-Rad Gene Pulser Plus). Be sure to set up controls with 10 ng pBR322 DNA (to gauge efficiency) and without DNA (to guard against contaminating DNA in the libraries).

3. Set up the Gene Pulser Plus to give a 2.5 kV pulse, using the 25 µF capacitor and the pulse controller set to 200 ohm.

4. Dry the cuvette with tissue and place in the electroporation chamber.

5. Pulse once. The registered time constant should be 4.5–5.0 msec.

6. Immediately add 1 ml fresh 2 × TY + 2% glucose to the cuvette, resuspend the cells, and transfer to a disposable culture tube.

7. Shake for 1 h at 37 °C to allow the cells to recover and to express antibiotic resistance.

8. Plate appropriate dilutions (in 2 × TY) on 90 mm 2 × TY agar plates containing 2% glucose and 100 µg/ml ampicillin (2 × TYAG). If using ligation mix and cells of known efficiency, plate aliquots containing at least 200 000 clones on 243 × 243 mm 2 × TYAG plates.

9. Grow overnight at 30 °C.

10. Harvest the repertoire the following day by flooding the plates with 2–10 ml 2 × TY and detach cells by scraping using a sterile spreader. Transfer cells to a sterile polypropylene tube and disperse clumps by vortexing. Add sterile glycerol to a final concentration of 15%, mix, and freeze 1 ml aliquots at − 70 °C. At this stage, freshly prepared aliquots of the library can be used to inoculate a culture to rescue phage and start selections as described in Chapter 1, *Protocol 8*.

8: Human antibodies by guided selection

From this repertoire, half-murine/half-human antibodies binding to the original antigen can be selected and analysed. For an extensive list of protocols for the rescue of phagemids, selection, and screening see Chapter 1, *Protocols 8–13*. After screening and identification of the best pairs (see also Section 4), the second shuffle can be performed as in *Protocol 5*.

3.2 Construction of fully human chain-shuffled repertoires by combining the selected human V_L genes with a human V_H repertoire

A number of selected V_L genes, which form binding pairs with the original murine V_H domain, are combined individually or in batch with a repertoire of human V_H genes. Care must be taken to ensure that the risk of contamination of the full human repertoire by half-human clones or the original murine antibody, is minimized. This is achieved by the use of tagged primers. The selected human V_L repertoire and linker can be reamplified using REVJH-NOXHO and FDTSEQ24TAG and assembled with a repertoire of human V_H genes amplified using PUCREVERSETAG and HUJHFOR. The primers FDTSEQ24TAG and PUCREVERSETAG have been appended with extra bases (the tag) which can be used as the site for annealing for tag-specific primers in the subsequent 'pull-through' PCR. After digestion with *Sfi*I/*Not*I, the pull-through product can be ligated into vector pCANTAB5myc (as before), or preferably, to minimize contamination of the murine V_H at later stages, into a different vector such as pCANTAB6 (ref. 11, also described in Chapter 1), which contains a polyhistidine tail just before the *myc* tag. PCR using a primer specific for the polyhistidine tag will discriminate between contaminating clones in pCANTAB5-*myc* (i.e. only half-human) and those in pCANTAB6 (fully human clones).

Protocol 5. Amplification of V_H and V_L + linker fragments from scFv repertoires to create full human repertoires

Equipment and reagents
As *Protocol 1*

Method
1. Set up the following PCR reaction mix:
   - H_2O 37 µl
   - 10 × Taq buffer   5.0 µl
   - 5 mM dNTP   1.5 µl
   - BACK primer[a]   2.5 µl
   - FOR primer[b]   2.5 µl
   - DNA[c]   1.0 µl
   - Taq DNA polymerase 0.5 µl

   Overlay with 2 drops mineral oil.

Protocol 5. *Continued*

2. Amplify by PCR using 25 cycles of 94°C for 1 min, 55°C for 1 min, 72°C for 2 min, followed by incubation at 72°C for 10 min. Check each PCR amplification with a 5 µl sample on a 2% gel.

3. Gel-purify the fragments on a 2% lmp (low melting point) agarose/TAE gel. Carefully excise the V_H (350 bp) and V_L/linker (420 bp) bands (using a fresh sterile scalpel or razor blade for each) and transfer each band to a separate sterile microcentrifuge tube. The fragments can then be purified from the agarose as in *Protocol 1*.

[a,b] Primer pairs (at 10 µM each)
Human V_H: PUCREVERSETAG; HUJHFOR
Human V_L + linker: REVJHNOXHO; FDTSEQ24TAG
[c] DNA is at 10 ng/µl and derived from the selected half human scFv fragments, or from the human scFv IgM-derived repertoire described in ref. 4.

The assembly reaction is performed in two stages. The first stage assembles the V_H and V_L/linker fragments together in the absence of the flanking primers. The product of this reaction is added to a second reaction with the 'pull-through' primers to amplify the assembled V_H–linker–V_L product. The use of tagged versions of PUCREVERSE and FDTSEQ24 allows for subsequent pull-through using REVERSETAG and FDTAG that anneal specifically to the tagged portion of the primers. This minimizes the chances of contamination of murine V_H or V_L in the full human repertoire. For this assembly use *Protocol 2*, but replace the oligonucleotide primers in step 3 with REVERSETAG and FDTAG. Gel-purify the assembled fragment (750 bp) on a 1% lmp (low melting point) agarose/TAE gel as in *Protocol 1* of this chapter and in *Protocol 7* of Chapter 1 for restriction enzyme digestion and gel purification. The fragment is cloned into a phagemid vector, and the repertoire used for rescue (to obtain display of the antibody) and selection on antigen (see Chapter 1).

4. Selecting half-human or completely human antibodies by display and enrichment steps

After having created antibody repertoires with V genes of half-murine, half-human origin (first shuffle), or of completely human origin (second shuffle), the antibody repertoires are subjected to rounds of growth, rescue (for display), and selection on antigen, to enrich the population for antigen binders (*Protocols 8–13* in Chapter 1). The success of converting rodent to human antibodies depends, to a large extent, on how the human binders are best identified and isolated, which in turn depends on the antigen. The concentration of antigen used in any of the selection procedures is an important factor,

8: Human antibodies by guided selection

and depends on whether the aim is to isolate all antibodies (in which case a high concentration, up to 1 mg/ml, is used), or to preferentially select for higher affinity antibodies (when a lower concentration is used, e.g. 2–10 µg/ml). Regardless of which of the selection protocols is used, the main objective is to be able to apply the same selection criteria to half or fully human versions as to the starting rodent antibody. It will usually be necessary to repeat any selection procedure two to five times before the population is significantly enriched.

The quickest and most convenient way of identifying binders from chain-shuffled repertoires at different stages of selection is by ELISA, using the methods described in Chapter 1 (*Protocols 9, 10, 11* and *13*). Not only does ELISA screening give a quantitative evaluation of positive clones in the repertoire (to determine whether sufficient rounds of selection have been carried out), but often the isolation of individual clones giving strong relative ELISA signals allows qualitative comparisons as well.

In certain cases, however, the ELISA signal obtained from soluble scFvs of half or fully human antibodies is too weak to analyse binding specificity and affinity in great detail. In these instances, sensitivity of detection in ELISA can be increased by analysing the clones as phage. Once cross-contamination with the parent rodent antibody is no longer a consideration (for example, when screening fully human chain-shuffled versions), it is feasible to grow clones in microtitre plates, induce soluble scFv, or rescue as phage (Chapter 1, *Protocols 9, 11*, and *13*) and perform a number of assays to determine both antigen-binding capacity and specificity. Individual clones can be rapidly identified by *Bst*NI fingerprint (*Protocol 6*, this chapter). Finally, the IC_{50} affinity value for the best clones can be determined and the off-rate component of the measured affinity determined if required (see Chapter 2). We have found these ELISA-based assays give results consistent not only with other methods of measuring affinity (using kinetic measurements in solution or in a biosensor), but even with affinities of whole recombinant immunoglobulins assembled from their scFvs. Alternatively, soluble antibody fragments can be analysed, based on *Protocols 11* and *13* of Chapter 1.

All fully human clones still positive for antigen binding after colony purification can be analysed further. A useful analysis to undertake at this point is to 'fingerprint' the PCR product of the scFv gene using a frequent cutting endonuclease such as *Bst*NI (*Protocol 6*). Another is to check to ensure that the human clones do not bind to a panel of irrelevant antigens, as undesirable cross-reactivity may have been introduced during the guided selection. The choice of antigens to use for this panel is up to the individual, but we generally use a panel of at least 10, to include haptens, purified proteins from different species, and antigens with structural similarities to the antigen of interest.

Protocol 6. Fingerprinting clones with the restriction endonuclease *Bst*NI

The primers used to generate material for fingerprinting have been described in *Protocol 2*. This method is only intended as a rough guide to diversity in a panel of clones, in between sequential rounds of selection, or as a guide to diversity between individual clones prior to sequencing.

Equipment and reagents
As *Protocol 1* with the following additional reagents:
- Nu Sieve agarose (Flowgen)
- 10 × buffer for *Bst*NI (NEB2, New England Biolabs)
- *Bst*NI enzyme (New England Biolabs)
- Acetylated BSA (10 mg/ml)

Method

1. Make up 50 µl PCR mixes containing:
   - H_2O 37.5 µl
   - 10 × Taq buffer   5.0 µl
   - 5 mM dNTPs   2.5 µl
   - FDTSEQ24 (10 µM)   2.5 µl
   - PUCREVERSE (10 µM)   2.5 µl

2. With a toothpick touch a single colony and bring in contact with the PCR reaction mix briefly. Throw away the toothpick.

3. Repeat for all clones. Overlay reactions with mineral oil and heat to 94°C for 5 min using the PCR block.

4. At the end of this incubation, add 0.2 µl Taq polymerase (5 U/µl) under the oil.

5. Cycle 25 times to amplify the fragments: 94°C for 1 min; 55°C for 1 min; 72°C for 1.5 min.

6. After the PCR, add the following underneath the mineral oil:
   - acetylated BSA (10 mg/ml)   1 µl
   - 10 × buffer   10 µl
   - H_2O to 100 µl
   - *Bst*NI (10 U/µl, New England Biolabs)   2 µl

7. Digest samples at 60°C for 2 h.

8. Remove an aliquot of the digested sample from underneath the oil. Analyse on a 4% (w/v) Nu Sieve agarose gel cast in TAE buffer containing 0.5 µg/ml ethidium bromide.

9. Compare the banding patterns of individual clones on a UV transilluminator.

References

1. Jones, P. T., Dear, P. H., Foote, J., Neuberger, M. S., and Winter, G. (1986). *Nature*, **321**, 522–5.
2. McCafferty, J., Griffiths, A. D., Winter, G., and Chiswell, D. J. (1990). *Nature*, **348**, 552–4.
3. Hoogenboom, H. R., Griffiths, A. D., Johnson, K. S., Chiswell, D. J., Hudson, P. and Winter, G. (1991). *Nucl. Acids Res.*, **19**, 4133–7.
4. Marks, J. D., Hoogenboom, H. R., Bonnert, T. P., McCafferty, J., Griffiths, A. D., and Winter, G. (1991). *J. Mol. Biol.*, **222**, 581–97.
5. Hoogenboom, H. R. and Winter, G. (1992). *J. Mol. Biol.*, **227**, 381–8.
6. Griffiths, A. D., Williams, S. C., Hartley, O., Tomlinson, I. M., Waterhouse, P., Crosby, W. L., Kontermann, R., Jones, P. T., Low, N. M., Allison, T. J., Prospero, T. D., Hoogenboom, H. R., Nissim, A., Cox, J. P. L., Harrison, J. L., Zaccolo, M., Gherardi, E., and Winter, G. (1994). *EMBO J.*, **13**, 3245–60.
7. Jespers, L. S., Roberts, A., Mahler, S. M., Winter, G., and Hoogenboom, H. R. (1994). *Bio/Technol.*, **12**, 899–903.
8. Figini, M., Marks, J. D., Winter, G., and Griffiths, A. D. (1994). *J. Mol. Biol.*, **239**, 68–78.
9. Marks, J. D., Griffiths, A. D., Malmqvist, M., Clackson, T. P., Bye, J. M., and Winter, G. (1992). *Bio/Technol.*, **10**, 779–83.
10. Clackson, T., Hoogenboom, H. R., Griffiths, A. D., and Winter, G. (1991). *Nature*, **352**, 624–8.
11. McCafferty, J., Fitzgerald, K. J., Earnshaw, J., Chiswell, D. J., Link, J., Smith, R., and Kenten, J. (1994). *Appl. Biochem. Biotechnol.*, **47**, 157–73.

9

Choosing and manipulating effector functions

INGER SANDLIE and TERJE E. MICHAELSEN

1. Choosing effector functions

The aim of antibody therapy is often target-cell destruction. The target cells may be microorganisms, infected cells, cancer cells, autoimmune T cells, or cells responsible for organ graft rejection and graft versus host disease. In all these cases it is crucial that the effector mechanisms of target-cell specific antibodies are elicited. Antibodies protect against bacterial infections by inducing bactericidal activity, i.e. complement-mediated bacteriolysis, or by inducing opsonophagocytosis mediated by neutrophils, monocytes, or macrophages. When experiments are carried out using antibodies against human target cells and with human serum as the source of complement, only a few of the antibody molecules induce good lysis. This is due to the presence of cell-surface molecules such as 'homologous restriction factor', 'decay accelerating factor', CD59, and 'membrane co-factor protein' (collectively called complement regulatory proteins or homologous complement restriction factors). They reduce the toxicity of the animal's own complement to the animal's own cells (1). On the other hand, when heterologous systems are employed, with human target cells combined with rat or mouse monoclonal antibodies and rabbit or guinea-pig complement, good lysis has often been observed. The results obtained with complement and target cells from the same donor, however, show that some antigens are good targets for cell-mediated lysis (CML), whereas most are not. The antigen CDw52, which is recognized by the CAMPATH-I family of antibodies (found on human lymphocytes and monocytes but which is absent from other blood cells including haemopoietic stem cells) is a good target for complement lysis (2).

While it is possible to analyse the interaction of antibodies with specific effector function molecules in *in vitro* systems, the results from these experiments are not sufficient to explain why some antibodies are effective *in vivo* and others are not; but clearly, differences in the *in vivo* biodistribution of antibodies, complement, and effector cells are crucial for the effect. Prediction of

Table 1. The biological activities of the human IgG subclasses[a]

	IgG1	IgG2	IgG3	IgG4
Complement activation				
Classical pathway	+++	+	+++	−
Alternative pathway	−	+	−	−
Fc receptor activation				
FcγRI	+++	−	+++	++
FcγRII, HR	+	−	+	−
FcγRII, LR	+	+	+	−
FcγRIII	+	−	+	−

[a] Adapted from ref. 17.

the therapeutic value of antibodies is difficult. The problem lies not only in determining the contribution of CML versus FcγR-related mechanisms in depleting target cells, but in our ability to predict the final outcome of the FcγR-dependent mechanisms, i.e. when a number of FcγR types on various effector cells in an animal are triggered simultaneously by immune complexes. Studies in *in vitro* systems employing effector cells transfected with individual FcγR genes and in animal models where specific mutations are introduced or gene disruption is performed to 'knock out' the function of the complement system or defined FcγRs may be necessary to clarify the relative importance of each component.

In a number of cases it may be desirable to utilize antibodies which lack one or more effector functions. The aim may be to block the activity of haemopoietic cells temporarily, allowing useful cellular functions to return once tolerance is established (3). There is also a potential advantage in the use of such antibodies to act as delivery agents, for example in radioimmunotherapy of tumours or as carriers of drugs to tumours (4). Sets of antibodies with identical V regions and different C regions, allow a direct comparison of properties both in *in vitro* and *in vivo* (5–11).

The biological activities of the four human subclasses are listed in *Table 1*.

2. Mediation of effector functions

In human blood and interstitial fluids the predominant antibody class is IgG, which is the major component of the secondary humoral immune response. Hence, this chapter will focus on this class of antibodies. The effector functions initiated by IgG are known as complement activation which can lead to cell-mediated lysis, antibody-dependent cell-mediated cytotoxicity (ADCC), and opsonophagocytic activity.

2.1 Complement activation and lysis

The classical complement activation pathway, leading to CML, is initiated by the binding of IgM or IgG to antigens on the surface of target cells, followed by multivalent interaction between the antibodies and the multi-subunit molecule C1q. C1q interacts with C1s–C1r–C1r–C1s, a Ca^{2+}-dependent tetrameric complex of two serine proteases. To activate C1r and C1s, C1q has to bind at least two of its subunits to the C_H2 domains of IgG or C_H3 domains of IgM. It is the presumed distortional change in the C1q structure upon binding that leads to activation of the C1 complex and further activation of the later component terminating in the assembly of C6–C9 to generate the membrane attack complex which causes lysis (12).

2.2 FcγR-mediated activities

ADCC and phagocytosis are mediated by effector cells that display Fcγ receptors (FcγR), binding the Fc part of IgG. For phagocytosis, complement receptors on effector cells also play a part (13). Humans have three groups of FcγRs: FcγRI (CD64), FcγRII (CD32), and FcγRIII (CD16). The FcγRI and FcγRII groups have three members (A, B, and C) and FcγRIII two (A and B). At least 12 isoforms are generated by differential splicing and utilization of different polyadenylation sites (14). Members of the FcγRI group are found on monocytes and macrophages and can be induced on neutrophils and eosinophils. FcγRIIA and C are found on monocytes, macrophages, and neutrophils, whereas FcγRIIB are found on B cells. Thus, members of the FcγRII class are the most broadly distributed FcγRs (15). FcγRIIA exists in two polymorphic forms (LR and HR) (16). FcγRIIIB is found on neutrophils and is linked to the surface by a glycosyl—phosphatidylinositol linkage. The FcγRIIIA, on the other hand, is a transmembrane protein found on natural killer (NK) cells, macrophages, and some T cells (14).

FcγRI is the only FcγR able to bind monomeric IgG with high affinity, unlike FcγRII and FcγRIII which bind only complexed IgG. However, in all cases, only the complexed IgG triggers biological functions by cross-linking FcγR on the effector cell membrane. Even if the FcγR are differentially expressed on different cell types, FcγRs of each group can mediate phagocytosis and/or ADCC under appropriate conditions (16). On immune complex binding, the FcγRs mediate intracellular signalling, resulting in endocytosis, enhancement of antigen presentation, secretion of lytic substances, mediator release, or phagocytosis depending on the effector cells. Phagocytosis may also be facilitated or triggered by complement activation, during which C3 and C4 split products are deposited on the bacterial surface. Since cells can constitutively express both FcγR (14) and complement receptors (17), a synergistic opsonic effect is achieved when the target is sensitized with both IgG and complement split products.

3. Measuring complement activation and lysis *in vitro*

3.1 The structural requirements for complement activation

Of the human IgGs, IgG1 and IgG3 are very effective at activating complement, IgG2 is only effective at high antigen concentrations, and IgG4 is ineffective (5, 9, 18, 19).

Protein engineering studies of mouse IgG2b suggest that three defined residues constitute the essential binding motif for C1q, Glu318, Lys320, and Lys322, and all human IgG isotypes contain this core C1q-binding motif in their C_H2 domain (20). However, the different human IgG isotypes show different abilities to bind C1q and activate complement. Naturally occurring IgG isotypes, with limited flexibility in the hinge region, are unable to bind C1q and thus do not activate complement, and it has been suggested that the Fab arms of such antibodies cover the C1q binding site on C_H2 (21). However, recent studies show that neither flexibility nor a 'spacer' between the Fabs and Fc is a prerequisite for complement activation, given that the heavy chains are connected with an S–S bond between C_H1 and C_H2 (22, 23). Rather, the C-terminal half of the C_H2 domain carries sites responsible for the isotype variation (24). Replacement of Ser331 with Pro improves complement activation of IgG4 (25, 26) to the level observed for IgG2, but not to the level of IgG1 and IgG3 (27). Other replacements are not found to have an effect.

Furthermore, it is crucial for complement activation that the IgG molecules are glycosylated. Both non-glycosylated human IgG1 and IgG3 antibodies have been shown to be deficient in complement activation (28). The presence of the C_H3 domain is also necessary for optimal complement activation as its removal reduces complement activation by 50% (29).

3.2 Comparing the IgG subclasses in complement activation and lysis

IgG1 has been reported to be superior to IgG3 in CML of target cells in some *in vitro* studies (5). In these studies, the matched set of antibodies have specificity for the hapten NP and its derivative NIP (5-iodo-4-hydroxy-3-nitrophenyl-acetyl). The hapten is introduced to the target cells (human red blood cells) as a lipid soluble molecule, NIP-kephalin. Whereas IgG3 binds C1q better than IgG1, IgG1 lyses cells more effectively than IgG3. Further studies have demonstrated that more C4b is deposited on the cell surface in the case of IgG1 (30). Neither IgG2 nor IgG4 show CML activity in these studies. In another series of experiments with anti-dansyl antibodies, IgG3 activates complement better than IgG1, whereas IgG2 and IgG4 show no activity (31). We, and others, have measured the effect of varying the antigen density as

9: Choosing and manipulating effector functions

well as the epitope patchiness on CML (18, 19). It has been found that whereas the IgG1 isotype is most effective at inducing CML at high concentrations of antigen, IgG3 is most effective at lower concentrations. In addition, IgG2 gives good lysis at very high concentrations of antigen. When comparing a matched set of antibodies recognizing the antigen CDw52 (CAMPATH-1) on human peripheral blood cells, IgG1 appears to be slightly more effective than IgG3, IgG2 also gives good lysis, whereas IgG4 shows no activity (9). This observation may indicate that the CDw52 antigen is equivalent to the NIP antigen at high concentrations.

In our experiments, BSA or Fab fragments are used as a carrier for NIP (32, 33). The hapten NIP is introduced to the target cells by direct NIP-labelling using NIP–CAP–O–Su or as rabbit anti-sheep red blood cell (SRBC)-Fab fragments labelled with NIP. Labelling BSA and Fab fragments with the succinimidyl ester NIP–CAP–O–Su is performed according to *Protocol 1*, and direct labelling of SRBC with NIP–CAP–O–Su as described in *Protocol 2*. In this way BSA and Fab fragments against SRBC as well as membrane surface proteins of SRBC are transformed to antigens for chimeric NIP-antibodies with human constant gamma-heavy chains.

Protocol 1. NIP/NP labelling of BSA and Fab

Reagents

- BSA
- Fab
- NIP–CAP–O–Su (Cambridge Research Biochem., cat. no. IT-03–18140)
- NP–CAP–O–Su (Cambridge Research Biochem., cat. no. IT-03–18040)
- Dimethylformamide (Pierce, cat. no 20673)
- 0.1 M NaHCO$_3$ (pH 8.5), 0.15 M NaCl
- 1 M ethanolamine (pH 8.5) (Sigma, cat. no. E-9508)
- PBS (pH 7.4) with 0.02% azide

Method

1. Dissolve BSA or Fab in 0.1 M NaHCO$_3$ (pH 8.5), 0.15 M NaCl, or dialyse for 16h at 4°C against this solution. The final concentration of the protein should be 5–40 mg/ml.

2. Dissolve NIP–CAP–O–Su or NP-CAP-O-Su at 1–20 mg/ml in dimethylformamide and use the solution immediately.

3. To obtain different haptenization ratios add 3 µl–400 µl of the NIP–CAP–O–Su or NP–CAP–O–Su solution to 1 ml protein. Keep at room temperature for 2 hours and then add 100 µl ethanolamine solution to block any unreacted NIP–CAP–O–Su or NP–CAP–O–Su. Dialyse against PBS with azide and store at 4°C.

4. Calculate the haptenization ratio by measuring absorbance at 280 nm and 405 nm of the haptenated conjugates and of solutions with known concentrations of NIP–CAP–O–Su or NP–CAP–O–Su. Also measure the absorbance of the protein solution at 280 nm. Calculate the

Protocol 1. *Continued*

concentration of NIP or NP by measuring absorbance at 405 nm and subtract their contribution to absorbance at 280 nm of the conjugates to get the concentration of protein in the conjugate. Use mol.wt of BSA: 70 000, Fab: 50 000, NIP–CAP–O–Su: 629, and NP–CAP–O–Su: 512 to calculate the molar ratio of NIP/NP and protein.

Protocol 2. Hapten and ^{51}Cr labelling of cells

Equipment and reagents

- NIP–CAP–O–Su (*Protocol 1*)
- NP–CAP–O–Su (*Protocol 1*)
- Dimethylformamide (*Protocol 1*)
- 0.1 M NaHCO$_3$ (pH 8.5), 0.15 M NaCl
- Sheep red blood cells (SRBC) (Local supplier)
- ACD anticoagulant
- PBS (pH 7.3) with 0.02% azide
- [^{51}Cr]Sodium chromate, 200–500 mCi/mg (Amersham, cat. no. CJS4)
- Veronal buffer (VB) (pH 7.2) containing 1% BSA, 5 mM CaCl$_2$, and 5 mM MgCl$_3$
- DMEM/Hepes (Gibco)
- fetal calf serum (Gibco)

Method 1, indirect method

1. To NIP-label cells for use in the CML assay (*Protocol 4*) wash SRBCs three times with PBS by centrifugation at 500–1000 *g* for 5 min. In three vials, mix 10 µl packed cells with 16 µl PBS and add 60 µCi [^{51}Cr]sodium chromate.

2. Leave the cells at 37 °C for 1 h and then add 2–13.5 µl of 1 mg/ml NIP–Fab conjugates (typically with 4–60 NIP residues per Fab molecule) to each vial and incubate for a further 1 h.

3. Wash the cells three times (as step 1) with VB and dilute to about 2×10^7 cells/ml.

4. For the ADCC assays (*Protocol 5*) mix the contents of three vials (each with 15 µl 3.5% SRBC) with about 100 µCi [^{51}Cr]sodium chromate and incubate for 1 h at 37 °C.

5. Add 2–13.5 µl of 0.1 mg/ml NIP–Fab conjugate (typically containing 4–60 NIP residues per Fab (molecule) and incubate for a further 1 h.

6. Wash the cells with DMEM/Hepes supplemented with 1% FCS (by centrifugation at 500–1000 *g* for 5 min) and adjust to 0.3×10^6 cells/ml.

Method 2, direct method

1. Wash ^{51}Cr-labelled SRBC (prepared as described in step 1 of the indirect method) three times with 0.1 M NaHCO$_3$, 0.15 M NaCl (pH 8.5) and dilute to a 5% suspension.

2. Mix 300 µl aliquots of this with 2–15 µl NIP–Cap–O–Su (2.5 mg/ml in dimethylformamide) and incubate at room temperature for 1 h.

9: Choosing and manipulating effector functions

3. Wash the cells three times and adjust to 3×10^7 cells/ml in VB for the CML assay, and to 0.3×10^6 cells/ml in DMEM/Hepes/1% FCS for the ADCC assay.

Other coupling procedures exist for active groups other than succimidyl ester, and once a hapten–carrier conjugate is made, or if the target naturally contains epitopes for the antibodies to be tested, the protocols for effector functions described below can be used. Complement activation and CML can be measured according to *Protocols 3* and *4*, respectively.

Protocol 3. Complement activation assay

Equipment and reagents

- Flat-bottomed microtitre plates designed for ELISA assays
- NIP/BSA (*Protocol 1*)
- NP/BSA (*Protocol 1*)
- Antibodies to be tested
- Rabbit anti-human C1q (Dako, cat. no. A136)
- Rabbit anti-human C3 (Dako, cat. no. A062)
- Rabbit anti-human C4 (Dako, cat. no. A065)
- Rabbit anti-human C5 (Dako, cat. no. A055)
- Human serum as complement source, stored at $-70\,°C$ (from local staff)
- Biotin-labelled sheep anti-rabbit IgG (produced locally)
- Streptavidin (produced locally)
- Biotin-labelled alkaline phosphatase (produced locally)
- Nitrophenylphosphate (NPP) (Sigma, cat. no. code 104–105)
- Microtitre plate washer
- Microtitre plate OD reader
- PBS (pH 7.4) with 0.05% Tween-20 (PBS-T)
- 0.1 M veronal buffer (pH 7.2) containing 5 mM $CaCl_2$, 5 mM $MgCl_3$, and 0.05% Tween-20 (VB-T)
- 0.1 m diethanolamine (pH 8.9)
- $MgCl_2$

Method

1. Coat microtitre plates with 100 or 150 μl 1–10 μg/ml NIP/BSA or NP/BSA dissolved in PBS (pH 7.4) with 0.02% azide overnight at 4°C. The plates can be stored for 1–4 months at 4°C when sealed with Scotch tape.
2. Wash the plates three times with PBS-T and once with distilled water.
3. Make serial dilutions of the test antibodies in VB-T and add to the plates. Add serum as the complement source at an optimal concentration (1:90–1:200) and incubate at 37°C for 2 hours.
4. Wash the plates three times with PBS-T.
5. Add rabbit anti-human C1q, rabbit anti-human C3, rabbit anti-human C4, or rabbit anti-human C5 at an optimal dilution (1:2000–1:10000) in PBS-T to parallel wells and incubate for 2 hours at 37°C.
6. Wash the plates three times with PBS-T.
7. Add optimal dilutions of a mixture of biotin-labelled sheep anti-rabbit

Protocol 3. *Continued*

IgG, streptavidin, and biotin labelled alkaline phosphatase in PBS–T and incubate for 2 hours at 37 °C.

8. Wash the plates three times with PBS–T.

9. Add a solution of 1 mg/ml NPP in 0.1 M diethanolamine (pH 8.9) with 5 mM $MgCl_2$. Incubate for 0.5–2 hours at 37 °C until an optimal absorbance is developed for reference antibodies.

10. Read the absorbance at 405 nm in a microtitre plate reader to evaluate the relative complement activation capacity of the different antibodies.

Protocol 4. Antibody-dependent complement-mediated lysis (CML)

Equipment and reagents

- Round-bottomed microtitre plates (Tek-nunc)
- Antibodies to be tested
- VB (*Protocol 2*) without Tween-20
- Hapten and ^{51}Cr-labelled target cells (*Protocol 2*)
- Human serum absorbed with SRBC
- Centrifuge for microtitre plates
- 0.1 M $CaCl_2$
- Harvesting device (Skatron)
- γ-counter and equipment and reagents

A. *Absorbing human serum with SRBC*

1. Add 0.75 ml 0.1 M EDTA, (pH 7.2) to 15 ml freshly drawn serum from a person with a low antibody level against SRBC.

2. Cool to 0 °C and add 3 ml washed, packed SRBC.

3. Incubate for 30 minutes and centrifuge for 10 minutes at 800 *g* and pipette off the serum.

4. Check for total depletion of antibodies against SRBC in the serum by the haemagglutination technique. If antibodies are still present, repeat steps 2 and 3.

5. Add 1.5 ml 0.1 M $CaCl_2$, aliquot, and store at −85 °C.

B. *Complement-mediated lysis*

1. Add 50 μl of a dilution of antibody (3 ng/ml–900 ng/ml) dissolved in VB into round-bottomed microtitre plates.

2. Add 50 μl haptenized and ^{51}Cr-labelled target cells, 2–3 × 10^7 cells/ml. Incubate for 10 min at room temperature, add 50 μl of complement (serum diluted 1:30 in VB, giving a final dilution of 1:90), and incubate for 1 h at 37 °C.

9: Choosing and manipulating effector functions

> 3. Centrifuge the plates at 300 *g* for 2 min, and collect the supernatants with a special harvesting device (Skatron). Measure the results in a gamma counter.
>
> 4. In each assay, incubate cells with 0.1% Tween-20 in water for total chromium release and replace antibody with VB alone for spontaneous release. Calculate lytic activity from the formula: % lysis = (test release − spontaneous release)/(total release − spontaneous release) × 100.

4. Measuring ADCC *in vitro*

4.1 The structural requirements for FcγR binding

Human IgG1 and IgG3 bind FcγRI with high affinity, IgG4 binds with a 10-fold lower affinity, and IgG2 shows no detectable binding. IgG1 and IgG3 also bind to the low affinity FcγRs, whereas IgG2 and IgG4 do not. The lower hinge region, encoded by the 5' end of the C_H2 exon, contains residues that are crucial for binding to all FcγRs (*Table 2*). Both IgG1 and IgG3 have the motif 234Leu–Leu–Gly–Gly237. IgG4 has a Leu/Phe interchange at residue 234, and IgG2 has 234Val–Ala–deletion–Gly237. Experiments suggest that the binding sites for FcγR on IgG are non-identical, but overlapping. However, the involvement of amino acids other than those in the lower hinge region may have to be taken into account since an FcγRIIA variant (LR) has been characterized, which binds to IgG2 as well as to IgG1 and IgG3 (16). Furthermore, it is crucial for optimal binding that the heavy chains are linked by a disulfide bridge between C_H1 and C_H2, as hingeless mutants show reduced binding (35). As for complement activation, the presence of carbohydrate linked to Asn297 is necessary (36).

4.2 Comparing the IgG subclasses in ADCC

In most *in vitro* studies, regardless of the type of effector cells employed, IgG1 is the most effective isotype in ADCC when matched sets of human IgG isotypes are compared. IgG3 also shows activity, but less than IgG1 (5, 7, 9). However, IgG3 has been reported to be as effective as IgG1 (37). Depending on the effector cells used, IgG4 shows no activity in some studies but good activity in others. IgG2 shows low or no activity in most studies (5, 7, 9).

Table 2. Sequence comparison of the lower hinge region of human IgG molecules (34)

	231	232	233	234	235	236	237	238
IgG1	Ala	Pro	Glu	Leu	Leu	Gly	Gly	Pro
IgG2	Ala	Pro	Pro	Val	Ala	—	Gly	Pro
IgG3	Ala	Pro	Glu	Leu	Leu	Gly	Gly	Pro
IgG4	Ala	Pro	Glu	Phe	Leu	Gly	Gly	Pro

When the experiments are performed with a larger panel of effector cells from different donors (the effectors being peripheral blood mononuclear cells activated with a mitogenic CD3 antibody and expanded with recombinant human IL-2), differences in the pattern of ADCC are seen in different individuals. In some individuals (three out of eight), only IgG1 gives good levels of ADCC. For other individuals, significant levels of ADCC are seen with all four IgG isotypes (10). Non-glycosylated IgG, may thus turn out to be a better choice than IgG4, if antibodies without effector functions are required. Non-glycosylated IgGs are obtained by growing the antibody-producing cells in the presence of tunicamycin or by mutating the sugar attachment site Asn297.

To study ADCC, target cells are hapten-labelled according to *Protocol 2* and ADCC activity measured according to *Protocol 5*.

Protocol 5. Antibody-dependent cell-mediated cytotoxicity (ADCC)

Equipment and reagents

- Round-bottomed microtitre plates (Nunc or Corning)
- Antibodies
- Vacutainers with ACD anticoagulant
- Effector cells: human peripheral mononuclear cells (MNC)
- 0.9% saline pH 7.3
- Hapten and ^{51}Cr-labelled target cells
- Centrifuge for microtitre plates
- Harvesting device (Skatron)

- γ-counter
- Lymphoprep (Nycomed)
- 50 ml, sterile, conical clinical centrifuge tubes (Costar, cat. no. 3252)
- Dulbecco modified Eagle medium (DMEM) (Gibco, cat. no. 041–91053M)
- Fetal calf serum (FCS) (Gibco, cat. no. 011–06290H)
- DMEM/Hepes/1% FCS (see *Protocol 2*)

A. Preparation of effector cells

1. Draw peripheral blood from healthy volunteers in ACD vacutainers.

2. Dilute the blood 1:1 with sterile, room temperature 0.9% saline solution.

3. Add 15 ml room temperature Lymphoprep to sterile 50 ml, conical centrifuge tubes.

4. Carefully pipette 25–35 ml 1:1 diluted blood over the Lymphoprep and centrifuge at 400 g for 20 minutes at room temperature (the brakes of the centrifuge must be off).

5. Pipette off the cell layer present at the interface between the Lymphoprep and the 1:1 diluted blood.

6. Wash the cells with DMEM/1% FCS and centrifuge at 300 g for 10 minutes at room temperature.

7. Adjust the cell concentration to 3×10^6 in DMEM/1% FCS.

B. Lysis of target cells

1. Add 50 µl of diluted antibody and 50 µl haptenized and ^{51}Cr-labelled target cells, 0.3×10^6, into round-bottomed microtitre plates. Incubate for 10 min at room temperature.

2. Add 50 µl effector cells, 3×10^6/ml. As effector cells use peripheral blood mononuclear cells (MNC) from a healthy donor suspended in DMEM/1% FCS and kept at 37°C, 5% CO_2, before use. Dilute all reactants in DMEM/Hepes/1% FCS.

3. Spin the contents of the microtitre plate at 300 g for 2 min, and incubate for 4 h at 37°C. Total and spontaneous release is achieved by adding 0.1% Tween-20 in water and replacing antibody with DMEM, respectively. Harvest supernatant and calculate lytic activity as described for the CML assay (Protocol 4).

5. Measuring phagocytosis and respiratory burst

In addition to ADCC mediated solely through Fc receptors, target cells may be destroyed by phagocytosis mediated through Fc and complement receptors. Antibodies bound to surface antigens on target cells may have opsonic activity, either directly through binding to Fc receptors on phagocytes or indirectly by their ability to activate the complement system thereby causing complement deposits on the target cell surface. These complement deposits, C3b and C3bi, may then bind to the complementary CR1 and CR3 receptors on the phagocytes, and thus enhance phagocytosis. Stimulation through both Fc and C receptors is reported to have a synergistic effect on the phagocytic process. Therefore, the capacity of each IgG subclass to work as an opsonin depends on its interaction with the particular FcγR and its ability to activate the complement cascade. There are two types of phagocytic cells involved in this process: polymorphonuclear leukocytes (PMN) and mononuclear leukocytes (i.e. monocytes and macrophages). PMN constitutively express FcγRIIa and FcγRIIIb. However, in an inflamed foci, gamma interferon (IFN-γ) will induce high-affinity FcγRI expression. Monocytes constitutively express FcγRI and FcγRIIa. Macrophages express all three groups of FcγRs. Both PMN and monocytes/macrophages express receptors for C3b/C3bi. However, little has been done to compare the contribution of each subclass to the phagocytic process. In some studies, IgG2 has proved to be an efficient opsonin provided complement is available (38, 39), whereas in many immunological textbooks it is classified as being devoid of effector function.

In our studies using the family of chimeric NIP antibodies, the IgG1, IgG3, and IgG4 subclasses were not markedly influenced by complement, whereas IgG2 gained high activity but was not as effective as the other subclasses (40). IgG3 internalized more target cells to PMN than IgG1 and thus mediated a higher respiratory burst signal (40). IgG2 required both high antigen concentration on the target cell and complement to induce phagocytosis and respiratory burst, indicating that this activity is solely mediated through complement receptors on the effector cells (40).

To measure phagocytosis and respiratory burst, the target cells are NIP-labelled according to *Protocol 2*. Phagocytosis and respiratory burst are then measured according to *Protocols 6* and *7*, respectively. When bacteria are used as the target cells in these assays, we have chosen the Gram-negative *Neisseria meningitidis*, however, we anticipate that other Gram-negative strains can also be used using the same procedures.

Protocol 6. Phagocytosis measured by flow cytometry

Equipment and reagents

- Freshly drawn venous blood from normal volunteers
- Heparin vacutainers
- 10 × lysis solution (LS): 83 mg/ml NH$_4$Cl, 10 mg/ml NaHCO$_3$, and 0.8 mg/ml EDTA (pH 6.8) (locally produced)
- 0.1 M NaHCO$_3$/0.15 M NaCl (pH 8.3)
- Hanks' balanced salt solution (HBSS) (Gibco, cat. no. 041–04020H) HBSS with 0.2% BSA (HBSS–BSA)
- U937 cell line (ATCC)
- Fluorescein-isothiocyanate isomer 1 (Becton Dickinson, cat. no. 12008)
- DMSO (Merck, cat. no. 2931)
- NIP- and FITC-labelled target cells (SRBC, yeast, or bacteria)
- Flow cytometer
- Human serum without antibodies against target cells

Method

1. Draw venous blood from normal volunteers into heparin vacutainers. Lyse the red cells with the LS (it takes 2–5 min).

2. Wash the leukocytes twice with HBSS–BSA by spinning at 250 *g* for 5–8 min at room temperature. Adjust the concentration to 5 × 10^6/ml and store at 4 °C until use.

3. When bacteria or yeast are used, fix them before they are NIP-labelled. Fix the cells in 70% ethanol for 24 hours at 20 °C, wash in HBSS, aliquot and store at −85 °C. Then NIP-label the cells as in *Method 2* of *Protocol 2*. Use ten times more bacteria or yeast cells compared to SRBC.

4. Wash the NIP-labelled cells and resuspend them in 0.1 M NaHCO$_3$/ 0.15 M NaCl (pH 8.3) and add 75 μl 1 mg/ml fluorescein-isothiocyanate isomer 1 in DMSO. Incubate for 1 h and wash the bacteria or yeast in HBSS containing 5 mg/ml BSA, aliquot and store at −85 °C.

5. When using NIP-labelled SRBC as target cells, label with FITC as described for bacteria and yeast in step 4. Wash the NIP-haptenized, FITC-labelled target cells twice in HBSS with 0.2% BSA and adjust the cell concentration to 2 × 10^8/ml.

6. Mix 5 μl aliquots of the target suspension with 50 μl of diluted test antibodies in U-bottomed microtitre plates. Incubate for 30 min at 20 °C.

9: Choosing and manipulating effector functions

7. When the phagocytosis is performed with PMN and complement, add 5 µl serum and incubate with shaking at 37 °C for 6 min. Then add 50 µl of the leukocytes prepared in steps 1 and 2, and incubate with continuous agitation at 37 °C for 6 min to complete the phagocytosis. Put the cells on ice until analysis on the flow cytometer.

8. Alternatively, the U937 cell line may be used as effector cells. In this case, the opsonized target cells have to be washed to remove any unbound IgG that will otherwise block FcγRI on U937. Mix U937 and opsonized target, spin the suspension down at 250 *g* for 2 min, and incubate in a humidified atmosphere at 37 °C for 30 min to complete the phagocytosis. Place the cells on ice until analysis on the flow cytometer.

Note: A detailed protocol for flow cytometer analyses is outside the scope of this chapter, but is fully described elsewhere (41). Briefly, the results are given either as mean fluorescence intensity (MFI) of all effector cells, or as per cent positive effector cells within the analysis region, i.e. per cent U937 or PMN that have phagocytosed the target cells. To discriminate between internalized and adherent target cells, add Trypan Blue to a final concentration of 0.5 mg/ml to quench the fluorescence from adherent cells.

Protocol 7. Respiratory burst measured by flow cytometry

Equipment and reagents

- Freshly drawn venous blood from normal volunteers
- Heparin vacutainers
- 10 × Lysis solution (LS): 83 mg/ml NH$_4$Cl, 10 mg/ml NaHCO$_3$, and 0.8 mg/ml EDTA (pH 6.8)
- HBSS with 0.2% BSA (HBSS–BSA) (*Protocol 6*)
- NIP- and FITC-labelled target cells (SRBC, yeast, or bacteria)
- 10 mg/ml dihydrorhodamine 123 (DHR) (Molecular Probes, cat. no. D632) in DMSO
- Flow cytometer
- Human serum without antibodies against target cells (*Protocol 4A*)

Method

1. Prepare effector cells according to *Protocol 6*, steps 1 and 2. NIP-labelled SRBC (*Protocol 2*) or NIP-labelled bacteria or yeast cells (*Protocol 6*, step 3) for use as target cells. The target cells are not FITC-labelled.

2. Mix aliquots of target cells with diluted antibody as in step 6, *Protocol 6*. When complement is included in the assay, add 5 µl serum to the mixture of target cells and antibodies and incubate with agitation at 37 °C for 6 min.

3. Add the leukocytes, and incubate with continuous agitation at 37 °C for 6 min to complete the respiratory burst. Then put the cells on ice until analysis on the flow cytometer.

Note: Flow cytometer analyses is fully described elsewhere (41). Briefly, the results are given either as mean fluorescence intensity (MFI) of all effector cells, or as per cent positive effector cells within the analysis region, i.e. per cent PMN that have undergone respiratory burst after internalizing the target cells.

6. Optimizing effector functions

Considering *in vitro* data, human IgG1 seems to be the most effective subclass for depletion of target human cells, and consequently, IgG1 is usually the subclass chosen when aiming to deplete target cells *in vivo*. However, in the future, mutant isotypes with defined changes which alter effector functions may well be employed. A few examples exist where mutations in individual isotypes improve effector functions *in vitro*.

CML increases up to 50-fold when induced by IgG3 mutants with a truncated hinge region (23). In these mutants, the amino terminal end of the hinge has been deleted, leaving a hinge of 15 amino acids. This is done by removing the three 5′ hinge exons in the γ3 heavy chain gene segment. The long hinge of IgG3 seems to down-regulate its potential in CML, and several IgG3 variants with a short hinge show an improved ability to induce CML, without a concomitant effect on FcγR-related mechanisms. The improvement is seen at both low and high concentrations of antigen on the target cells.

In addition, IgG1 dimers with improved activity in CML have been constructed through genetic engineering (42). A Ser/Cys444 substitution in C_H3 led to dimerization, and half of the secreted IgG1 was detected as covalent dimeric immunoglobulin $(H_2L_2)_2$ linked 'tail to tail'. A 200-fold increase in CML compared with monomeric IgG1 was obtained with dimeric IgG1. Neither of these IgG variants have been tested in *in vivo* studies. When the lower hinge sequence 233Pro–Val–Ala–deletion–Gly237 of IgG2 is substituted with the corresponding sequence from IgG1, namely 233Glu–Leu–Leu–Gly–Gly237, a 4-fold improvement in FcγRI binding is observed, such that the IgG2-variant binds FcγRI better than IgG1 (43). This mutant has neither been tested in *in vitro* nor *in vivo* ADCC assays.

A proportion of human IgG4 is secreted as half molecules (HL) from antibody-producing cells (5), indicating that hinge–disulfide bridges between heavy chains are formed less efficiently than for the other subclasses. The amino acid sequence in the hinge region varies among isotypes and species, but in humans it typically contains the sequence Cys–Pro–X–Cys–Pro. IgG4 contains the motif Cys–Pro–Ser–Cys–Pro. A mutant with a single amino acid substitution (from Ser to Pro) in this motif leads to the production of only covalently linked H_2L_2 antibodies (44). The hinge modification also affects the biodistribution of the chimeric anti-carcinoma antibody cB72.3(γ4P) which has this particular heavy chain (44). Compared to the corresponding antibody with a normal γ4 heavy chain, the biodistribution of the mutant more closely resembles that observed for IgG1. This implies that a higher proportion of the mutant antibodies are distributed to the tumour in nude mice with a carcinoma xenograft. Less is distributed to the kidney, so the tumour/kidney ratio is significantly improved. It is suggested that the non-disulfide bridged form of IgG4 is easily broken down and cleared via the kidney.

9: *Choosing and manipulating effector functions*

In summary, simple mutations introduced in antibody genes by site-directed mutagenesis have been shown to improve effector functions. Both increased CML and FcγRI binding have been observed, as well as improved biodistribution.

Acknowledgements

We would like to thank Ingunn B. Rasmussen, Ole Henrik Brekke, and Audun Aase for their helpful comments. This work was supported by grants from The Norwegian Research Council and The Norwegian Cancer Association.

References

1. Lachmann, P. J. (1991). *Immunol. Today*, **12**, 312.
2. Bindon, C. I., Hale, G., and Waldmann, H. (1988). *Eur. J. Immunol.*, **18**, 1507.
3. Waldmann, H. (1989). *Annu. Rev. Immunol.*, **7**, 407.
4. Adair, J. R. (1992). *Immunol. Rev.*, **130**, 5.
5. Bruggemann, M., Williams, G. T., Bindon, C. I., Clark, M. R., Walker, M. R., Jefferis, R., Waldmann, H., and Neuberger, M. S. (1987). *J. Exp. Med.*, **161**, 1351.
6. Shaw, D. R., Khazaeli, M. B., and LoBuglio, A. F. (1988). *J. Natl Cancer Inst.*, **80**, 1553.
7. Steplewski, Z., Sun, L. K., Shearman, C. W., Ghrayeb, J., Daddona, P., and Koprowski, H. (1988). *Proc. Natl Acad. Sci. USA*, **85**, 4852.
8. Dangl, J. L., Wensel, T. G., Morrison, S. L., Stryer, L., Herzenberg, L. A., and Oi, V. T. (1988). *EMBO J.*, **7**, 1989.
9. Riechmann, L., Clark, M., Waldmann, H., and Winter, G. (1988). *Nature*, **332**, 323.
10. Greenwood, J., Clark, M., and Waldmann, H. (1993). *Eur. J. Immunol.*, **5**, 1098.
11. Jefferis, R. and Lund, J. (1993). In *Protein engineering of antibody molecules for prophylactic and therapeutic applications in man* (ed. M. Clark), pp. 115–26. Academic Titles, Nottingham.
12. Burton D. and Woof, J. M. (1992). *Adv. Immunol.*, **51**, 1.
13. Frank, M. M. and Fries, L. F. (1991) *Immunol. Today*, **12**, 322.
14. van de Winkel, J. G. J. and Capel, P. J. A. (1993). *Immunol. Today*, **14**, 215.
15. Ravetch, J. V. (1994). *Cell*, **78**, 553.
16. van de Winkel, J. G. J. and Anderson, C. (1991). *J. Leukocy. Biol.*, **49**, 511.
17. Ross, G. D. (1989). *Curr. Opin. Immunol.*, **2**, 50.
18. Michaelsen, T. E., Garred, P., and Aase, A. (1991). *Eur. J. Immunol.*, **21**, 11.
19. Valim, Y. M. L. and Lachmann, P. J. (1991). *Clin. Exp. Immunol.*, **84**, 1.
20. Duncan, A. R. and Winter, G. (1988). *Nature*, **332**, 738.
21. Brekke, O. H., Michaelsen, T. E., and Sandlie, I. (1995). *Immunol. Today*, **16**, 85.
22. Shopes, R. (1993). *Mol. Immunol.*, **30**, 603.
23. Brekke, O. H., Michaelsen, T. E., Sandin, R., and Sandlie, I. (1993). *Nature*, **363**, 628.
24. Tao, M. H., Canfield, S. M., and Morrison, S. I. (1991). *J. Exp. Med.*, **173**, 1025.
25. Tao, M. H., Smith, R. I., and Morrison, S. L. (1993). *J. Exp. Med.*, **178**, 661.
26. Xu, Y., Oomens, R., and Klein, M. H. (1994). *J. Biol. Chem.*, **269**, 3469.

27. Brekke, O. H., Michaelsen, T. E., Aase, A., Sandin, R., and Sandlie, I. (1994). *Eur. J. Immunol.*, **24**, 2542.
28. Tao, M. H. and Morrison, S. L. (1989). *J. Immunol.*, **143**, 2595.
29. Utsumi, S. M., Okada, K., Udaka, and Amano, T. (1985). *Mol. Immunol.*, **22**, 811.
30. Bindon, C. I., Hale, G., Bruggemenn, M., and Waldmann, H. (1988). *J. Exp. Med.*, **168**, 127.
31. Morrison, S. L., Smith, R. I. F., and Wright, A. (1994). *The Immunologist*, **2**, 119.
32. Michaelsen, T. E., Aase, A., Westbye, C., and Sandlie, I. (1990). *Scand. J. Immunol.* **32**, 517.
33. Aase, A. and Michaelsen, T. E. (1991). *J. Immunol. Methods*, **136**, 185.
34. Kabat, E. A., Wu, T. T., Reid-Miller, M., Perry, H. M., and Gottesman, K. S. (1987). *Sequences of proteins of immunological interest.* Fourth edition. US Department of Health and Human Services, Bethesda, Maryland.
35. Michaelsen, T. E., Brekke, O. H., Aase, A., Sandin, R., Bremnes, B., and Sandlie, I. (1994). *Proc. Natl Acad. Sci. USA*, **91**, 9243.
36. Pound, J., Lund, J., and Jefferis, R. (1993). *Mol. Immunol.*, **30**, 233.
37. Michaelsen, T. E., Aase, A., Norderhaug, L., and Sandlie, I. (1992). *Mol. Immunol.*, **29**, 319.
38. Amir, J., Scott, M. G., Nahm, M. H., and Granoff, D. M. (1990). *J. Infect. Dis.*, **162**, 163.
39. Sawada, S., Kawamura, T., and Masuho, Y. (1987). *J. Gen. Microbiol.* **133**, 3581.
40. Aase, A. and Michaelsen, T. E, (1994). *Scand. J. Immunol.*, **39**, 581.
41. Aase, A., Sandlie, I., Norderhaug, L., Brekke, O. H., and Michaelsen, T. E. (1993). *Eur. J. Immunol.*, **23**, 1546.
42. Shopes, R. (1992). *J. Immunol.*, **148**, 2918.
43. Chappel, M. S., Isemann, D. E., Everett, M., Xu, Y. Y., Dorrington, K. J., and Klein, M. H. (1991). *Proc. Natl Acad. Sci. USA*, **88**, 9036.
44. Angal, S., King, D. J., Bodmer, M. W., Turner, A., Lawson, A. D. G., Roberts, G., Pedley, B., and Adair, J. R. (1993). *Mol. Immunol.*, **30**, 105.

10

Producing antibodies in *Escherichia coli:* from PCR to fermentation

ANDREAS PLÜCKTHUN, ANKE KREBBER, CLAUS KREBBER,
UWE HORN, UWE KNÜPFER, ROLF WENDEROTH, LARS NIEBA,
KARL PROBA, and DIETER RIESENBERG

1. Introduction

The expression of antibody fragments in *Escherichia coli* has become a widely used technique for several reasons; it is, in principle, an easily accessible methodology for anyone used to growing *E. coli*, and it can be advantageously combined with phage-display technology which also makes use of *E. coli* (1–3).

Nevertheless, successes in obtaining large amounts of protein by simply placing genes encoding antibody fragments behind a signal sequence and a strong promoter have been rather mixed. This may at first seem surprising to those molecular biologists who have previously expressed large amounts of their favourite functional protein in *E. coli*, using readily available commercial plasmids. The reason one has to write a chapter about this technology at all lies almost entirely in the protein folding problem and in the individuality of antibodies. The successful expression of high amounts of antibody fragments in *E. coli* is thus an exercise in understanding gene regulation, cell physiology, and, most importantly, protein folding to a sufficient degree to maximize the amount of functional antibody protein.

Therefore, it is the purpose of this chapter to describe some of the problems frequently observed as well as giving possible solutions. In particular, it will be shown in detail how an extremely simple benchtop fermentation of *E. coli* can be used as a more efficient alternative to the traditional shake-flask cultures, and how more sophisticated instrumentation can be used to yield up to several *grams* of functional antibody protein per litre of culture, an amount which generally should overcome any worries that *E. coli* is an unsatisfactory host for antibody production.

The emerging topic, irreversibly linked with expression, is protein folding. The whole point of antibody expression is obviously to obtain active, antigen-binding material. In this context, two related questions are of importance.

The first is a general one, namely in which exact molecular and genetic format to put the variable domains together to obtain a correct binding site. The second question is how to improve the folding properties of the antibody molecule, which has been shown to be the most important yield limiting factor (see below). Fortunately, it has now become possible, at least in some cases, to engineer antibodies to improve folding, and this will be discussed below.

We must base this chapter on a fairly detailed investigation of a relatively small number of recombinant antibodies, about a dozen different ones, which in our laboratory are studied in various formats each with numerous mutants (binding to proteins, peptides, oligosaccharides, hydrophobic or hydrophilic haptens, containing either human or murine frameworks). Even when combining the published literature as much as possible, the database of well-characterized recombinant antibodies is still tiny compared to the 10^{12} possible antibodies; and for every rule emerging there just may exist an exceptional antibody sequence which ignores this rule.

The main part of this chapter will discuss the strategies and factors important for obtaining soluble antibody, after a few candidate antibodies have been identified from phage display, or after cloning from a hybridoma cell line, or having synthesized the genes. We will first discuss the pros and cons of converting the antibody genes into various fragments for expression, such as scFv, Fv, dsFv, Fab, and fusion proteins (*Figure 1*). This will be followed by an overview of the expression strategies available and suitable vectors. We will then discuss progress in protein engineering techniques to improve expression, by eliminating the tendency of the molecule to aggregate, and finally we give procedures to produce antibody fragments in small, medium, and large scale, up to gram amounts.

2. Cloning of antibody variable genes

The first step in an antibody project involves obtaining the antibody genes. We can distinguish four starting scenarios. First, if a hybridoma cell is available, the mRNA must be isolated, and the antibody genes amplified by PCR (*Protocol 1*, see also Chapter 1). Even though only one antibody sequence should, in principle, be encoded by the hybridoma this is not always the case and it can still be advantageous to enrich the correct antibody sequence by phage display. It is not at all uncommon to have additional rearranged antibody sequences present, either from the myeloma fusion partner or from aberrant allelic exclusion in the B-cell fusion partner (4).

Second, a technically similar situation arises when antibodies are to be isolated from an immunized mouse. This involves isolating the spleen and from it all splenic mRNA, which is then amplified by PCR and assembled for phage display. After selecting binding clones, a few candidate antibodies are obtained ready for expression. We will not, however, discuss phage display

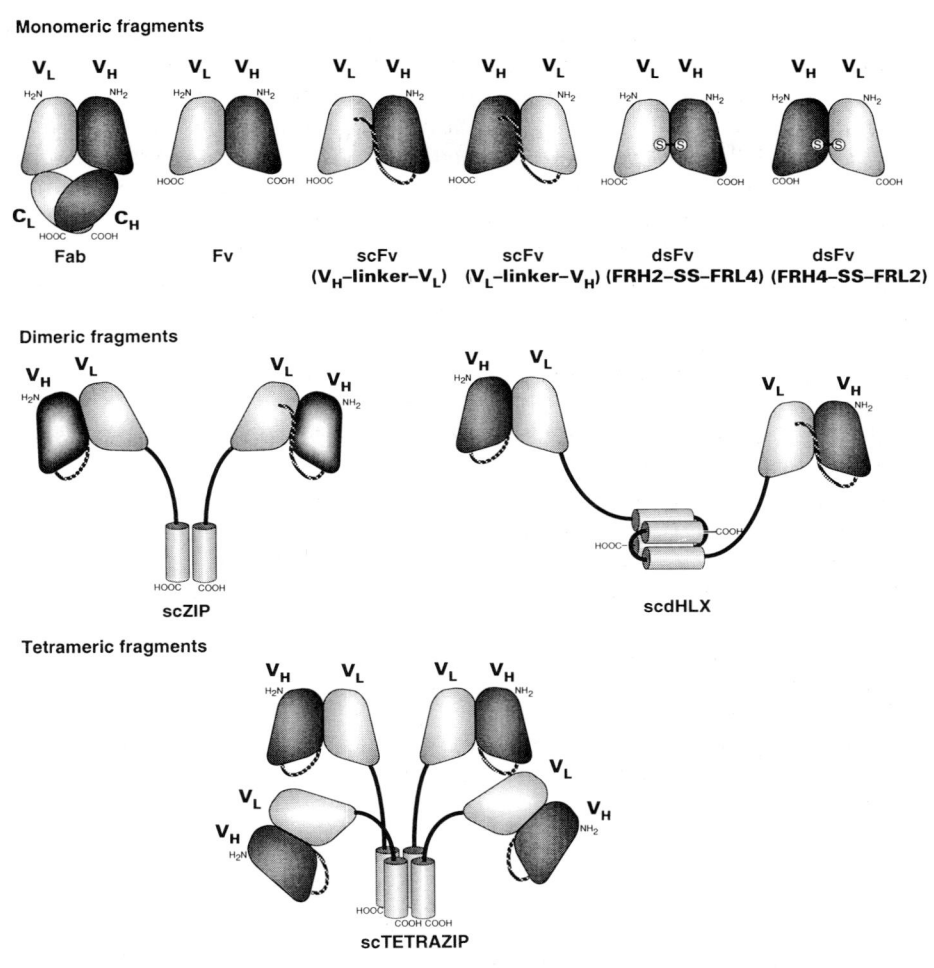

10: Producing antibodies in E. coli: from PCR to fermentation

Figure 1. Antibody fragments suitable for functional high-level expression in *E. coli*. For details on the cloning see ref 16; for details on the miniantibodies see Pack *et al.* (31–33); for details on the dsFv fragments see refs 25–27.

or library construction in much detail here, as this is done elsewhere in this volume (Chapters 1, 2, and 8).

Advances have been made in constructing large antibody libraries from germline sequences and synthetic D and J segments (3). In this third scenario, again binding clones need to be selected, usually by phage panning, and a few candidates will then be ready for expression.

The fourth, frequent scenario is that a known antibody sequence needs to be modified. A typical example would be 'humanizing' an antibody, which has shown promise in preclinical mouse experiments, for human therapy (5).

Other examples may include the re-engineering of the variable domains to give higher stability, better folding (see below) or for speciality functions, such as a metal binding site. In this case, a total gene synthesis is often the fastest way. In a number of cases, we have used modifications of the PCR-based assembly method of Prodromou and Pearl (6). The single most important parameter for a speedy gene synthesis by this method is the quality of the synthetic oligonucleotides. These need to be free of one-base deletions (see general comment on *Protocol 1*). Correct clones can be enriched by one round of phage panning, although this actually amounts to curing the symptom.

Protocol 1 describes a procedure for assembling the genes of an scFv fragment, which is a very useful first step in many antibody engineering projects. The scFv fragment can then be easily converted to any other fragment shown in *Figure 1* and engineered further, as discussed throughout the text. The initial procedure of gene assembly can be carried out in the same way both for first cloning into a phage-display vector or directly in an expression vector. Note, however, that optimized display and expression vectors have rather different requirements (see below).

Protocol 1. Assembly of an scFv fragment from a hybridoma

Equipment and reagents

- $1-5 \times 10^6$ cells from a growing or frozen hybridoma culture
- PCR primers (*Table 1*) and plasmids (*Figure 3*)
- Helper phage (e.g. VCS Stratagene)
- F+, supE, recA− strain (e.g. XL1 Stratagene)
- Anti M13-HRP conjugate (Pharmacia Biotech)
- PEG 6000
- DEPC-treated water
- 10 × Amplitaq PCR buffer: 100 mM Tris-HCl, pH 8.3 at 25°C, 500 mM KCl, 15 mM $MgCl_2$, 0.001% gelatin (w/v)

- Standard molecular biology equipment and reagents to:
  —purify mRNA (Pharmacia QuickPrep mRNA Purification Kit)
  —perform a cDNA synthesis reaction (Pharmacia Biotech First-Strand cDNA Synthesis Kit)
  —perform PCR reactions (Chapter 1, *Protocol 3*)
  —cut and gel purify DNA (Chapter 1, *Protocol 7*)
  —ligate and transform DNA (Chapter 8, *Protocols 3 and 4*)
  —grow bacteria and phages (Chapter 1, *Protocols 8–11*)
  —perform an ELISA (Chapter 1, *Protocol 13*)

A. *Synthesis of cDNA*

1. Take $1-5 \times 10^6$ cells from a growing or frozen hybridoma culture and perform an mRNA preparation as described in the Pharmacia Quick-Prep mRNA Purification Kit.[a]

2. Ethanol precipitate the purified mRNA which is present in a volume of 0.75 ml elution buffer as follows:

    prepare 2 aliquots: each containing:
    - mRNA   0.32 ml
    - ethanol (chilled to −20°C)   0.8 ml

10: Producing antibodies in E. coli: from PCR to fermentation

- glycogen (10 mg/ml)   10 µl
- 2.5 M K-acetate, pH 5.0   32 µl

3. Collect the precipitated mRNA from one aliquot by centrifuging at 16 000 g, 4°C for 30 min.[b]
4. Wash with 1 ml 90% ethanol and dissolve in 20 µl H$_2$O (diethylpyrocarbonate-treated).
5. The mRNA solution is now ready for cDNA synthesis. Add 1 µl cDNA synthesis primer[c] and heat the solution to 65°C for 10 min.
6. Centrifuge the sample as before and add 1 µl 200 mM DTT and 11 µl Bulk First-Strand Reaction Mix.[d]
7. Incubate at 37°C for 1 h.

B. *Preparation of PCR products*

1. Use 5 µl of the completed first-strand cDNA reaction for PCR amplification of V_L and V_H.

   PCR conditions[e]:
   - cDNA   5 µl
   - dNTPs (10mM each)   1 µl
   - AmpliTaq 10 × PCR Buffer   5 µl
   - 1 µl LB primer mix (100 pmol/µl) or HB primer mix (100 pmol/µl), respectively   1 µl
   - 1 µl LF primer mix (100 pmol/µl) or HF primer mix (100 pmol/µl), respectively   1 µl
   - H$_2$O   37 µl

2. Add wax or mineral oil and heat for 5 min to 92°C.
3. Add 2.5 units AmpliTaq and perform the following 7 cycles: 1 min at 92°C, 30 sec at 63°C, 50 sec at 58°C, and 1 min at 72°C. Followed by 23 cycles:[f] 1 min at 92°C, 1 min at 63°C, and 1 min at 72°C.
4. Gel-purify the V_L and V_H genes and determine the DNA concentration of both chains.[g]

C. *Assembly of scFv gene*

1. Use equimolar amounts of both domains (approximately 10 ng) for the assembly PCR[h]:
   - V_L 10 ng
   - V_H 10 ng
   - 10 × PCR buffer   5 µl
   - dNTPs (10 mM each)   1 µl
   - MgSO$_4$ optimized for the enzyme and the template
   - H$_2$O to an end volume of   50 µl

2. Add wax or mineral oil and heat for 5 min to 92°C.

Protocol 1. *Continued*

3. Add 1–2 units polymerase and perform 2 cycles: 1 min at 92 °C, 30 sec at 63 °C, 50 sec at 58 °C, and 1 min at 72 °C.
4. Add 1 μl scfor and scback primer mix[e] (each primer 50 pmol) and run 5 additional cycles as described in step 3.
5. Continue with 23 cycles: 1 min at 92 °C, 1 min at 63 °C, and 1 min at 72 °C.
6. Gel-purify the scFv antibody gene.

D. *Digestion and cloning of scFv gene*

1. Perform a *Sfi*I digest for 3–4 hours at 50 °C (overlay with mineral oil).
2. Purify the digested fragment and ligate it into *Sfi*I digested pAK100 or pAK200 vector (molar ratio vector to insert 1.5:1).
   - vector DNA   200 ng
   - scFv gene fragment   20 ng
   - T4 ligase   1 unit
   - 10 × ligase buffer with 10 mM ATP   2 μl
   - add H$_2$O to a final volume of   20 μl

   Incubate overnight at 16 °C.
3. Transform 5–10 μl of the ligation reaction into competent XL1Blue (Stratagene) cells.
4. Plate on 2 × YT, 1% glucose, Cam (30 μg/ml) plates.

E. *Screening for binders*

1. Pick 10 colonies and grow them separately in 2 ml 2 × YT, 1% glucose, Cam (30 μg/ml) until they reach an OD_{600} = 0.5.
2. Add 2 ml 2 × YT, 1% glucose, Cam (30 μg/ml), 1 mM IPTG, 10^{10} p.f.u. VCS helper phage (Stratagene) and grow overnight or 6 h at 37 °C.[j]
3. Centrifuge the culture. Take 1.6 ml supernatant and mix it with 0.4 ml 20% PEG 6000, 2.5 M NaCl in a 2.2 ml Eppendorf cap in order to precipitate the phages.
4. Incubate on ice for 15 min and spin at 16 000 *g*, 4 °C for 20 min.
5. Dissolve the phage pellet in 400 μl PBS containing 2% skimmed milk powder and use 100 μl phage solution per well in an ELISA assay to distinguish functional scFv antibody-displaying phages from those which display a non-functional or non-productive antibody fragment. (See Chapter 1, *Protocol 13*.)

10: Producing antibodies in E. coli: from PCR to fermentation

6. If possible include an ELISA control that shows that free antigen is able to compete with bound antigen for phage binding to distinguish non-specific 'sticky' phage from specifically binding phage (see Chapters 2 and 8). In principle the same ELISA protocol which was used for the hybridoma screening procedure should be used. Up to 10^7 bound phages can be detected with an anti M13-HRP conjugate (Pharmacia Biotech) which is a component of the RPAS (Recombinant Phage Antibody System) Detection Module. To minimize background problems, it is highly recommended that 2% skimmed milk powder is included in *all* incubation steps.[j]

[a] According to the manufacturer, this kit can be used for up to 5×10^7 cells, but in order to obtain pure mRNA it is highly recommended that the oligo(dT)–cellulose column is not overloaded. It is not the quantity of mRNA but rather the quality that is most important in generating recombinant scFv antibodies. The use of 5×10^6 cells typically yields 10–20 µg of mRNA.

[b] mRNA can be stored in ethanol at –20°C for several months.

[c] Use 0.2 µg/µl pd(N)$_6$ or d(T)$_{18}$ or 30 pmol/µl of a specific primer mix (*Table 1*). As far as subsequent PCR reactions are concerned, all three versions are successful; however, for the heavy chain pd(N)$_6$ or specific primers are superior over d(T)$_{18}$, whereas for the light chain no differences can be observed. In order to decide which kind of specific primers should be used it is of advantage to perform an isotype determination with the hybridoma of interest.

[d] All solutions needed for first strand cDNA synthesis, except the specific primer mix which is listed in *Table 1*, come as part of the Pharmacia Biotech First-Strand cDNA Synthesis Kit.

[e] Primers for first PCR: LB (light back), HB (heavy back), LF (light forward), and HF (heavy forward) are primer mixes listed in *Table 1*.
Primers for the second (assembly) PCR: scback, scfor are listed in *Table 1*.

[f] Do not elongate for 10 min at 72°C at the end since AmpliTaq adds an additional A at the 3' end of the PCR product in a high fraction of the molecules. Alternatively, use a linker assembly strategy where the additional A matches to the template strand. The use of proof-reading polymerases for the first PCR may be an alternative, but it was found that Taq works more generally, whereas proof-reading polymerases usually require more optimization, e.g. they may require MgSO$_4$ titrations. This can be a problem when only limited template is available. An alternative might be the Expand™ High Fidelity PCR System of Boehringer Mannheim which uses a mixture of Taq (non proof-reading) and Pwo (proof-reading) DNA polymerase in order to combine greater fidelity and higher yields with low error rates.

[g] Using the listed primer mix, the expected lengths of V_L and V_H PCR products are between 375–402 bp and 386–440 bp, respectively (depending on the CDR lengths of the antibody fragment).

[h] A number of proof-reading polymerases like Pwo (Boehringer Mannheim), Pfu (Stratagene), and Vent (New England Biolabs) have worked successfully in our hands. For unknown reasons, different enzymes worked best in different cases. If problems occur it is therefore worthwhile testing several different enzymes, in addition to carrying out MgSO$_4$ titrations.

[i] For some scFvs growth at a lower temperature after infection may be necessary. The phage titre after overnight incubation is in the range of 10^{11} to 5×10^{11} c.f.u. per ml supernatant.

[j] If no functional clone shows up in ELISA of single clones, perform one round of phage panning in order to enrich the functional binders. Problems can occur if the quality of the oligonucleotide primers is not satisfactory (see (m) in General comments below) or if the hybridoma transcribes more than one functional or even non-functional heavy or light chain variable region gene (4). It was found that several kappa chain secreting hybridomas, where X63Ag8.653 myeloma cells were used as a fusion partner, are able to transcribe a functional lambda chain which competes with the kappa V_L gene for in-frame scFv antibody assembly (S. Bornhauser, personal communication). Therefore, it is highly recommended that any lambda chain primer in the PCR reactions is left out if the isotyping indicates that the hybridoma of interest secretes a kappa light chain.

Table 1. Primers for assembling mouse scFv fragments in the orientation V$_L$–linker–V$_H$ with the vector systems of *Figure 3*

Primer VH back:

	5' (Gly$_4$Ser)-linker *Bam*HI V$_H$ 3'	Degeneracy (d)	μl Mix
HB1	**ggcggcggcggctcc** ggtggtggt<u>ggatcc</u> GA<u>K</u>GT<u>R</u>MAGCTTCAGGAGTC	8	4
HB2	**ggcggcggcggctcc** ggtggtggt<u>ggatcc</u> GAGGT<u>B</u>CAGCT<u>B</u>CAGCAGTC	9	4
HB3	**ggcggcggcggctcc** ggtggtggt<u>ggatcc</u> CAGGTGCAGCTGAAG<u>S</u><u>A</u>STC	4	3
HB4	**ggcggcggcggctcc** ggtggtggt<u>ggatcc</u> GAGGTCCA<u>R</u>CTGCAACA<u>R</u>TC	8	4
HB5	**ggcggcggcggctcc** ggtggtggt<u>ggatcc</u> CAGGT<u>Y</u>CAGCT<u>B</u>CAGCA<u>R</u>TC	12	7
HB6	**ggcggcggcggctcc** ggtggtggt<u>ggatcc</u> CAGGT<u>Y</u>CA<u>R</u>CTGCAGCAGTC	4	2
HB7	**ggcggcggcggctcc** ggtggtggt<u>ggatcc</u> CAGGTCCACGTGAAGCAGTC	1	1
HB8	**ggcggcggcggctcc** ggtggtggt<u>ggatcc</u> GAGGTGAA<u>S</u><u>S</u>TGGTGGAATC	4	2
HB9	**ggcggcggcggctcc** ggtggtggt<u>ggatcc</u> GA<u>V</u>GTGA<u>W</u>G<u>Y</u>TGGTGGAGTC	12	5
HB10	**ggcggcggcggctcc** ggtggtggt<u>ggatcc</u> GAGGTGCAG<u>S</u><u>K</u>GGTGGAGTC	4	2
HB11	**ggcggcggcggctcc** ggtggtggt<u>ggatcc</u> GA<u>K</u>GTGCA<u>M</u>CTGGTGGAGTC	4	2
HB12	**ggcggcggcggctcc** ggtggtggt<u>ggatcc</u> GAGGTGAAGCTGATGGA<u>R</u>TC	2	2
HB13	**ggcggcggcggctcc** ggtggtggt<u>ggatcc</u> GAGGTGCA<u>R</u>CTTGTTGAGTC	2	1
HB14	**ggcggcggcggctcc** ggtggtggt<u>ggatcc</u> GA<u>R</u>GT<u>R</u>AAGCTTCTCGAGTC	4	2
HB15	**ggcggcggcggctcc** ggtggtggt<u>ggatcc</u> GAAGTGAA<u>R</u>STTGAGGAGTC	4	2
HB16	**ggcggcggcggctcc** ggtggtggt<u>ggatcc</u> CAGGTTACTCT<u>R</u>AAAG<u>W</u>GT<u>S</u>TG	8	5
HB17	**ggcggcggcggctcc** ggtggtggt<u>ggatcc</u> CAGGTCCAACT<u>V</u>CAGCA<u>R</u>CC	6	3.5
HB18	**ggcggcggcggctcc** ggtggtggt<u>ggatcc</u> GATGTGAACTTGGAAGTGTC	1	0.7
HB19	**ggcggcggcggctcc** ggtggtggt<u>ggatcc</u> GAGGTGAAGGTCATCGAGTC	1	0.7

Primer VH for:

5' *Eco*RI 3'
scfor   ggaattc<u>ggccc</u>ccgag

	5' *Eco*RI *Sfi*I V$_H$ 3'	
HF1	ggaattc<u>ggccc</u> **ccg**<u>aggc</u>C GAGGAAACGGTGACCGTGGT	1
HF2	ggaattc<u>ggccc</u> **ccg**<u>aggc</u>C GAGGAGACTGTGAGAGTGGT	1
HF3	ggaattc<u>ggccc</u> **ccg**<u>aggc</u>C GCAGAGACAGTGACCAGAGT	1
HF4	ggaattc<u>ggccc</u> **ccg**<u>aggc</u>C GAGGAGACGGTGACTGAGGT	1

Specific primers for c-DNA synthesis:

CLκ   ACTGGATGGTGG
CLλ   ACTCTTCTCCACA
IgA   GGTGGTTATATCC
IgM   CTGATACCCTGG
IgG+E RCTGGACAGGG

Primer VL back:

5' *Sfi*I FLAG 3'
scback ttactcgc<u>ggccc</u>**agc**c<u>ggcc</u>atggcggactacaaaG

	5' FLAG V$_L$ 3'	(d)	μl Mix
LB1	gccatggcg*gactacaaa*GA Y ATCCAGCTGACTCAGCC	2	1
LB2	gccatggcg*gactacaaa*GA Y ATTGTTCTC<u>W</u>CCCAGTC	4	2
LB3	gccatggcg*gactacaaa*GA Y ATTGTG<u>M</u><u>T</u><u>M</u>ACTCAGTC	12	5
LB4	gccatggcg*gactacaaa*GA Y ATTGTG<u>Y</u><u>T</u>RACACAGTC	8	3.5
LB5	gccatggcg*gactacaaa*GA Y ATTGT<u>R</u>ATGAC<u>M</u>CAGTC	8	4
LB6	gccatggcg*gactacaaa*GA Y ATT<u>M</u>AGAT<u>R</u>A<u>M</u>CCAGTC	16	7

210

10: Producing antibodies in E. coli: from PCR to fermentation

Table 1. Continued

LB7	gccatggcg*gactacaaa*GAYATTCAGATGAYDCAGTC	12	6
LB8	gccatggcg*gactacaaa*GAYATYCAGATGACACAGAC	4	1.5
LB9	gccatggcg*gactacaaa*GAYATTGTTCTCAWCCAGTC	4	2
LB10	gccatggcg*gactacaaa*GAYATTGWGCTSACCCAATC	8	3.5
LB11	gccatggcg*gactacaaa*GAYATTSTRATGACCCARTC	16	8
LB12	gccatggcg*gactacaaa*GAYRTTKTGATGACCCARAC	24	8
LB13	gccatggcg*gactacaaa*GAYATTGTGATGACBCAGKC	12	6
LB14	gccatggcg*gactacaaa*GAYATTGTGATAACYCAGGA	4	2
LB15	gccatggcg*gactacaaa*GAYATTGTGATGACCCAGWT	4	2
LB16	gccatggcg*gactacaaa*GAYATTGTGATGACACAACC	2	1
LB17	gccatggcg*gactacaaa*GAYATTTTGCTGACTCAGTC	2	1
λB	gccatggcg*gactacaaa*GATGCTGTTGTACTCAGGAATC	1	1

Primer V$_L$ for:

	5' (Gly$_4$Ser)-linker	V$_L$ kappa 3'	
LF1	**ggagccgccgccgcc**	(agaaccaccaccacc)$_2$ ACGTTTGATTTCCAGCTTGG	1
LF2	**ggagccgccgccgcc**	(agaaccaccaccacc)$_2$ ACGTTTTATTTCCAGCTTGG	1
LF4	**ggagccgccgccgcc**	(agaaccaccaccacc)$_2$ ACGTTTTATTTCCAACTTTG	1
LF5	**ggagccgccgccgcc**	(agaaccaccaccacc)$_2$ ACGTTTCAGCTCCAGCTTGG	1
		V$_L$ lambda	
λF	**ggagccgccgccgcc**	(agaaccaccaccacc)$_2$ ACCTAGGACAGTCAGTTTGG	0.25

In this nomenclature, 'back' refers to 'toward the 3' end of the gene' and 'for' to 'toward the 5' end of the antibody gene'. The sequences are given using the IUPAC nomenclature of mixed bases (shown in underlined capital letters), with a column listing the degeneracy encoded in each primer and the volume (microlitres) to be used to set up the PCR mix (*see Protocol 1*).

The LB1–LB17 series encodes a length of 20 bases complementary to the mature mouse antibody κ coding sequence (in capital letters). Underlined is the preceding sequence which encodes the shortened FLAG sequence (14). Since the FLAG uses the N-terminal Asp of the mature antibody (encoded by GAY, in italics), only three additional amino acids are necessary. The FLAG codons are then preceded by the codons specifying the end of the signal sequence. The λB primer for mouse lambda chains is constructed analogously.

The 'V$_L$-for' primers hybridize with the J-elements (capital letters) and encode three repeats of the Gly$_4$Ser sequence, the terminal one of which has a very different codon usage so that wrong overlap is minimized.

The 'V$_H$-back' primers encode the other part of the linker, overlapping with 'V$_L$-for' in the sequence shown in bold. The hybridization is within the region given in capital letters.

The 'V$_H$-for' primers hybridize within the J$_H$ region.

The final assembly of the scFv gene is carried out with scback and scfor as described in *Protocol 1*.

K = T, G; R = A, G; M = A, C; B = C, G, T; S = G, C; Y = T, C; V = A, C, G; W = A, T.

General comments on primary PCR and assembly (Protocol 1)

Protocol 1 has the following features, which we believe to be significant improvements:

(a) The scFv fragment is assembled from only two pieces, not three. Therefore, only two fragments have to be matched in concentration.

(b) The linker is usually chosen to be 20 amino acids long for V$_L$–linker–V$_H$ assemblies, to avoid dimerization or aggregation of scFv fragments (7–10).

(c) To avoid any wrong overlap during assembly PCR, the (Gly$_4$Ser) repeats are encoded by different codons.

(d) Polymerases with proof-reading capacity are used whenever possible.

(e) The set of mouse primers (*Table 1*) has been optimized, incorporating all presently known mouse sequences and combining previous primer sets (11–13).

(f) The scFv encodes a convenient shortened version of the FLAG peptide, which introduces only three additional amino acids at the N-terminus (14). This way, the scFv can be detected in any fusion protein.

(g) *Sfi*I is used as the universal cloning site for directional cloning. The enzyme *Sfi*I has a number of remarkable advantages. It recognizes eight bases, interrupted by five non-recognized nucleotides (GGCC-NNNNNGGCC), and sites are therefore very rare. By choosing two *different* sticky ends, directional cloning is possible. Furthermore, avoiding symmetry in the sticky ends, self-dimerization of either insert or vector becomes impossible. Finally, *Sfi*I has the interesting property that it always cuts two sites at once, and therefore single-cut plasmids do not occur as intermediates (cutting plasmids with single *Sfi*I sites requires two plasmid molecules) (15).

(h) To further test and improve ligation efficiency, the recipient vector is used with a tetracycline resistance (2101 bp) cassette between the two different *Sfi*I sites. The loss of *tet* resistance can easily monitor the successful cutting and ligation.

(i) The procedure has been successfully used for library cloning, e.g. from immunized mice in an analogous way.

(j) The procedure as detailed here describes the assembly of an scFv fragment via PCR cloning into the two different *Sfi*I sites. Since compatible vector sets are available for expression as soluble fragments or for phage display (16), the procedure can be carried out identically in both cases.

(k) Compatible vector sets are available which allow conversion of the scFv fragment to any fragment shown in *Figure 1*. This is described in more detail in Ge *et al.* (16). It may be advantageous to carry out a preliminary characterization of the candidate antibodies in the scFv format, as this is fairly general.

(l) Vector features are listed in Section 3.2.1.

(m) Using gene synthesis, e.g. by the PCR-based method of Prodromou and Pearl (6), one or both domains can be synthesized completely. The procedure can then be followed from *Protocol 1C*. As indicated in the text, the quality of each of the oligonucleotides is most decisive. Since the PCR-based method randomly amplifies single molecules, single-base deletions present in any one of the oligonucleotides can be amplified into the final product. While the 'capping' used in the usual oligonucleotide synthesis protocol (17) tries to minimize this problem, it can never be eliminated completely. For such projects we recommend that the oligonucleotide supplier is carefully chosen and that documentation on stepwise yields for each oligonucleotide is requested. A low total yield is usually a warning sign that the 'full-length' oligonucleotide pool contains a significant proportion of molecules with random single-base deletions. Gel purification can ameliorate, but not solve, the basic problem, which lies in poor synthesis quality.

(n) Similar precautions are appropriate in primer synthesis for library projects. The use of mixed bases in particular, poses a problem for some oligonucleotide suppliers. Bottle changes on the synthesizer may introduce moisture in the system, leading to lower overall yields, with concomitant base deletions. Any sequence absent from a complex primer mixture will obviously decrease the functional library size. Thus, for complex library projects, the oligonucleotide supplier should be chosen with care.

2.1 Choice of antibody format

An important question for any expression project is which format the antibodies are to be expressed in. A similar question can of course also be posed in the design of a phage library. Since the conversion from one format into the other is straightforward, one need not be constrained by the availability of a particular library.

The antibody structure is separated into an antigen-binding part (the Fab fragment) and into an effector part (the Fc fragment), connected by a hinge region (Chapter 9, ref. 18). There is no evidence, to the best of our knowledge, that these two parts interact with each other in the final IgG molecule or during the folding. Therefore, the recombinant Fab fragment is most likely a precise replica of the same Fab fragment in the context of the whole antibody. The only exception is that some subclasses of antibodies are glycosylated in the C_H1 domain (18), and occasionally, CDRs code for an adventitious glycosylation site, usually unwanted anyway, which of course will not be reproduced by bacterial hosts.

The Fab fragment has no effector function, and any desired biological function of the recombinant antibody, other than antigen recognition, must therefore be engineered into it—most advantageously in the form of fusion proteins. The type of partners which can be fused is limited only by imagination, as enzymes, cytokines, metal binding domains, and receptor ligands have all been used, fused to either Fab or single-chain Fv fragments (see below) (1, 19, 20). The glycosylation of the Fc part is necessary for its function, both in complement activation as well as in antibody-dependent cellular cytotoxicity (ADCC) (21), and so the expression of whole antibodies in *E. coli* would not be useful for biological function, even if they were expressed with significant yields (22). For the second function of the Fc part, namely to provide bivalency, there are simpler solutions which have been shown to work well in *E. coli* (see below).

For *E. coli* (2), the choice is therefore between Fab fragments and Fv fragments and their derivatives (single-chain Fv fragments (scFv) (23–25) and disulfide-linked Fv fragments (dsFv) (see *Figure 1*)) (25–27). All other fragments (single domains, CDR peptides) are not faithful representations of the original antibody-binding site and therefore of interest only in special situations and will not be discussed further.

Fab fragments make use of the natural stabilizing effect of the constant domains C_H1 and C_L. These constant domains not only increase the size of the interface between the light and the heavy chains, but also protect the interface at the 'bottom' of the Fv fragment against partial denaturation and aggregation. Another advantage is the non-covalent tight association of the two chains, leading to a more facile combinatorial approach to libraries (3). However, there is a price to be paid. In a number of examples, the corresponding Fab fragment was expressed at lower levels than the Fv or scFv fragment because of increased folding problems (2). Apparently problems with aggregation-prone intermediates become exacerbated when these are coupled to another domain, such as the constant domains. Fab fragments may thus be thermodynamically favoured, but kinetically (during folding) disfavoured, compared to Fv derivatives. This observation is not general, however, as some Fv or scFv can be proteolytically labile, such that the achievable levels of functional protein for the Fab fragment could be similar or even better. While folding mutations found to improve Fv fragment folding (see Section 4) (28) were also found to improve Fab fragment folding, all Fabs remained at a lower level than the Fv derivatives in these experiments.

The Fv fragment is only part of a natural protein assembly. It has not been optimized by evolution to stay associated as a V_L–V_H heterodimer, since normally, the constant domains help in achieving this. Therefore, it is not surprising to find a large variation in the stability of domain association, since the hypervariable loops contribute a significant part to the V_H/V_L domain interface (18). Because of this, as yet, unpredictable association behaviour, pure Fv fragments are only rarely used.

Most popular has been a variant of the Fv fragment, in which the two chains are coupled genetically to give the so-called single-chain Fv (scFv) fragment (23–25). This can be done in two orientations, V_H–linker–V_L or V_L–linker–V_H. The large number of antibodies converted to these formats allow us to make some comments about the properties of such molecules. While the antibody has an approximate pseudo two-fold molecular axis of symmetry, the distance between the C-terminus of V_L and the N-terminus of V_H is around 39–43 Å (29, 30), while the distance between the C-terminus of V_H and the N-terminus of V_L is around 32–34 Å (29, 30). To obtain similar molecular properties, a V_L–V_H linker must obviously be longer than a V_H–V_L linker.

If the linker is too short, the molecule prefers to dimerize or multimerize: one V_H domain of one scFv pairs with a V_L domain of a different scFv molecule. This behaviour has been investigated as a function of linker length (7, 8). Since several sites often remain functional in such oligomers, some avidity effects can be observed, although this is somewhat unpredictable and depends on the primary sequence of the antibody (7–9). The 'miniantibody' format (31–33) shown in *Figure 1* may be a more general solution to this problem. While these molecules have minimal oligomerization domains, many alternative fusions, which lead to dimers or multimers, are of course conceivable and some have been made in the laboratory. This phenomenon of scFv aggregation has also been exploited for constructing heterodimers ('diabodies') (10). It appears that a 15–20 amino acid linker is usually appropriate for obtaining largely monomeric V_H–linker–V_L molecules.

We recommend, therefore, the use of 15 or 20 amino acid linkers in the orientation V_H–linker–V_L and 20 to 25 amino acid linkers in the orientation V_L–linker–V_H. The latter orientation has advantages because of the ease with which a minimal FLAG-tag of only three extra amino acids can be added for detection (14, see below). These design principles should also be used in creating phage libraries, and if a linker of repeating sequences is used, e.g. the $(Gly_4Ser)_n$ repeats, the similarity of the DNA sequences between the repeats has to be minimized to avoid deletions through recombination or during PCR assembly (16).

The advantage of the scFv strategy is that it secures an equimolar mixture of V_H and V_L, and makes the V_H–V_L association concentration independent, but it keeps V_H and V_L in a rather loose (and possibly mobile) complex. Another method for obtaining covalent linkage of V_L and V_H is by designing disulfide bonds between them (25–27). Framework positions have been identified which will be useful for many antibodies, although probably not for all, because of the variability in relative orientation of V_H and V_L. There may not be a single solution to a globally stabilizing disulfide bond, but a number of bonds are generally promising. In direct comparisons (25), the functional expression of disulfide-containing Fv fragments was lower than the same fragment without the disulfide bond or the scFv fragment, but this disadvantage may be offset by ending up with a very stable molecule (25–27).

Future approaches will most likely combine several of these strategies, and also incorporate mutations useful for minimizing aggregation problems.

3. Expression strategies

3.1 Overview

In general, one strives to produce a protein in the native state. The only case when alternative expression methods can become attractive, is if the native expression does not yield sufficient material. We stress again that, all other things being properly designed, the primary sequence of the antibody is a most decisive factor (28). We will analyse potential reasons for this, provide possible solutions to some problems, and compare alternative techniques.

All antibody expression strategies have to take account of the fact that a crucial intramolecular disulfide bond stabilizes each of the immunoglobulin domains (1, 2, 34). This disulfide must be present in the final product, and it must be formed either by making use of the periplasmic disulfide-forming machinery or during *in vitro* folding, with thiol/disulfide couples catalysing disulfide bond formation. We will first present an overview of the four basic antibody production techniques in *E. coli* (*Figure 2*), before discussing the two native strategies in more detail.

The first strategy (35, 36) consists of secreting the recombinant antibody to the periplasm. Today we know that this allows use to be made of the periplasmic disulfide-forming machinery of DsbA, DsbB, and DsbC and possibly other factors (37). Using this strategy, all monovalent and multivalent types of antibodies in *Figure 1* have been successfully expressed (reviewed in refs 1, 2). The main attraction is the ease of handling and the direct compatibility with the phage display format (3). Using high cell-density fermentation (see Section 5.3), yields of up to several grams per litre of functional antibody can be obtained, at least for antibodies with a reasonable folding behaviour. The disadvantage of this strategy is, however, that some particular antibody sequences can give relatively poor folding yields, whose molecular causes are only starting to be unravelled (see Section 4). Furthermore, some antibodies show sensitivity to proteases.

The second strategy is a direct companion of the first. It is to isolate the insoluble, periplasmic material (which has failed to fold or has even been deliberately accumulated, by carrying out antibody secretion at high temperature with strong promoters). This material must be refolded *in vitro* (38–40).

The third strategy is to express the antibody fragment without a signal sequence in the cytoplasm of special strains so that disulfide bonds can be formed there (41, 42). This strategy is only possible with *E. coli* strains, which have been made deficient in thioredoxin reductase (TrxB).

The fourth strategy is finally to produce cytoplasmic inclusion bodies (16, 29, 30, 43, 44). Antibody inclusion bodies are not fundamentally different from any other inclusion bodies and, thus, many guidelines from general inclusion body production can be followed (45). As in the second strategy, the crucial step is high-yield refolding *in vitro*.

An important trend which we have observed in a number of well-studied examples is a strong correlation between the *in vitro* and *in vivo* folding behaviour. This means that the same molecules which gave poor *in vivo* folding yields also lead to most of the aggregation by-products *in vitro*. Therefore, it appears that re-engineering the molecules for improved folding may be useful for *any* expression strategy.

While this chapter concentrates on native expression of antibody fragments, detailed protocols to refold antibodies from inclusion bodies have been given elsewhere (16, 29, 30, 43, 44). It should be emphasized that some

10: Producing antibodies in E. coli: *from PCR to fermentation*

empirical optimization of the refolding procedure will be necessary, because of the individuality of antibody fragments. General considerations are given in references 16, 29, 30, 45.

3.2 Secretion

The secretion of proteins to the periplasm is a natural process in *E. coli* (46), whose dependence on a signal sequence is well known. Much less understood are the requirements on the mature protein. Different signal sequences, by having different nucleotide sequences, may affect the translation initiation, and thus the rate of protein translation and transport, ultimately influencing the aggregation process accompanying protein folding. The choice of signal sequence is mostly governed by technical issues such as the ease with which rare 8-base restriction sites can be engineered into the sequence without disturbing its function. The *pelB* signal (from *Erwinia cavotovora* pectate lyase) has been found to be useful in this respect by a number of investigators, and because of its somewhat poorer translation initiation (probably related to the nucleotide sequence following the start codon) (A. Krebber, unpublished) may be particularly advantageous for phage display. In comparison, the signal from the *E. coli* outer membrane protein A (*ompA*) appears to be translated more efficiently and thus may be better suited for expression vectors than phage display. The weaker expression of *pelB* signals can be overcome by using optimized upstream regions, as we have done in new generation vectors (see below).

While the periplasm contains the machinery to form disulfides, no general periplasmic chaperone has been identified up to now (47). Thus, the recombinant proteins may be more or less 'on their own', and may have to be engineered for efficient folding (see Section 4). Several periplasmic proteases have been discovered (47, 48), but it is not clear whether there may be more, and the various available multiple deletion strains (48) have not been thoroughly evaluated at the time of writing. It should be pointed out, however, that the degradation is frequently a *symptom* (being a consequence of poor folding, since misfolded material gets degraded) and not a *cause* of poor expression (49).

3.2.1 Secretion vectors

Why is it necessary to give any thought to this topic at all, when so many vectors for the production of a multitude of different proteins have been described? The main reason is the stress imposed on *E. coli* by secreting the antibody. As discussed elsewhere in more detail, the cell attempts to minimize the stress by getting rid of the plasmid, or in severe cases by eliminating the genes by plasmid rearrangement or mutations. It is the residual expression of the protein which causes all the problems and must be carefully controlled to just the right level. This makes it immediately clear why different antibodies may have different requirements for the level of background expression they

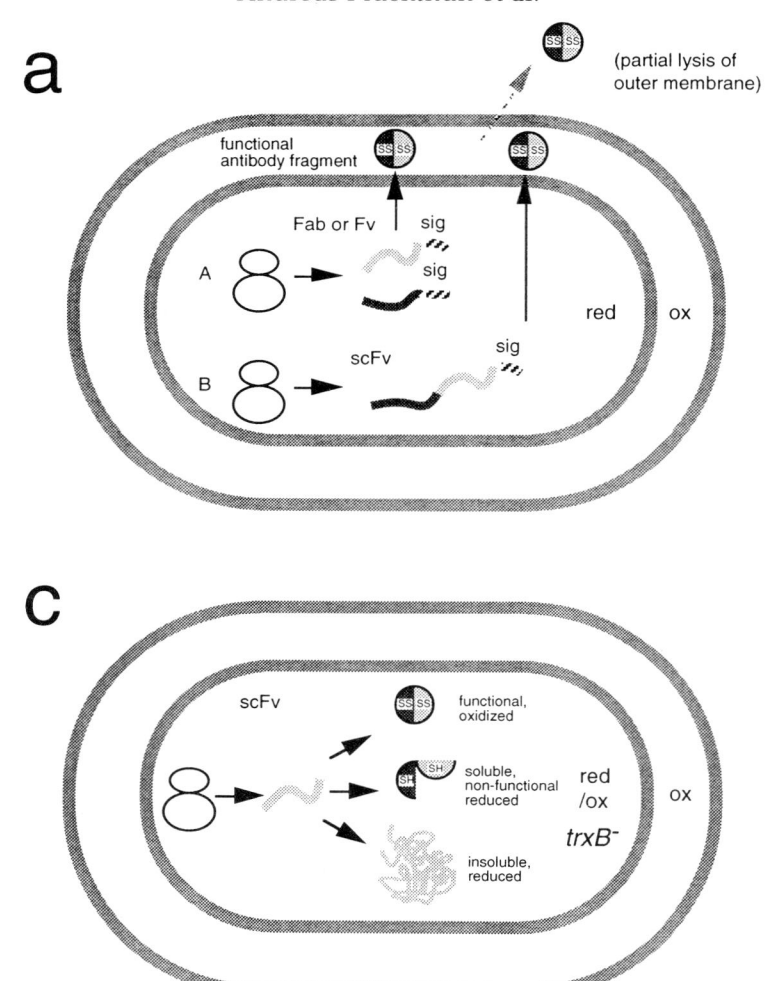

Figure 2. Expression strategies for antibody fragments in *E. coli.* (a) Expression by secretion. This has been demonstrated for all the fragments in *Figure 1*, and is symbolized here for a two-chain fragment (Fv or Fab) (A) and a single-chain fragment (B). As discussed in the text, this strategy can lead to very high amounts of functional protein and is compatible with the fermentation conditions in Section 5. The side reactions are insoluble periplasmic protein (see (b)) and a leakiness of the outer membrane, which depends on the growth conditions and the primary sequence of the antibody. (b) Expression as insoluble periplasmic protein with subsequent *in vitro* refolding. This strategy is an attempt to rescue the protein unable to fold in strategy (a). To maximize

can tolerate. The design of the expression vector must therefore optimize the vector stability, minimize the promoter leak-rate, and optimize (not maximize) the expression level in the induced state.

We will discuss the problem using the example of *lac*-promoter based plas-

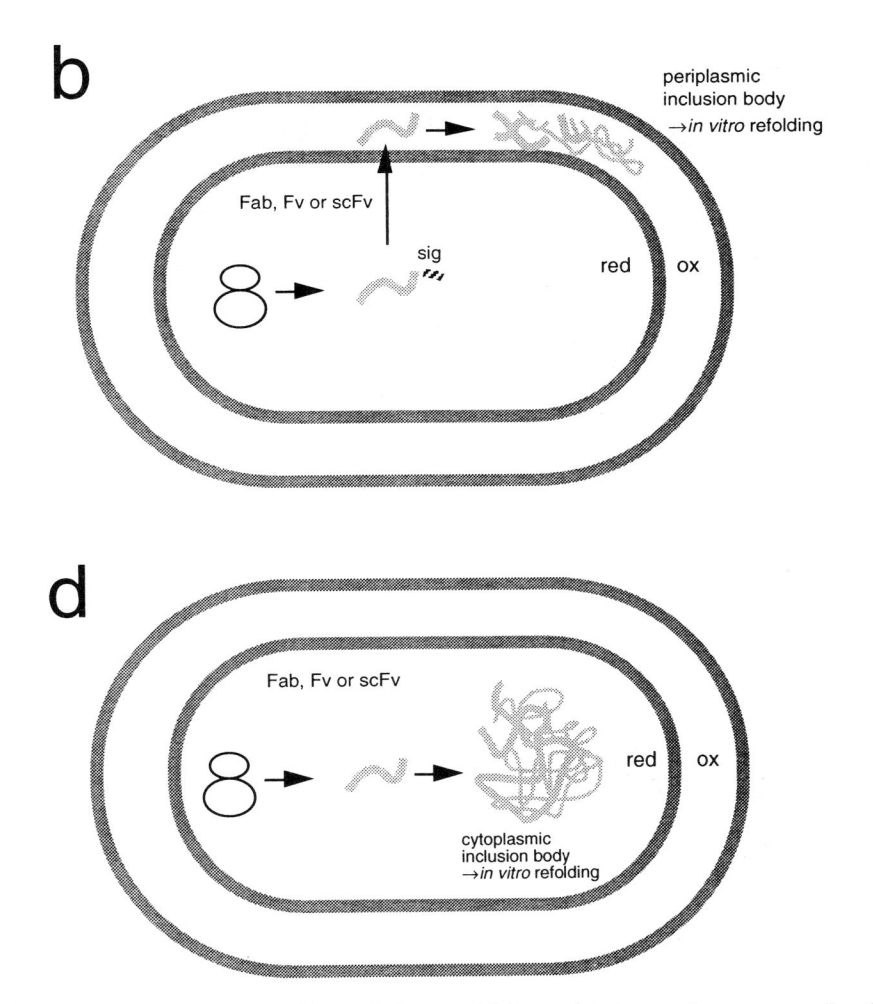

10: Producing antibodies in E. coli: from PCR to fermentation

b

periplasmic inclusion body
→*in vitro* refolding

Fab, Fv or scFv

sig

red   ox

d

Fab, Fv or scFv

red   ox

cytoplasmic inclusion body
→*in vitro* refolding

the yield, this strategy is usually carried out at high temperatures, since aggregation is desired. (c) Functional expression in the cytoplasm. This is only possible under conditions where disulfide formation is favoured, such as in *trxB⁻* strains. The antibody fragment is partitioning between an insoluble and, of course, non-functional form, a reduced, soluble but non-functional form, and the correctly oxidized and functional form. The *trxB⁻* mutation increases the percentage of the latter form, but not of the soluble, non-functional form. (d) Expression as cytoplasmic inclusion bodies with subsequent *in vitro* refolding. To maximize the yield, this strategy is usually carried out at high temperatures, since aggregation is desired.

mids, although other promoters have also been used successfully (1). Any system can be used which adheres to the following rules. The expression system has to be inducible at room temperature. The stress on the host cell (indirectly linked to incorrectly folded and aggregated antibody protein) greatly increases at higher temperature. Thus, the optimal expression temperature is around 26 °C. While one can use temperature inducible systems (involving a brief shift to 42 °C), this appears to still give more aggregation than keeping the cells at 26 °C continuously (50). In the fermenter, the temperature jumps upset the balance of the cells, and are not advantageous. Thus, a chemical inducer is preferred together with a plasmid-encoded repressor. The system with *lac*I as the repressor gene, IPTG as the inducer, and the *lac* operator fulfils this need. Undoubtedly, other similar systems can be used in an analogous fashion.

Most important is the tight control of the promoter/operator system. The 'leakage' of expression is responsible for all the unfavourable effects such as periplasmic leakage, plasmid loss, and rearrangement, because of the relatively long time cells spend in pre-cultures, or the main culture before induction. In the *lac*-based system, arranged precisely as in the natural *lac* operon, two sources for background expression exist. The first is a read-through from the *lac*I repressor gene, upstream of the promoter, which becomes *worse* when its promoter strength is increased (the *lac*I^q variant) indicating that it is not the amount of repressor protein which is limiting, but that the *lac*I message continues into the antibody gene (A. Krebber, C. Wülfing and A. Plückthun, unpublished; see also ref. 51). Consequently, a strong terminator upstream of the *lac* promoter is essential for tight repression. The second source of background expression is leakage of the promoter itself. This is most efficiently minimized by the presence of the natural CAP site (52), leading to glucose repression.

In general, the promoter must have a 'window' as wide as possible (defined as the ratio of induced over non-induced transcription) (53). The more toxic the product is, the more advantageous it becomes to lower both the non-induced as well as the induced levels by factors which affect both levels equally: promoter strength and translation initiation. Since for such toxic protein products, only a fraction of the protein may fold correctly anyway, a high expression level is not that crucial. From a more pragmatic point of view, there may never be a universal vector, but, using a terminator upstream of the promoter in all cases, several strength variants of the promoter (e.g. *lac*, *lac*UV5, *tac*) combined with inducer titration can be an easy solution for elucidating the optimal levels for a given protein (*Figure 3*).

We believe that for phage display, a strong promoter is of no importance and actually a burden. In contrast, given the enormous stress fusion proteins of antibodies with *gene3* impose on the cell, a low background level before induction is of utmost importance, or losses of clones from the library or gene deletions will quickly accumulate. Thus, phage display and soluble expression have rather different optima, the latter requiring much stronger transcription and translation. Thus the use of phage-display vectors for soluble expression, e.g. making use of a suppressible stop codon in the antibody–*gene3* fusion protein, can be used for characterization but not for efficient production.

The antibody cassette should be modular in nature. One solution is given in the pAK and compatible pIG series (16) of vectors (*Figure 3*). This way, the initial product can easily be converted into a number of fusion proteins or miniantibody formats (16) (*Figure 4*). In the new vectors, the PCR product can be cut with only one enzyme, an 8-base cutter (*Sfi*I), which because of its unspecified three nucleotides in the overlap, allows directional cloning, if both sites are different. This enzyme has the interesting property that it only makes double-cuts in the plasmid, thus further facilitating library cloning (15) (see *Protocol 1* and note g (p. 212)).

Particularly important are tag sequences to follow and purify the antibody conveniently. We prefer to have a shortened FLAG (14) requiring only three extra amino acids when put in front of the light chain, or four on the heavy chain. Purification can be carried out with FLAG sequences, using an anti-FLAG column, but is most conveniently carried out with the histidine tag (54) in combination with immobilized metal ion affinity chromatography (IMAC) (see below). The his tag can be detected by an antibody (P. Lindner, unpublished), or it can be preceded by another tag, the myc tag (55).

Such insertion cassettes also allow fusion to other protein domains. We have used dimerization and tetramerization devices to obtain dimeric or multimeric antibodies conveniently with these vectors (31–33). These devices are amphiphathic helices, linked to an scFv fragment via a flexible hinge (*Figure 1*). For easy purification, a his-tail can be attached as well. For convenience, *gene3* cassettes are also available to make phage-display vectors and expression vectors compatible (16).

Outside of the expression cassette, the vector must code for an antibiotic resistance. In phage display, the expression of the antibiotic resistance gene must not be too high, or low phage titre will be observed, perhaps because of interference between the single-strand production and transcription. Chloramphenicol (Cam) and ampicillin (Amp) resistance are satisfactory, while kanamycin (Kan) resistance appears to give somewhat lower phage titres in a direct comparison of w.t. phages (S. Spada, C. Krebber, and A. Plückthun, unpublished). In fermentation, antibiotics, notably ampicillin, are quickly degraded by enzymes from lysed cells so that plasmid maintenance is not possible by antibiotics alone. Thus, additional so-called post-segregational killing mechanisms such as the *hok/sok* system (56) have been employed by us.

If it is to be used for phage display and/or site-directed mutagenesis, the plasmid needs an f1-origin of replication (57). To replicate as a double-stranded plasmid, the plasmid also needs another origin, usually derived from *ColE1*. The pUC-derived variant is temperature sensitive: it has a low copy number (lower than pBR322) at room temperature and a high copy number at 37°C (58).

3.3 Functional antibodies from the cytoplasm

The *E. coli* cytoplasm is 'reducing', which means that under normal circumstances the equilibrium between reduced and oxidized cysteine is on the

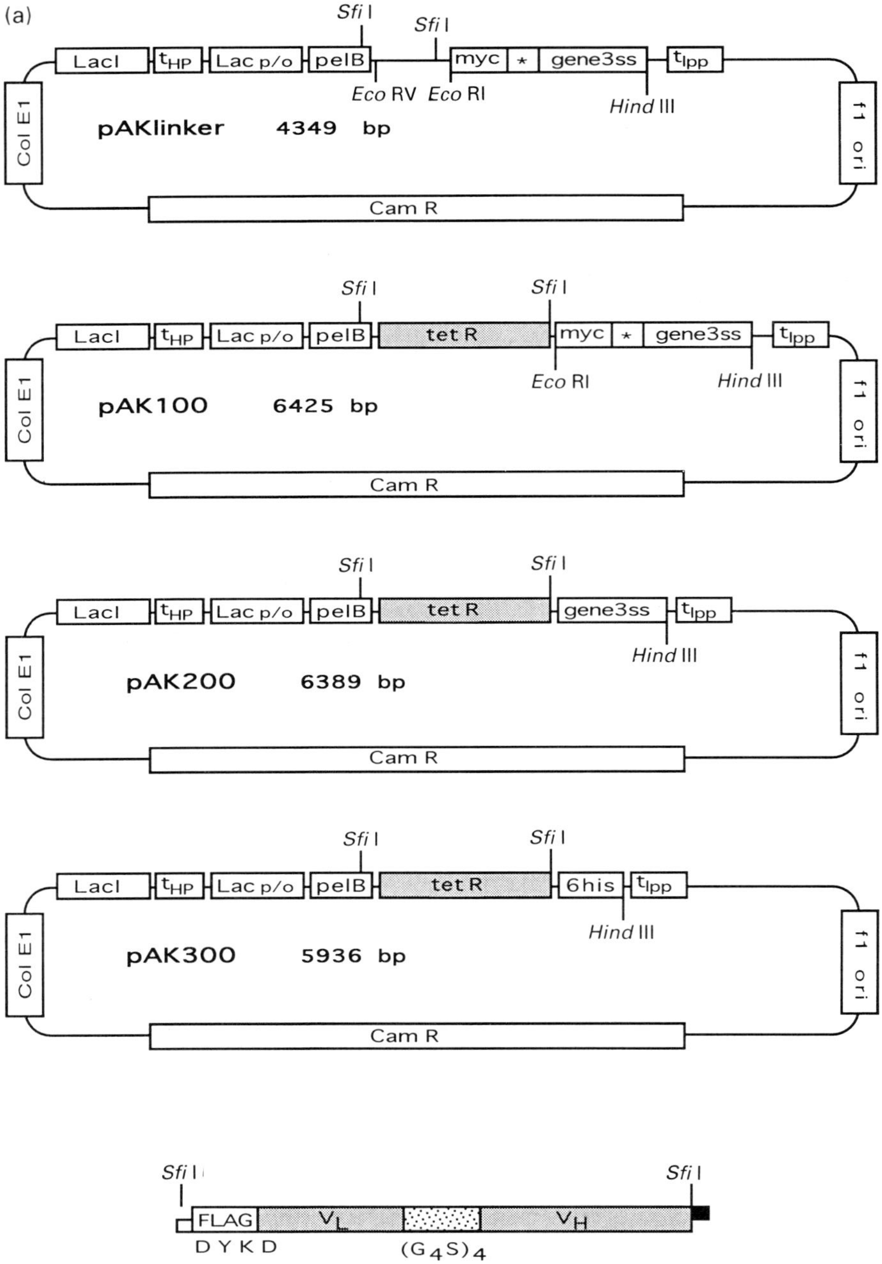
(a)
Sfi I
Sfi I
LacI
tHP
Lac p/o
pelB
myc
*
gene3ss
tlpp
Eco RV
Eco RI
Hind III
Col E1
pAKlinker 4349 bp
f1 ori
Cam R
Sfi I
Sfi I
LacI
tHP
Lac p/o
pelB
tet R
myc
*
gene3ss
tlpp
Eco RI
Hind III
Col E1
pAK100 6425 bp
f1 ori
Cam R
Sfi I
Sfi I
LacI
tHP
Lac p/o
pelB
tet R
gene3ss
tlpp
Hind III
Col E1
pAK200 6389 bp
f1 ori
Cam R
Sfi I
Sfi I
LacI
tHP
Lac p/o
pelB
tet R
6his
tlpp
Hind III
Col E1
pAK300 5936 bp
f1 ori
Cam R
Sfi I
Sfi I
FLAG
VL
VH
DYKD
(G4S)4

10: Producing antibodies in E. coli: from PCR to fermentation

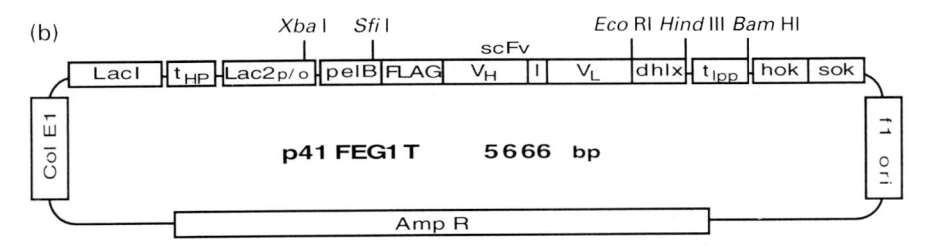

(b)

```
            Xba I    Sfi I                      Eco RI  Hind III  Bam HI
                            scFv
  ┌─────┬──────┬───────┬────┬────┬──┬──┬────┬────┬────┬────┐
  │ LacI│ t_HP │Lac2p/o│pelB│FLAG│V_H│ * │V_L │dhlx│t_lpp│hok│sok│
  └─────┴──────┴───────┴────┴────┴──┴──┴────┴────┴────┴────┘

              p41 FEG1 T        5666 bp

  Col E1                                                    f1 ori
            ┌──────────────────────────────────┐
            │             Amp R                │
            └──────────────────────────────────┘
```

Figure 3. Vectors and cloning strategies. (a) Vectors useful for the initial cloning of antibody fragments using the strategy outlined in *Protocol 1* (A. Krebber and A. Plückthun, unpublished). The vectors pAK100, pAK200, and pAK300 contain a *tet* resistance cassette (*tetA* and *tetR*; 2101 bp, shaded) to facilitate the monitoring of *Sfi*I cutting, both by gel electrophoresis and by religating and subsequent plating on tetracyline plates. Note that both *Sfi*I sites are different (see text), indicated by the different shadings of the sticky ends at the scFv insert at the bottom. The pAKlinker vector only contains a short DNA linker. The expression cassette is shown on the top of each vector. It comprises the *lacI* repressor gene, a strong terminator, t_{HP} (51), the *lac* promoter/operator, and the *pelB* leader sequence, which has been modified to contain an *Sfi*I site. After ligation, the antibody fragment is fused in-frame to *gene3* (pAKlinker, pAK100, pAK200) or to a his-tail for purification (pAK300) (54). The in-frame fusion using pAK100 to *gene3* first leads into a myc-tag (55) as a detection handle, in addition to the short 3-amino acid FLAG at the N-terminus (14). The asterisk represents an amber codon. Depending on the strain used it is possible to switch between soluble expression of scFvs (by using non-suppressor strains like JM83) and expression of scFv *gene3* fusions (by using suppressor strains like XL1). The *gene3* portion (denoted ss for super-short) starts at position 250 in the precursor protein, thus avoiding extraordinarily long glycine linkers and, most importantly, any unpaired cysteine of g3p. The two origins for phage replication and plasmid replication are as usual (details in ref. 16).

All fusion partners, including the helices for multimerization shown in *Figure 1*, can be added as *Eco*RI–*Hin*dIII fragments in pAKlinker or pAK100 (for details see ref. 16).

The chloramphenicol cassette (*Cam R*) is originally derived from pACYC184, but its expression strength has been adapted by randomizing the promoter and selecting clones with optimal growth and selection properties (C. Krebber and A. Plückthun, unpublished).

(b) Example for a high-level expression derivative of the pAK-vectors as used in high cell-density fermentation (see Section 5) (U. Horn, A. Krebber, D. Reisenberg, and A. Plückthun, unpublished). The vectors use a much stronger upstream region (denoted as *lac2 p/o* derived from a T7 upstream region), a different resistance (which is of no importance other than for transformation, since the selective pressure is not sufficient anyway for plasmid maintenance), and as a post-segregational killing mechanism, the *hok/sok* system (56). The particular example shown is that of a miniantibody.

(c) Two-step assembly strategy of scFv fragments as detailed in *Protocol 1*. The primer mixes are found in *Table 1*.

See Figure 3c over page.

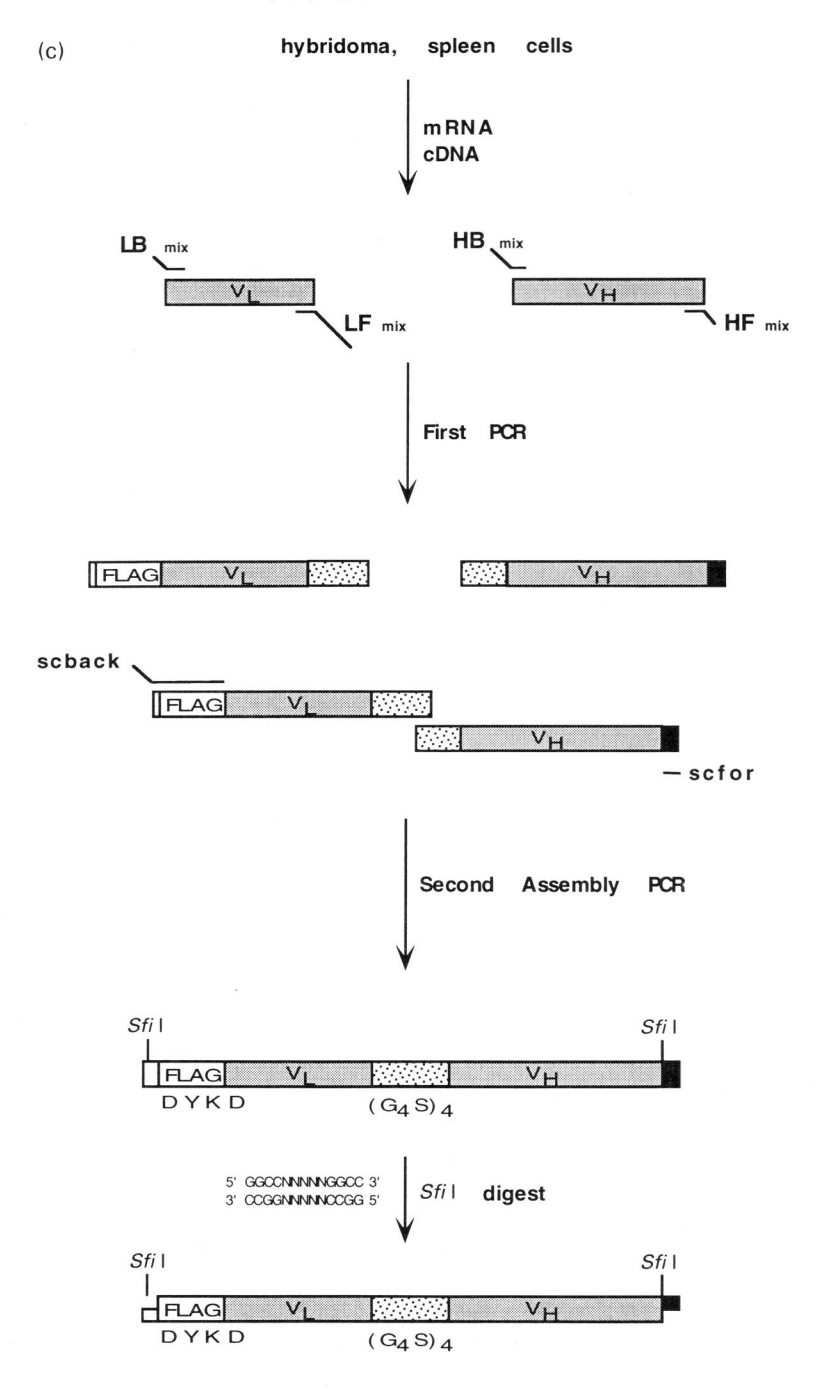

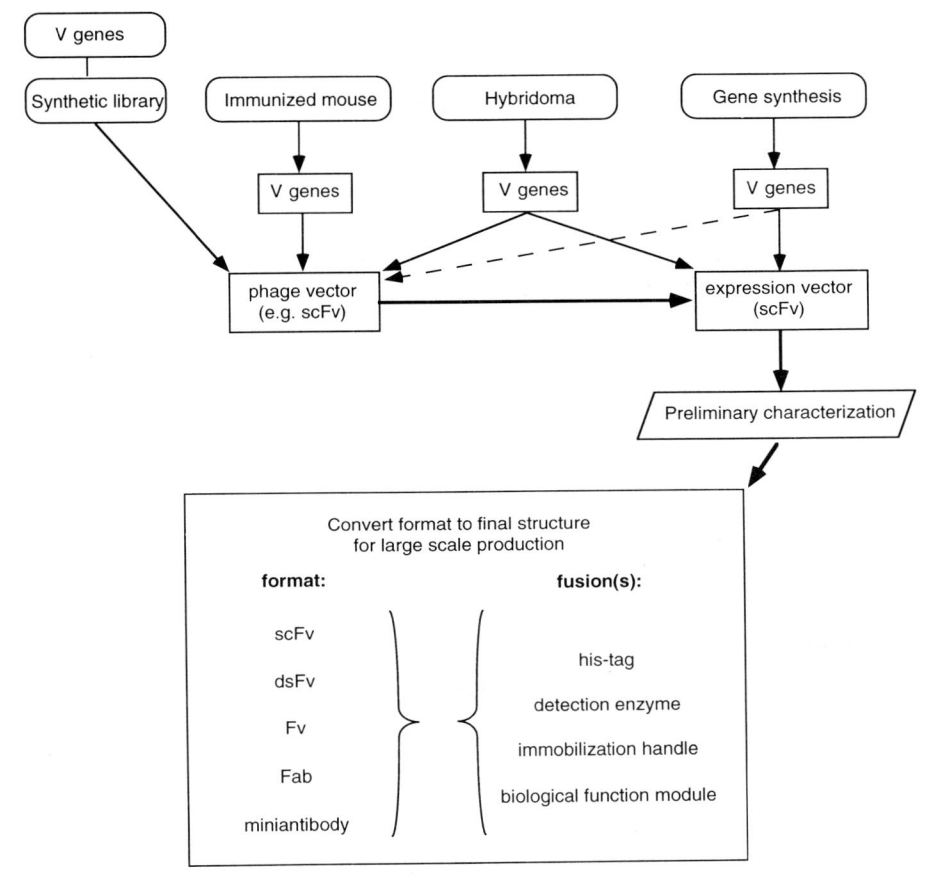

10: Producing antibodies in E. coli: from PCR to fermentation

Figure 4. Schematic cloning, conversion, and expression strategies. Four different starting scenarios are shown, leading to the cloning of the V-genes in either a phage vector or directly into an expression vector. While not strictly required, it is often advantageous to have a streamlined strategy by first cloning an scFv fragment and then to convert the best candidate to the exact fragment and/or fusion protein desired. For details, see also ref. 16.

reduced side. In order for disulfide bonds to form, two conditions must be met: first the thermodynamic equilibrium must allow it, and second, there must be an efficient kinetic mechanism available to form disulfides, i.e. reagents with which the disulfides formation can be catalysed.

E. coli depends on reduced cysteines not only in a number of cytoplasmic enzymes, but also in some proteins such as thioredoxin (Trx) and glutaredoxin (Grx), which play a crucial role as co-factors in the biosynthesis of deoxyribose, cysteine, and the reduction of methionine sulfoxide back to methionine (59). To keep Trx and Grx reduced, NADPH is used by two different enzymes, thioredoxin reductase and glutathione reductase. If thioredoxin reductase (TrxB) activity is diminished, cytoplasmic disulfides can form (42), but the mechanism by which this occurs is presently still unclear. It is not known whether molecular oxygen acts directly as the oxidant, or whether oxidation exploits accumulating oxidized glutathione, or other factors.

Strains deficient in TrxB appear to grow normally under standard laboratory conditions, indicating that only a slight perturbation of the disulfide metabolism is taking place. It was found that significant amounts of functional scFv were only obtained in the presence of this $trxB^-$ mutation (41). The active scFv correlated with the amount of *oxidized* material detectable by gel electrophoresis, but *not* with the amount of soluble scFv, which was constant and higher than the active amount. This clearly shows that there is soluble, inactive (and probably reduced) scFv formed under these conditions, in addition to the soluble oxidized, active scFv. Increasing the promoter strength can dramatically increase the amount of soluble and insoluble scFv—but the amount of active, disulfide containing scFv does not seem to change. These results suggest that cytoplasmic oxidation is the limiting reaction for functional expression, at least under the experimental conditions tested (41).

The suitability of this strategy for fragments other than scFv and the influence of antibody folding mutants (see below), the effect of co-expression of cytoplasmic folding factors, and the influence of particular protease-deficient strains still remains to be thoroughly evaluated.

3.3.1 Vectors for cytoplasmic soluble expression

From the much more limited experience with cytoplasmic, functional expression in $trxB^-$ strains, it appears that there is much less stress imposed on the cells than in the periplasmic expression strategy. Thus, the tightness of the regulation does not appear to be so critical. Both a standard *lac p/o*-based system as well as a T7-based system (16, 41, 43) gave satisfactory results. For the T7-based system, a BL21(DE3) strain, containing the T7 polymerase gene in the chromosome (60) and the $trxB^-$ mutation, was constructed (41).

3.3.2 Vectors for cytoplasmic inclusion body formation

For the preparation of cytoplasmic inclusion bodies, the T7-based system is particulary useful. In this case, it is crucial to grow the cells at 37°C, and not at lower temperature, for favouring inclusion body accumulation, and a normal $trxB^+$ w.t. strain harbouring the T7 polymerase gene can be used. For both strategies (Sections 3.3.1 and 3.3.2) the same vector can be used (16).

4. Improving expression: the influence of the sequence on expression and folding

4.1 Analysis and long-term solutions

The primary sequence of the antibody is emerging as a most decisive factor in determining the yield of functional protein and many anecdotal observations have pointed in this direction. Two antibodies have been compared in the form of the Fv fragments, scFv fragments, and Fab fragments for soluble periplasmatic expression in *E. coli* (28) and a similar relative functional expression ratio was found for all formats. Thus the very high yield for one of the antibodies (61) was shown to be due to the fact that this particular antibody sequence gives practically no insoluble protein at all, even at 37°C.

Clearly, it would be desirable to transfer these spurious favourable properties to *any* antibody. In a series of loop-grafting experiments, it was found (G. Wall, H. Bothmann, S. Jung, K. Bauer, A. Knappik, and A. Plückthun, unpublished) that these effects partially reside in the framework, partially in the CDR loops. It has been possible, based on sequence comparisons and analysis of point mutants, to engineer the antibody framework to achieve better folding antibodies (28). Although this has allowed an insight into the folding problem, the problem is far from being solved.

This mutational analysis has shown that there are two separable, but related phenomena, occurring in most antibody fragments, albeit to widely varying degrees. Upon secreting the antibody to the periplasm, the cells become leaky and eventually lyse. The exact mechanism remains unknown, but it does depend on external factors such as the medium. For instance, it does not happen in the fermentation discussed in Section 5. Furthermore, a significant portion of the protein may end up in insoluble protein, which has its signal precisely removed and thus must have seen signal peptidase. This is most likely a *periplasmic* aggregate. Both phenomena can be suppressed by *different* mutations, yet the mutations act synergistically and both may constitute folding mutations on the same folding pathway. An *in vitro* analysis has shown that the improved folding yield is also seen during the oxidative folding reaction *in vitro* under similar conditions as used preparatively for renaturation from inclusion bodies (28). This means that such mutations are also extremely useful if the ultimate production strategy is the refolding *in vitro*. It was also found that these mutations do not affect folding kinetics or thermodynamics but the aggregation reaction.

Unfortunately, our understanding has not yet advanced to the point where it is obvious how to improve routinely any given sequence other than to introduce any of the mutations so far found useful. Yet, a few points have emerged:

(a) The mutations found useful so far are all in turns, and may constitute subdomains involved in late folding steps *in vitro* and *in vivo*. Most likely, the favourable amino acids slow down the aggregation step itself.

(b) Several mutations, while not improving the soluble/insoluble ratios, are useful simply because the cells can be grown to much higher densities without lysis.

(c) The residues most frequently found in the database at framework positions are not necessarily the best for folding and expression, indicating that the mammalian cell may have more efficient chaperoning mechanisms, allowing it to minimize or even ignore the problems. However, a detailed analysis of these mutations in mammalian cells is still outstanding.

(d) Simple CDR grafting to superior frameworks has variable effects, depending on how much of this problem is contributed by the CDRs themselves.

(e) There is no obvious correlation between thermodynamic stability and folding efficiency (soluble expression yield), provided a minimum stability of an average scFv fragment (of probably a few kcal) is reached.

(f) While the usual problem is protein folding and *not* secretion, there are exceptions. Some antibody frameworks appear to encode too many positive charges in the sequence following the signal sequence, leading to the accumulation of precursor (62).

The ultimate solution for obtaining superior folding antibodies will come from framework engineering, starting from the best-known natural variants. Before this will be achieved, the merits of some more immediate approaches must be discussed.

4.2 Short-term solutions

Since the expression of antibodies on phage goes through an intermediate stage in which the hybrid protein is anchored to the inner bacterial membrane, protein folding of the phage-bound antibody also occurs in the periplasm and probably follows the same, or at least similar, constraints. There is anecdotal evidence from various laboratories that sequences with severe folding problems may be preferentially lost from libraries. It is unclear, however, how efficient this selection is. Usually, phage display is carried out at 37 °C and then only molecules with at least some tolerable properties at this temperature will be selected. Whether this effect is actually desired (i.e all selected antibodies should be at least mediocre folders) or not (the library may become very small and restricted and the very best intrinsic binders, or the ones that recognize the desired epitope, may all be lost) depends on the particular problem at hand. Clearly, the long-term solution will be synthetic libraries in which *all* members have good folding properties.

Another obvious series of experiments is the effect of overexpressing molecular chaperones. The protein-folding reaction in the secretion strategy occurs after transport through the membrane, in the bacterial periplasm. At the time of writing, no general periplasmic bacterial chaperone had been identified (47), and it is generally assumed that there is no ATP in the periplasm. On the other hand, two periplasmic prolyl *cis–trans* isomerases (Sur A (63) and Rot A (64)) and several proteins involved in disulfide formation (DsbA, DsbB, DsbC) (37) have been characterized (47). Furthermore, numerous *E. coli* proteins in cytoplasmic folding have been identified

10: Producing antibodies in E. coli: from PCR to fermentation

(65), as have factors involved with bacterial transport. Up to now, however, no dramatic, unequivocal, and *general* positive effect of overexpressing any of these factors on periplasmic antibody expression has emerged (66), to the best of our knowledge (see, for example, ref. 67), the only exception being on T-cell receptor fragments (49). The effect of overexpressing cytoplasmic chaperones during *cytoplasmic* functional expression (see Section 3.3) in *trxB*$^-$ strains has not yet been thoroughly evaluated.

Since the periplasm is connected to the 'outside world' by pores which can be traversed by small molecules, it is tempting to add compounds to the growing culture which are believed to improve folding by preventing aggregation. Sucrose has been reported to be successful in this respect (68, 69), but the effect is clearly not general. With other antibodies tested, no effect whatsoever was seen (L. Nieba and A. Plückthun, unpublished). The same holds true for betaine and sorbitol (70), for which no positive effect whatsoever was detected (L. Nieba and A. Plückthun, unpublished), at least for the antibody fragments tested.

One variable which is almost universally found to be useful, is low temperature. Its most significant effect may be on favourable partitioning of folding intermediates to the folded state and not to aggregates.

In summary, these results point to a great diversity of effects in protein folding among natural antibody sequences, which are clearly determined as much by the details of the sequence as by the common antibody domain topology. Antibodies differ enormously in pI, thermal stability, tendency to aggregate, or surface properties. It is an unsolved question whether the large differences in productivity seen in different hybridomas are also at least partially related to protein folding.

An antibody format must be chosen in which the two domains can assemble in a thermodynamically stable manner, discussed elsewhere in this chapter. We believe that a further analysis of the sequence requirement may provide the most dramatic general solution to the problem.

5. Growth and fermentation

Antibody expression can be carried out on various scales. We will first describe the simplest, small-scale shake-flask culture experiment, then an extremely simple benchtop fermentor—which may supersede the traditional shake-flask cultures in the 2–5 litre-range—and finally fermentation in a stirred tank reactor with more sophisticated controls for high cell-density cultivation with which gram per litre amounts of functional antibody fragments have been obtained. The procedures given are for functional secretion strategies (see Section 3.2).

5.1 Cultivation in standard shake flasks

For shake-flask cultures, complex media are usually used, since the cell density is not a major concern. While LB medium has been frequently used, somewhat higher cell densities have been obtained with super broth (SB) medium (see below). For most antibodies (but depending on the primary sequence, see Section 4), leakiness of the outer membrane sets in shortly after induction, often ending in complete cell lysis, thus limiting the duration of expression. With SB medium, the onset of leakiness can be somewhat retarded, thereby prolonging the useful time for induction, although the mechanism by which this occurs is still unclear. Antibodies with improved folding properties (naturally occurring or obtained by engineering) can be induced for much longer times at room temperature, often even overnight.

Protocol 2. Shake-flask culture

Equipment and reagents

- 500 ml and 5 litre flasks
- Shaker
- LB agar: aqueous solution of 10 g/litre tryptone, 5 g/litre yeast extract, 5 g/litre NaCl, and 15g/litre agar. Sterilize by autoclaving
- 1 M IPTG
- SB (super broth) medium: for 1 litre SB medium mix the following aqueous solutions, which have been autoclaved separately: 950 ml SB base (20 g/litre tryptone, 10 g/litre yeast extract, 10 g/litre NaCl), 33 ml 1.5 M K_2HPO_4, 5 ml 1 M $MgSO_4$, 10 ml/50% glucose (500 g/litre), and 1 ml of antibiotic stock solution (1000 ×)
- 1000 × antibiotic: 100 mg/ml ampicillin in H_2O or 25 mg/ml chloramphenicol in ethanol

Method

1. Transform the bacterial strain of choice (see text) with the expression plasmid. Plate out on LB agar plates, containing 0.5% glucose (5 g/litre)[a] and the appropriate antibiotic. Incubate overnight at 37 °C then store the plates at 4 °C.

2. Inoculate 50 ml of SB medium (in a 500 ml flask) with a single bacterial colony and shake overnight at 25 °C.[b]

3. Inoculate 1 litre of SB medium (in a 5 litre flask) with the preculture and shake at 25 °C at 150–200 r.p.m., depending on the type of shaker.

4. Induce the expression at OD_{550} = 0.5–1.0, by adding 0.25 ml 1 M IPTG.[c] Continue shaking until the culture reaches the stationary phase.[d]

5. Harvest the cells by centrifugation (8000 g, 10 min, 4 °C). Continue with cell disruption and protein purification (see Section 6).

[a] Addition of glucose is optional. However, background expression is reduced if working with the *lac* promoter system in which the *CAP* site is present, and bacterial growth is usually improved.
[b] Fermentation temperature depends on the stability of the expressed antibody fragment. In cases of an exceptionally stable fragment, fermentation can be carried out at 37 °C. However, for most antibody fragments, 25 °C is preferred.
[c] For use with promoter systems using the *lac* operator.
[d] Depending on the strain and properties of the antibody fragment expressed, this will be within 4–6 hours after induction and the final OD_{550} will be 3–6.

5.2 Cultivation in benchtop flasks to medium cell-densities (≤ 5 g/litre)

After the first tests on a recombinant antibody, which may be carried out in shake flasks, larger amounts are often needed, usually 5–50 mg. Typical examples are structural studies using crystallography or NMR, biophysical studies, or animal studies. This range is frequently not easily accessible by shake-flask culture, and a very simple methodology is needed such that several protein mutants can be simultaneously prepared in a short time. For this purpose, we developed a simple benchtop-cultivation flask (*Figure 5*), which is capable of achieving significantly higher cell-densities than shake flasks. Protocols are presented for the use of (a) SB medium (similar to shake-flask culture, see above) and (b) a defined glucose–mineral salt medium (Glu–MM), which is especially useful for the production of labelled proteins and in general for the production of recombinant proteins under well-defined conditions. This medium avoids the sequential consumption of various constituents of the medium, as is the case when using SB medium. To obtain very high cell densities, more sophisticated equipment is needed (see Section 5.3) and only glucose–mineral salt medium should be used.

In general, specific productivities (amount of antibody per cell) appear roughly comparable between the different cultivation systems. However, the controlled growth conditions in the benchtop and high cell-density

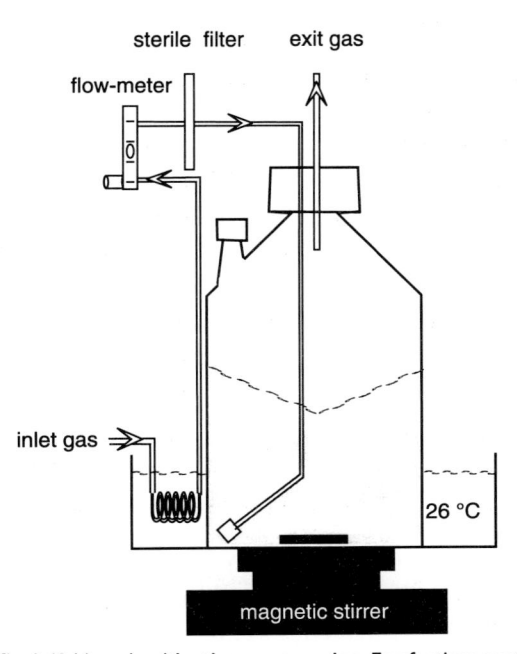

Figure 5. Benchtop-flask (2 L) and cultivation accessories. For further explanations see text.

fermenters lead to lower cell leakage and thus a lower loss of protein to the medium. With the high cell-density fermentation described in *Protocol 6*, up to 4 g active bivalent miniantibody per litre of *E. coli* have been achieved (Riesenberg *et al.*, unpublished observations).

To reach high volumetric yields with simple equipment, one has to analyse the critical factors which limit bacterial growth. (a) A high-yield coefficient (the mass of cells per mass of nutrients) of the medium is necessary, meaning that enough of all nutrients must be present to allow growth of cells to reasonable densities. There is a maximal tolerable concentration, however, which means that if higher cell densities are required, feeding of substrates to the growing culture is necessary (see Section 5.3). Therefore, the simple equipment used here precludes the densities which are achievable in a controlled fermentation. (b) Acid production, notably acetate, of the cells has to be avoided since it poisons the cells and stops growth. Acid production occurs because the glycolysis and tricarboxylic acid cycle are running faster than the respiratory electron–transport chain. To minimize acid production, the culture has to be optimally aerated, and to minimize the effect of the acids it has to be maximally buffered. Since the equipment was designed to be as simple as possible, no pH titration as in high cell-density cultivation (HCDC) is used (see Section 5.3).

The benchtop-flask cultivation of *E. coli* allows exponential growth to a dry biomass of ~5 g/litre in a very simple experimental set-up (*Figure 5*). In addition to a standard bottle, only a waterbath, magnetic stirrer, and pressurized air are necessary. A pH controller is not needed, due to the high buffering capacity of the SB and the Glu–MM medium. The buffering capacity cannot be increased further because higher concentrations of the buffer components ($Na_2PO_4.2H_2O$ and KH_2PO_4) would inhibit growth considerably. Intensive aeration is necessary. The yield coefficient, Y, is the ratio of mass of cells (X) for a given mass of nutrient and can be calculated from growth and substrate consumption. For glucose, $Y_{X/glu}$ is approx. 0.45, for the sole nitrogen source, NH_4Cl, $Y_{X/N}$ is approx. 6 in Glu–MM.

The glucose–mineral salt medium can also be used for complete labelling of recombinant proteins with [^{13}C]glucose and/or [^{15}N]NH_4Cl, as required for NMR studies. Since the amounts of labelled substrates can be calculated from their known yield coefficients (71), an estimation of an almost complete consumption of the labelled substrates via their yield coefficients is possible. Thus, the costs for the labelled compounds can be kept to a minimum. For example, about 50 mg completely labelled recombinant protein can be obtained from one benchtop-flask cultivation, using 1 litre of medium with 5 g final dry biomass/litre, even if the recombinant protein comprises only 1% of the synthesized biomass.

Protocol 3 was successfully applied to *E. coli* RV308 (Section 4.3.1) and *E. coli* BL21 (DE3) containing various expression plasmids. The strain RV308 accumulates the metabolic by-product acetate only at concentrations which are not inhibitory for growth. Acetate production should not exceed 4 g/litre.

10: Producing antibodies in E. coli: *from PCR to fermentation*

Before using other strains, the kinetics of growth, the kinetics of glucose and NH₄Cl consumption, as well as the kinetics of acetate formation, should be determined from samples taken over time and analysed using the assay kits (Section 5.3.4).

Protocol 4 uses the complex SB medium similar to that employed in the shake-flask procedure (*Protocol 2*), and has been applied for the expression of different scFv fragments in *E. coli* JM83, but it should also be useful for other strains.

Protocol 3. Cultivation in 2-litre benchtop flasks, using glucose–mineral salt medium

Equipment and reagents

- The experimental set-up is illustrated in *Figure 5*.
  The benchtop flask is a *2 litre DURAN laboratory bottle with DIN thread* (cat. no. 2180163) fitted with a *screw-cap system with Drechsel head* (cat. no. 2570401); both from Schott Glaswerke.
  For safety reasons, the air pressure should be controlled by a *reducing valve* to ≤ 0.5 bar (Druckregler, 1.6 bar from Ludi & Co., AG).
  The inlet air is temperature-equilibrated by letting it pass through a *copper coil* placed in the waterbath.
  A standard *air-flow meter* with a maximum setting of 20 litres per minute is installed to guarantee a constant flow of air (WISAG).
  Air is sterilized with a *0.2 μm PTFE filter* (Sartorius, model MIDISART 2000).
  A *glass tube with a porous sparger* at the end (pore size 2) used for generating air bubbles,[a] is connected via *silicon tubing* to the inlet rod of the cap. (Porous sparger supplied by Schott Glaswerke, cat. no. 2585732; note that this appears in the catalogue as 'microfilter candle with narrow tube'.)
  An ordinary *stir bar magnet* (about 4 cm length by about 5 mm diameter) is driven by a *standard magnetic stirrer* (IKAMAG REO).
  The benchtop flask is placed in a *waterbath* and this is then placed on top of the magnetic stirrer with *appropriate support* (e.g. lab. jacks, wooden blocks, etc.).
- 500 ml Erlenmeyer flasks

- Preculture medium (500 ml): dissolve the following in 400 ml water, 17 g Na₂HPO₄.2H₂O, 6 g KH₂PO₄, 250 mg NaCl, 2.9 g NH₄Cl, 5 ml iron(III) citrate hydrate (stock at 6 g/L), 50 μl H₃BO₃ (stock at 3 g/100 ml), 50 μl MnCl₂.4H₂O (stock at 15 g/100 ml), 50 μl EDTA.2H₂O (stock at 8.4 g/100 ml), 50 μl CuCl₂.2H₂O (stock at 1.5 g/100 ml), 50 μl Na₂MoO₄.2H₂O (stock at 2.5 g/100 ml), 50 μl CoCl₂.6H₂O (stock at 2.5 g/100 ml), 1 ml Zn (CH₃COO)₂.2H₂O stock at 0.4 g/100 ml).
  Make up to 500 ml and mix, distribute 100 ml to each of five 500 ml Erlenmeyer flasks, and autoclave; pH should be 6.9.
  After autoclaving, to each 100 ml aliquot, add 2 ml 50% glucose (final concentration 10 g/L) and 0.25 ml 24% MgSO₄.7H₂O.
- Main culture medium (1 litre): dissolve the following in 800 ml water: 34 g Na₂HPO₄.2H₂O, 12 g KH₂PO₄, 500 mg NaCl, 5.8 g NH₄Cl, 10 ml iron (III) citrate hydrate (stock at 6 g/L), 100 μl H₃BO₃ (stock at 3 g/100 ml), 100 μl MnCl₂.4H₂O (stock at 15 g/100 ml), 100 μl EDTA·2H₂O (stock at 8.4 g/100 ml), 100 μl CuCl₂.2H₂O (stock at 1.5 g/100 ml), 100 μl Na₂MoO₄.2H₂O (stock at 2.5 g/100 ml), 100 μl CoCl₂.6H₂O (stock at 2.5 g/100 ml), 2 ml Zn (CH₃COO)₂·2H₂O stock at 0.4 g/100 ml). Make up to 1 litre and mix, transfer to 2-litre flask, and autoclave; pH should be 6.9.
  After autoclaving, add 40 ml 50% glucose (final concentration 20 g/L), 2.5 ml 24% MgSO₄·7H₂O, and 1 drop of antifoaming agent Ucolub N115 (from Fragol Industrieschmierstoff GmbH).
- *E. coli* RV308 cultures (grown on LB agar, see *Protocol 2*)

A. Precultures

1. Collect *E. coli* RV308 from Petri-dish (grown overnight on LB agar at 26 °C) with 5 ml preculture medium, suspend and vortex.

233

Protocol 3. *Continued*

2. Inoculate each 100 ml preculture aliquot with 0.8 ml of this cell suspension, incubate in a shaker (200 r.p.m. at 26°C). Five flasks will give enough cells for inoculating the main culture.

B. *Main culture*

1. *Inoculation*: collect exponentially growing precultures by centrifugation (e.g. 450 ml for 5 min at 5500 r.p.m., resuspend in main culture medium and transfer to 1 litre main culture medium in 'benchtop' flask equilibrated at 26°C (initial OD_{550} = 1). OD_{550} measurements are given for a Pharmacia Novaspec II.

2. *Cultivation conditions*: temperature[b] T = 26°C (waterbath); agitation 1300 r.p.m. (using magnetic stirring[c] aeration: use aeration rate of about 6 litres of air per min until OD_{550} reaches about 6, then increase to about 15 litres air per min until the end; induction: add 0.25 ml 1 M IPTG[d] at an OD = 2.

3. Harvest cells by centrifugation (8000 g, 10 min, 4°C). Continue with cell disruption and protein purification (see Section 6).

[a] It is important that only very small air bubbles are released from the porous part of the sparger in order to achieve good aeration, i.e. a high transfer rate of oxygen from the gas phase into the liquid phase.
[b] Use temperature-equilibrated air (e.g. passed through copper tubing in the waterbath).
[c] To avoid oxygen limitation, the culture should be grown only to OD_{550} ~ 14 which corresponds to about 5 g dry biomass/litre.
[d] In case of promoter systems using the *lac* operator.

Protocol 4. Cultivation in 2 litre benchtop-flasks, using SB medium

Equipment and reagents

- Benchtop-flask (*Figure 5, Protocol 3*).
- LB agar and SB medium (see *Protocol 2*)
- Antifoam agent, Ucolub N115 (see *Protocol 3*)

Method

1. Follow steps 1 and 2 from *Protocol 2*.

2. Inoculate 1.5 litres of SB medium, containing 0.5 ml antifoam reagent, with the preculture and incubate at 25°C.[a] Set aeration to 6 litres/min and stir with maximal speed.[b]

3. Induce expression at OD_{550} = 1.5–2.0, by adding 0.375 ml 1 M IPTG.[c] Continue incubation until culture reaches the stationary phase.[d]

10: Producing antibodies in E. coli: *from PCR to fermentation*

4. Harvest cells by centrifugation (8000 *g*, 10 min, 4°C). Continue with cell disruption and protein purification (see Section 6).

ᵃ See *Protocol 2*, note *ᵇ*.
ᵇ The speed, which still allows proper rotation of the stirbar, depends on the size of the stirbar, the type of magnetic stirrer, and the distance between the stirrer and stirbar.
ᶜ In case of promoter systems using the *lac* operator.
ᵈ Time to stationary phase depends on the strain and properties of the expressed antibody fragment. This will be within 6–8 hours after induction and the final OD_{550} will be about 15.

5.3 Cultivation of *E. coli* in a 10 litre fermenter to high cell-densities (~ 100 g/litre)

We will first describe some peculiarities of high cell-density cultivations (HCDCs). The main problems related to these cultures are (for reviews see refs 72–74):

(a) limitation of and/or inhibition by substrates

(b) limited O_2 supply

(c) formation of inhibitory metabolic by-products (mainly acetate)

(d) evolution of high amounts of CO_2

(e) generation of heat.

Generally, HCDCs consist of a batch phase and a subsequent fed-batch phase to avoid growth inhibition, by high initial amounts of substrates. In developing glucose–mineral salt media, one has to consider the maximum concentration of substrates at which they become inhibitory for growth:

- 50 g per litre for glucose
- 3 g per litre for ammonia
- 10 g per litre for phosphate
- 8.7 g per litre for magnesium
- 0.8 g per litre for molybdenum
- 44 mg per litre for boron
- 4.2 mg per litre for copper
- 68 mg per litre for manganese
- 0.5 mg per litre for cobalt
- 38 mg per litre for zinc
- 1.15 g per litre for iron(II)

Due to the high-yield coefficients of the substrates containing the above-mentioned elements, HCDC media can be designed which contain sufficient amounts for the whole fermentation of all nutrients except glucose, magnesium sulfate, and ammonia. These components must be added in the fed-batch phase. Magnesium sulfate and glucose can be added together in one feeding solution. The second feeding solution contains ammonia which has a dual function: first it serves as a nitrogen source; second it is used to adjust the pH. To avoid significant precipitation of $Mg(NH_4)PO_4$, the concentration of Mg (e.g. as $MgSO_4 \cdot 7H_2O$) should be low in the initial medium for the batch phase. Additional Mg can be added later on via the feeding solution together with glucose.

There are several ways to meet the increasing oxygen demand of growing *E. coli* cultures, such as increasing the speed of the stirrer, increasing the aeration rate, increasing the molar fraction of oxygen in the gas used for aeration via a gas-mixing station, decreasing the temperature, and decreasing the specific growth rate of the culture. Controls for growth at reduced specific growth rates are also appropriate for reducing the formation of metabolic by-products, for reducing the rate of heat production, and for lowering the rate of formation of CO_2, which are all detrimental for growth over their respective threshold levels.

The formation of growth-inhibitory metabolic by-products, especially acetate (above 5 g/litre), is a serious problem in wild-type strains. Acetate formation is suppressed in glucose–mineral salt media during aerobic growth at a specific growth rate, $\mu < 0.2\ h^{-1}$, which can be achieved by limiting the supply of glucose in the fed-batch phase. Some strains exist with reduced acetate production (see also below).

5.3.1 Strains for HCDC

We use the following host strains for HCDC:

(a) *E. coli* K12 TG1 (DSM 6056): ((*lac-pro*), *sup*E, *thi*E, *hsd*D5/F'*tra*D36, *pro*A⁺B⁺, *lac*I^q, *lac*ZΔM15). This strain produces significant amounts of acetate in glucose–mineral salt media and has to be forced to grow, via reduced glucose feeding, to high cell densities. In mineral salt media with feeding of glycerol as the carbon source, the acetate formation is much lower than with feeding of glucose.

(b) *E. coli* K12 SK1590: (*gal thi sbc*B15 *end*A *hsd*R4) (75). This thiamine auxotrophic strain behaves in glucose–mineral salt media like *E. coli* TG1.

(c) *E. coli* K12 RV308 (ATCC#31608): (*lac*74-*gal* ISII:OP308 *str*A). Only small (not growth inhibitory) amounts of acetate accumulate during limited or unlimited growth in glucose–mineral salt medium. The mutations responsible are not yet known.

5.3.2 Fermenter and accessories

For HCDCs, stirred tank reactors should contain the usual controls for pH, pressure, temperature, and antifoam dosage control, closed-loop controls for pO_2 with stirrer speed or gas-flow ratio (O_2, air) as controller output variables. It is a good idea to place the fermenter on a balance. In addition, an

10: Producing antibodies in E. coli: from PCR to fermentation

exponential feeding of substrate according to a time schedule must be possible for limiting the supply of a carbon source (e.g. glucose, see *Protocol 5*). An on-line measurement and control of the carbon source (e.g. glucose) must be available for unlimited growth to high cell-densities (*Protocol 6*).

Figure 6 shows the experimental set-up we use for HCDCs. The 10 litre fermenter BIOSTAT-ED10 (B. Braun Biotech International) serves as the stirred tank reactor. Accessories include a digital measurement and control unit (DCU) for process control and data monitoring, a multi-fermenter control system (MFCS) for process management, data monitoring, and data storage. The gas-flow ratio controller is used to supply air or air enriched with pure oxygen. The bioreactor exhaust gas stream is analysed with an Uras 10E analyser for carbon dioxide and a Magnos 6G analyser for oxygen (Hartmann & Braun). The fermenter sits on a balance. Feeding devices consist of substrate reservoirs (standard glass bottles), balances (Sartorius), and peristaltic pumps (Watson Marlow). The accessories also include equipment for the generation of exponential profiles for substrate feeding, which includes the dosage controller YFC 02Z and the balance I 8100P-**D from Sartorius and the peristaltic pump WM 5034U/55RPM from Watson Marlow.

The system for on-line monitoring and control of the glucose concentration is shown in *Figure 7* and described in more detail elsewhere (76). A home-made by-pass serves for sampling. The culture is moved at 100 ml per minute through the by-pass with a peristaltic pump (503S, Watson Marlow) and a glass T-piece. After various dilutions (total dilution 1:54), including addition of the metabolism inhibitor NaN_3 (0.01% final concentration) and degassing the modified sample, it is moved into the flow injection analyser (FIA). A commercial FIA (FIAstar 5020 Analyzer, TECATOR) is used for the on-line measurement of glucose. The glucose sensor, based on glucose oxidase (GOD) (Medingen) is built in a 37°C thermostat (FIAstar 5101, TECATOR). GOD converts glucose and oxygen into gluconic acid and hydrogen peroxide. Signals from the electrode are used for calculating the actual glucose concentration. It takes 53 seconds to transport the sample from the fermenter to the glucose sensor, and an additional 60 seconds are needed for measurement. Hence, the total cycle time for glucose analysis amounts to about 2 minutes. This period for data acquisition is sufficiently short, even at high cell-densities. The control and feeding device is also shown in *Figure 7*. Via a serially FIA-connected laptop an analogue signal is sent to the glucose controller in the digital control unit. This controller determines glucose feeding with a peristaltic pump (503U/55 r.p.m., Watson Marlow). Glucose is usually controlled at a level of 1.5 g/litre.

5.3.3 Fed-batch strategies for the cultivation of *E. coli* at various specific growth rates

There are many strategies for cultivating *E. coli* to high cell-densities. The vast majority of the HCDC-literature describes processes characterized by a

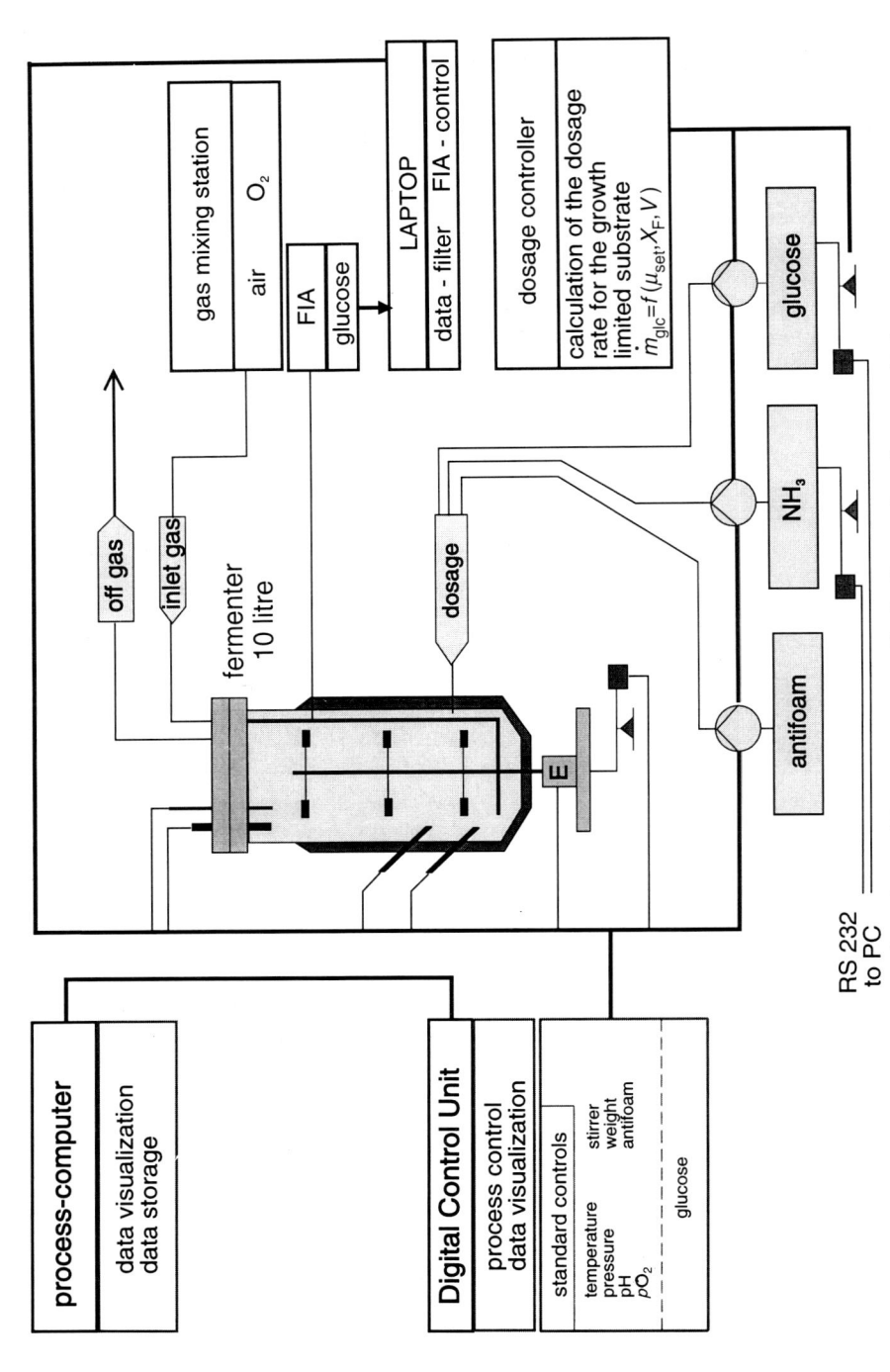

Figure 6. Experimental fermenter set-up. For further explanations see text.

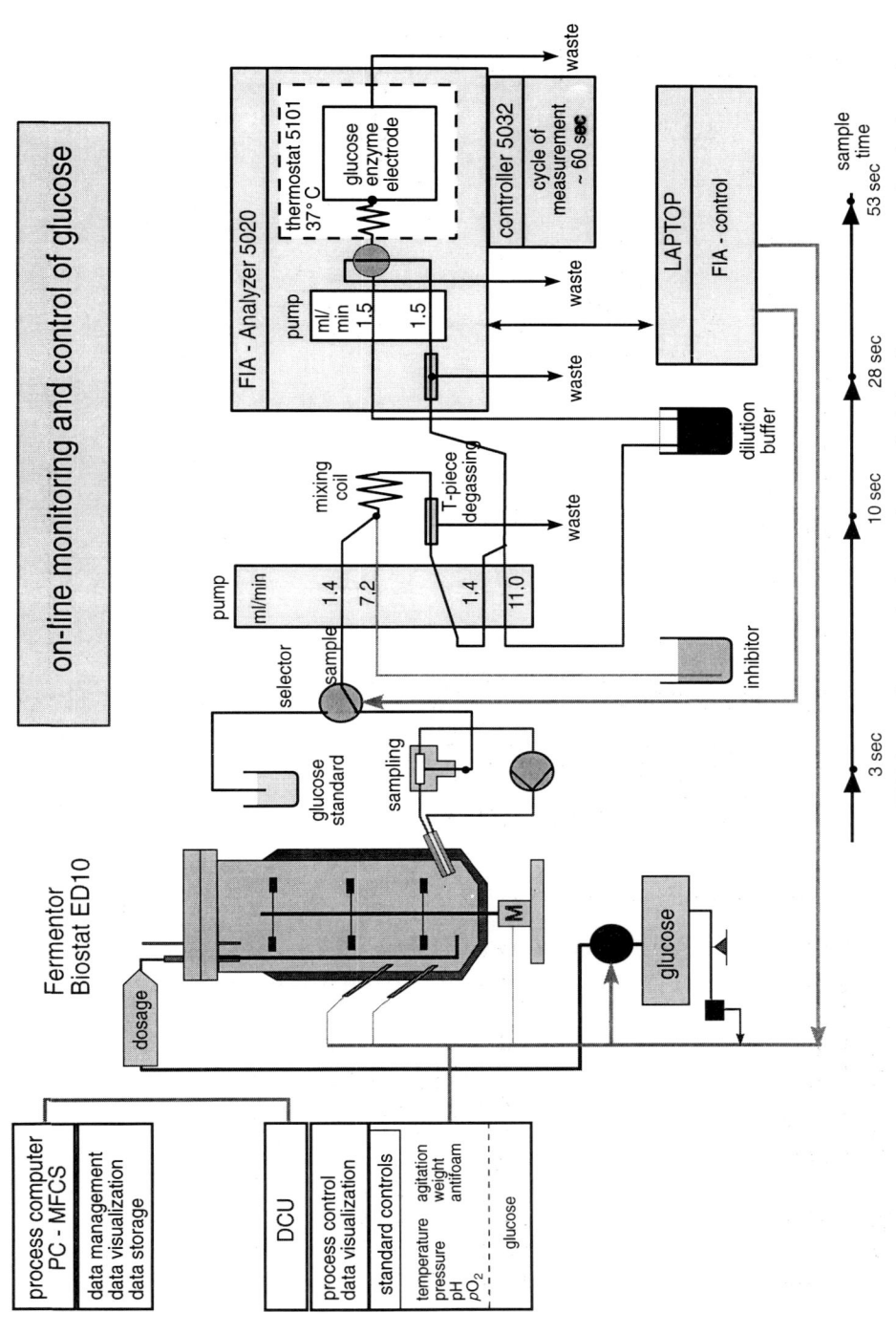

Figure 7. Fermenter coupled flow-injection analyser (FIA) for on-line monitoring and control of the glucose concentration in the culture. For further explanations see text.

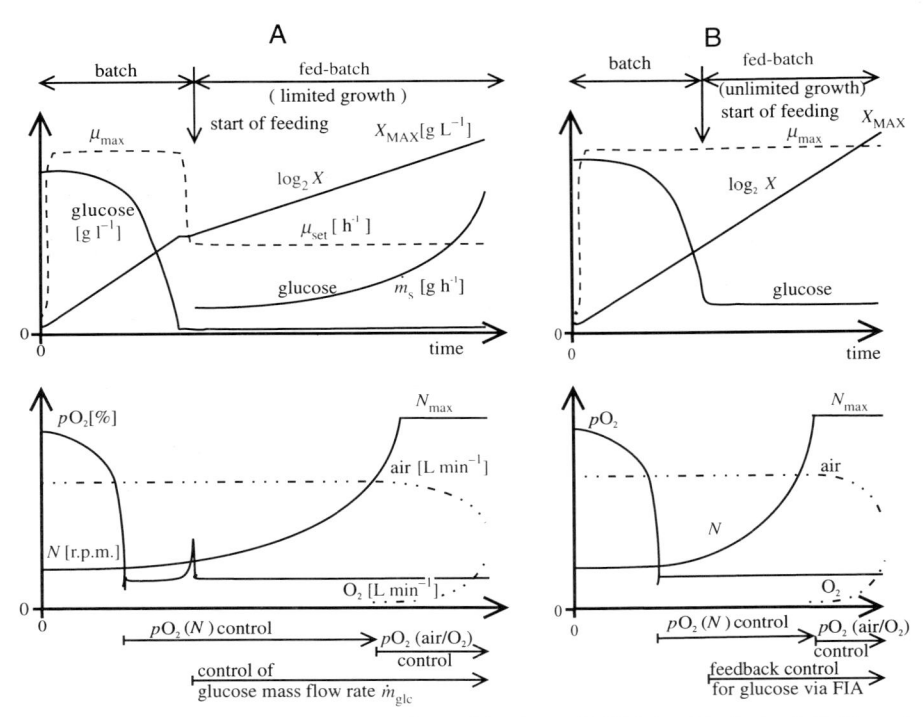

Figure 8. Principles of high cell-density cultivations (HCDCs) with limited (left) and unlimited (right) growth in the fed-batch phase. A, HCDC-type 1: glucose feeding such that a submaximal growth-rate results ($\mu_{\text{fed-batch}} < \mu_{\text{max}}$). B, HCDC-type 2: controlled supply of glucose resulting in non-limited growth ($\mu_{\text{fed-batch}} = \mu_{\text{max}}$).

limited supply of carbon sources (77–81). In these, the cells are forced to grow at reduced specific growth rates μ, with the advantage of only a slow and small accumulation of the metabolic by-product acetate. Few reports describe the use of strains or mutants with reduced acetate formation under aerobic conditions. But, in these cases, the feeding of substrates has only been manually controlled, until now (32, 82, 83). *Figure 8* illustrates the two main principles of HCDCs. The first type is characterized by limited growth in the fed-batch phase ($\mu_{\text{set}} < \mu_{\text{max}}$). The concentration of glucose in the culture is always nearly zero due to its limited supply and its immediate consumption (*Protocol 5*) (*Figure 8*). The second type is characterized by a controlled supply of glucose, so that the concentration of glucose is always above zero (e.g. grams per litre). Thus, this type of HCDCs allows unlimited growth during the fed-batch phase ($\mu_{\text{fed-batch}} = \mu_{\text{max}}$). Only strains with no, or drastically reduced, acetate formation (82–86) can follow type 2 to high cell-densities.

Protocol 5. HCDC—type 1[a]: feeding of glucose limiting growth ($\mu_{\text{fed-batch}} < \mu_{\text{max}}$)

Equipment and reagents

- See *Figure 6*, the glucose-FIA and the gas-mixing station are not necessary
- Preculture medium (1 litre): to 800 ml H$_2$O add 8.6 g Na$_2$HPO$_4$·2H$_2$O, 3 g KH$_2$PO$_4$, 500 mg NaCl, 1 g NH$_4$Cl, 10 ml iron(III) citrate hydrate (stock at 6 g/L), 100 μl H$_3$BO$_3$ (stock at 3 g/100 ml), 100 μl MnCl$_2$·4H$_2$O (stock at 15 g/100 ml), 100 μl EDTA·2H$_2$O (stock at 8.4 g/100 ml), 100 μl CuCl$_2$·2H$_2$O (stock at 1.5 g/100 ml), 100 μl Na$_2$MoO$_4$·2H$_2$O (stock at 2.5 g/100 ml), 100 μl CoCl$_2$·6H$_2$O (stock at 2.5 g/100 ml), 2 ml Zn(CH$_3$COO)$_2$·2H$_2$O (stock at 0.4 g/100 ml). Make up to 1 litre and mix, aliquot 100 ml into each of five 500 ml Erlenmeyer flasks, and autoclave; pH should be 6.9.

  After autoclaving, add the following sterilized solutions separately to each 100 ml aliquot: 2 ml 50% glucose (final concentration 10 g/L) and 0.25 ml 24% MgSO$_4$·7H$_2$O (final concentration 0.6 g/L).

- Main culture medium (for 10-litre fermenter, volume 8 litres): dissolve the following in 6 litres H$_2$O, 132.8 g KH$_2$PO$_4$, 32 g (NH$_4$)$_2$HPO$_4$, 17 g citric acid, 100 ml iron(III) citrate hydrate (stock at 6 g/L), 1 ml H$_3$BO$_3$ (stock at 3 g/100 ml), 1 ml MnCl$_2$·4H$_2$O (stock at 15 g/100 ml), 1 ml EDTA·2H$_2$O (stock at 8.4 g/100 ml), 1 ml CuCl$_2$·2H$_2$O (stock at 1.5 g/100 ml), 1 ml Na$_2$MoO$_4$·2H$_2$O (stock at 2.5 g/100 ml), 1 ml CoCl$_2$·6H$_2$O (stock at 2.5 g/100 ml), 20 ml Zn(CH$_3$COO)$_2$·2H$_2$O (stock at 0.4 g/100 ml). Make up to 7 litres (1 litre comes from the inoculum), mix, and autoclave; pH should be 6.9.

- After autoclaving, add the following sterilized solutions separately: 200 g glucose in 500 ml water (final concentration 25 g/L) and 12 g MgSO$_4$·7H$_2$O in 50 ml water (final concentration 1.5 g/L). Adjust pH to 6.8 with ~73 ml 25% (v/v) NH$_3$. Make up to 8 litres.
- 25% NH$_3$ (v/v) (for adjusting pH)
- Feeding solution[b] (glucose–magnesium sulfate solution): dissolve 750 g glucose in 600 ml water (gives ~1070 ml solution), and autoclave. Dissolve 22.2 g MgSO$_4$·7H$_2$O in 50 ml water, and autoclave. These two solutions are mixed and fed together.
- LB medium (*Protocol 2*, without agar)
- Culture grown on LB agar (*Protocol 2*)
- 50 ml, 500 ml, 1 L, and 10 L flasks

A. Precultures[c]

1. Use some colonies from a Petri-dish (grown overnight on LB agar at 26°C) to inoculate 10 ml liquid LB medium in a small flask; shake for 5 h (200 r.p.m., 26°C); transfer about 1 ml culture to 100 ml medium in 500 ml flasks for preculture 1 and incubate overnight as above; use about 10 ml for inoculating 100 ml of medium in 500 ml flasks for preculture 2 (9 flasks altogether) and incubate for several hours as above to obtain enough biomass to start the main culture (8 litres) at OD_{550} ~0.2.

B. Main culture in the BIOSTAT ED10™ fermenter[d]

1. **Conditions**:
   (a) Operate the fermenter for the production of miniantibodies at a temperature of 26°C,[e] a pressure of 0.15 MPa, a pH of 6.8,[f] and a gas-flow rate of 10 L min^{-1}.

Protocol 5. *Continued*

    (b) Suppress foaming by controlled supply of an antifoam reagent.[g]

    (c) Control the pO_2 to be $\geq 20\%$ of saturation.[h]

    (d) Set the initial agitation[i] to about 300 r.p.m.

2. **Batch phase**:

    (a) Inoculate the fermenter with an exponentially grown preculture.

    (b) Take samples for off-line analysis to follow growth and substrate consumption.

    (c) Estimate, in advance, the time for the end of the batch phase on the basis of the kinetics of OD_{550} and glucose, including the yield coefficient for glucose.

    (d) Carefully watch the sudden rise in pO_2 after the complete consumption of initial glucose.

    (e) Let the culture starve for some minutes to allow the metabolization of the excreted acetate.

3. **Fed-batch phase**:

    (a) Calculate the initial glucose mass flow rate (see below).

    (b) Start the glucose feeding.

    (c) Take samples as above.

    (d) Induce[j] product (e.g. miniantibody) formation at the desired biomass concentration.

    (e) Calculate the glucose mass flow rate ($\dot{m}_{glc}$) (see also nomenclature for symbols and explanations at the end of this chapter):

$$\dot{m}_{glc}(t) = Q_{glc,F} \cdot m_c(t) \cdot s^{-1} \cdot \exp[\mu_{set}(t - t_F)],$$

$$\text{with } Q_{glc,F} = [(\mu_{set} \cdot Y_{X/glc}^{-1}) + m_E] \cdot X_F.$$

$Q_{glc,F}$ is the initial glucose mass flow rate at the start of the feeding. Calculate the biomass at the start of feeding from the measured optical density (OD_{550}) of the culture knowing the conversion coefficient between OD_{550} and X. Since $\mu_{set} = 0.2\ h^{-1}$, $Y_{X/glc} \sim 0.45$, $m_E \sim 0.025\ g\ g^{-1}\ h^{-1}$ are fixed values, $Q_{glc,F}$ can also be calculated. Measure the weight of the culture continuously. The actual glucose mass flow rate $\dot{m}_{glc}(t)$ is continuously calculated by the process computer and transferred to the glucose feeding pump for glucose feeding.

---

[a] The process strategy for HCDC based on time-dependent exponential substrate feeding according to refs. 79–81, 87.

[b] The composition and preparation of media and feeding solution are essentially those described by Riesenberg *et al.* (77, 78).

[c] An attractive alternative is the use of glycerol stocks with high cell titre (about 10 g dry biomass per litre) for direct inoculation of the medium for the main culture. Riesenberg et al. (88) have described the preparation, long-term storage, and application of the special E. coli glycerol stocks. The inoculation ratio is 1:300, i.e. about 35 ml stock are sufficient for 10 litres to begin the fermentation with a reasonable amount of biomass. In glucose–mineral salt media the consumption of glycerol is suppressed until glucose, has been completely consumed.

[d] If only a fermenter which does not sit on a balance is available, quasi-exponential glucose feeding can be realized using the dosage controller (see *Figure 6*) alone. In that case, one makes the simplification m_c = const. (see above).

[e] This temperature is optimal for formation of miniantibodies during HCDC according to Pack et al. (32).

[f] Aqueous ammonia (25% or less concentrated) serves for adjustment of the pH and as the nitrogen source for growth.

[g] Ucolub N115 is an appropriate antifoam agent during HCDC of E. coli RV308 producing miniantibodies and other recombinant proteins. It can be purchased from Fragol Industrieschmierstoff GmbH.

[h] Two closed-loop controls for pO_2 operate throughout the whole fermentation: first pO_2 (agitation) and second pO_2 (gas-flow ratio). The gas is air or air-enriched with pure oxygen.

[i] The initial speed of the stirrer should be low. This guarantees that the pO_2 (agitation)-control starts in the batch-phase after the pO_2 reaches the set-value of 20% saturation. This enables a sudden rise in the pO_2 to high saturation values after the complete consumption of the initially added glucose. Thus, the end of the batch phase is clearly indicated.

[j] We used a *lac p/o*-expression-system to induce antibody formation (32). The addition of 1 mM IPTG (final concentration) at cell densities up to 70 g dry biomass per litre is sufficient for induction.

Protocol 6. HCDC[a]—type 2: feeding of glucose during non-limiting growth ($\mu_{\text{fed-batch}} = \mu_{\text{max}}$)

Equipment and reagents
- See *Figures 6* and *7*. The flow injection analysis (FIA) for on-line measurement and control of glucose is necessary. The fermenter balance is not needed.
- Preparation of media and feeding solution: see *Protocol 5*

Method

1. Precultures: see *Protocol 5*.
2. Carry out the main culture in the BIOSTAT-ED10™ fermenter.
3. For initial growth conditions see *Protocol 5*.
4. Inoculation and sampling of the batch phase are the same as in *Protocol 5*. Follow the decrease of glucose concentration in the culture, start the glucose-FIA after the glucose concentration has fallen below 8 g/L, control the glucose concentration at 1.5 g/L.[a]
5. For the fed-batch phase, let the glucose-FIA operate until the end of the fermentation, induce product (e.g. miniantibody) formation at the desired biomass concentration according to *Protocol 5*.

Protocol 6. *Continued*

[a] HCDC-type 2 is distinguished from HCDC-type 1 with respect to the pO_2 kinetics during the transition from the batch phase to the fed-batch phase. Since the glucose-FIA is started before the initial glucose is completely consumed, no intermediate increase in pO_2 occurs (see Figures 8A and B). Nevertheless, the controls for maintaining pO_2 at 20% saturation are the same for both HCDC-types. Due to the higher specific growth rate of the cells growing in the fed-batch phase of HCDC-type 2, the duration of HCDC-type 2 is considerably shorter than HCDC-type 1. It must be stressed, however, that only strains with significant reduced acetate accumulation (e.g. *E. coli* RV308) can be cultivated at μ_{max} until high biomass concentrations (~100 g L^{a-1}).

5.3.4 Off-line analysis

Glucose is analysed using the glucose analyser ESAT 6660 (Medingen) or the glucose test kit No. 716251 from Boehringer Mannheim. Ammonia nitrogen is determined in the usual way using the Kjeldahl method or the ammonia test kit No. 1112732 from Boehringer Mannheim. Acetate is determined using the test kit No. 148261 from Boehringer Mannheim. Other tests are available (89). Determination of cell density and biomass: cell density is measured with Novaspec II (Pharmacia) as optical density (OD) in a 1 cm lightpath cuvette at 550 nm either directly or after dilution of the culture with 0.9% NaCl solution. Cell dry weight X (g L^{-1}) is calculated on the basis of a calibration curve (e.g. for *E. coli* RV308 according to $X = 0.36\ OD_{550}$).

6. Antibody purification

Almost all biochemical or biomedical work will have to be carried out with purified antibody fragments, be it for an evaluation of binding properties or the measurement of biological effects. However, the degree of purity required can vary depending on the experiment. We describe here a general and extremly fast technique, which should permit the purification of antibody fragments to high purity. It is particularly important that this technology is scaleable over a very wide range.

6.1 General considerations

Since antibody fragments vary widely in surface composition and isoelectric points, there can be no generic purification scheme based on classical approaches, such as ion-exchange chromatography. While recombinant antibodies *can*, of course, be purified with these techniques, a new procedure would have to be worked out for every antibody. This will only be worthwhile if very large amounts of the same fragment are required at very high purity—e.g. for a clinical study.

The framework of the variable domains is itself not constant enough to act as a generic affinity purification scheme (90). Often antibodies are prepared as fusions for purification. These fusion can be very small peptide tags at either

10: Producing antibodies in E. coli: *from PCR to fermentation*

end of the domain or fusions to other protein domains, including of course the constant domains.

Most fusions in use today, notably peptides, are themselves recognized by another protein which needs to be immobilized (91). However, affinity columns with a protein ligand are fairly impractical on large scales: they are expensive and sometimes difficult to regenerate if used on raw extracts. Furthermore, the conditions which disrupt the intermolecular forces are the same conditions which destabilize the intramolecular forces within the antibody and immobilized ligand, and either protein may denature after applying the elution conditions, unless a competing ligand is available (reviewed in ref. 91).

6.2 Immobilized metal ion-affinity chromatography (IMAC)

We have previously described (54) the methodology of oligo-histidine tails with IMAC (92) in purifying antibody fragments in a one-step procedure in the native state. Upon frequent use of this method with a variety of antibodies we have discovered, however, that not every fragment can be purified to the same degree of purity in a single step. Our working assumption is that difficulties may be related to the formation of small, soluble aggregates of the antibody, leaving only few his-tails per monomer available. Furthermore, association with foreign molecules may explain the presence of more contaminants with some recombinant antibodies than others. Finally, low expression rates may simply introduce more *E. coli* background, when normalized to the same amount of recombinant protein.

Thus, it is sometimes necessary to include a second purification step after IMAC. Most convenient is ion-exchange chromatography. However, in traditional chromatography, this would be rather laborious: the eluant from standard condition IMAC, high in salt and imidazole, would first have to be dialysed before ion-exchange chromatography. Therefore, we have now developed a method with two columns directly in-line, which can perform both steps together in about 15 min (*Figure 9, Protocol 7*) (C. Krebber, L. Nieba, and A. Plückthun, unpublished).

IMAC relies on the metal-chelating ability of juxtaposed histidine residues (92). In order for this to be selective over any ion-exchange effects with the immobilized metal, high salt has to be included. In the procedure detailed below, the sample is loaded on to the IMAC column at pH 7.0 or 8.0 in high salt. Next, the buffer is changed to low salt and a gradient of imidazole is started. Imidazole, being a competitive ligand for the metal, elutes the protein under very mild conditions. The relevant fractions are then directly pumped on to the second column, which is usually an ion-exchange column, although other columns can also be used.

Depending on the ion-exchange column used (see below), the protein will bind because of the low salt. Ion-exchange chromatography is barely influenced by imidazole, if its pH is adjusted with acetic acid. Then, a salt gradient can be started to purify the antibody fragment directly. The whole procedure with both columns will take as little as 15 min.

The ion-exchange column can be chosen with the following criteria in mind. Most *E. coli* proteins have isolectric points around 4–6 (93), and are thus negatively charged at pH 7.0. If the antibody fragment, whose isolectric point is almost invariably known from its sequence at this stage, has a predicted positive net charge at pH 7.0, it will often be in the run-through of an anion exchanger, and it would be bound almost by itself by a cation exchanger. Since very acidic isoelectric points of antibodies are rare, most fragments are easy to purify away from the remaining *E. coli* proteins at this stage.

Annoying contaminants are co-purified proteases of *E. coli* (under native and denaturing conditions (54)), which may degrade some fragments upon extended storage. In our hands, all low molecular weight protease inhibitors investigated so far have been disappointing, and thus the most important precaution against proteases is great care during the purification steps, including rigorous regeneration of the columns.

Two commercial resins for IMAC are available from different manufacturers: iminodiacetic acid (IDA) and nitrilo-triacetic acid (NTA). Both have been used successfully for antibody purification (54). For perfusion chromatography, special materials are necessary (94). Only IDA is currently available for IMAC perfusion chromatography and therefore the 2-step method has been worked out on this material with immobilized nickel. Nevertheless zinc (sometimes found to be a better ligand for IDA (54)) might work as well as nickel does. In standard chromatography we have found that in general Ni–NTA is the favoured material for scFv purification.

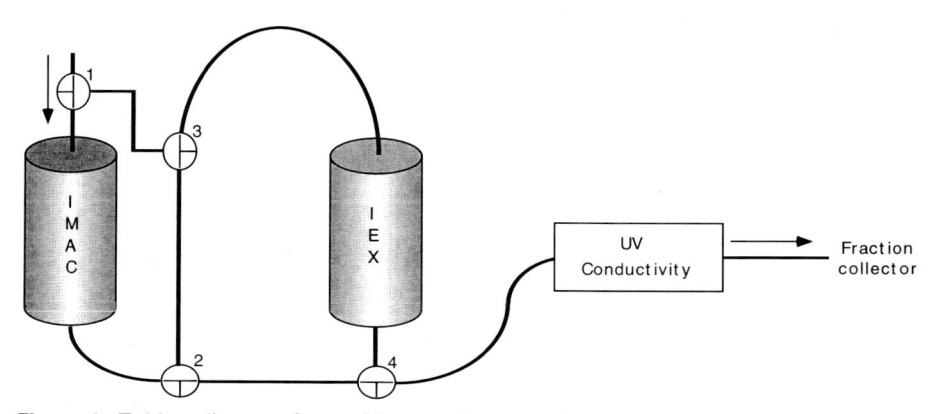

Figure 9. Tubing diagram for rapid two-column purification of antibody fragments. At the beginning of the chromatography, all flow is through IMAC (valve positions as shown). Upon antibody elution, the flow is redirected to the ion-exchange column (IEX), by turning valves 2 and 4. The adsorbed protein is then eluted with a new gradient, by turning valves 1 and 3.

10: Producing antibodies in E. coli: *from PCR to fermentation*

While the following procedure is described for the BioCAD system (Perseptive Inc.), it should be straightforward to adapt to any computerized chromatography system, or even manual chromatography, if the tubing diagram in *Figure 9* is followed.

Protocol 7. Purification of a single-chain Fv fragment with histidine tail by rapid two-column chromatography

This protocol is given for 5–10 g wet weight of *E. coli* cells (2 L of shake-flask culture).

Equipment and reagents

- Appropriate expression system to produce his-tagged antibody fragments
- Extraction buffer: 20 mM Hepes, 0.5 M NaCl adjusted to pH 7 with NaOH
- DNaseI (Boehringer Mannheim) 1 mg/ml in 50 mM Tris–Cl, 10 mM $MgCl_2$, pH 7.5
- Automated LC-System: BioCAD60 workstation with dual channel variable wavelength UV/visible detector, semipreparative flow cell (PerSeptive Biosystems), fraction collector Advantec SF-2120 (Toyo Roshi International)
- Columns: POROS20 MC/M 4.6 mm/100 mm (metal chelate)
  POROS20 HQ/M 4.6 mm/100 mm (anion exchange)
  POROS20 HS/M 4.6 mm/100 mm (cation exchange)
  (all PerSeptive Biosystems)
- Cell disruptor: French Press (Aminco)
- Centrifuge (Sorvall SS-34)
- Metal ion stock solution: 0.5 M $NiCl_2$
- Stock buffer system used on the BioCAD workstation:
  —100 mM MHA adjusted to pH 4.5 with HCl (33 mM Mes, 33 mM Hepes, 33 mM sodium acetate)
  —100 mM MHA adjusted to pH 7.5 with NaOH (33 mM Mes, 33 mM Hepes, 33 mM sodium acetate)
  —100 mM imidazole adjusted to pH 7 with acetic acid
  —3 M NaCl stock solution
  —distilled water
- SDS–PAGE equipment and reagents

Note: This method can be carried out with columns and chromatographic equipment from other manufacturers as long as the general principles laid out in the text and *Figure 9* are followed.

Method

1. Resuspend the cell pellet in 15–20 ml Hepes extraction buffer.
2. Disrupt the cells in a French Press.
3. Add DNaseI to a final concentration of 10 μg/ml.
4. Centrifuge the suspension (Sorvall SS-34, 48 200 g, 4°C, 30 min) and carefully collect the supernatant.
5. Filter the solution (0.22 μm, use filter with low protein binding properties).
6. Load the filtrate on to an Ni–IDA Poros column (1.66 ml) (pre-loaded with 3 ml 0.5 M $NiCl_2$) pre-equilibrated with 20 mM MHA-buffer (6.6 mM Mes, 6.6 mM Hepes, 6.6 mM Na-acetate), 0.5 M NaCl, pH 7.0. The flow rate should be 6 column volumes (CV) per minute.

Protocol 7. *Continued*

7. After loading the sample, wash the column with 20 mM MHA-buffer (pH 7.0) containing 0.5 M NaCl until the baseline is reached.
8. A washing step with 5 mM imidazole, 0.5 M NaCl, pH 7.0 is then applied for 5 CV, followed by a 5 mM imidazole washing step **without** salt.[a]
9. Perform the elution either using an imidazole gradient from 5 to 80 mM imidazole (pH 7.0) (**no salt**) (10 CV) or a step elution with 80 mM imidazole (pH 7.0) (**no salt**) (6 CV).[a]
10. Use the BioCAD workstation to allow loading of the IDA-elution directly on to a second column, without collecting the samples.[b,c,d]
11. This column can be either an anion-exchange or a cation-exchange column.[b] After automatically loading the imidazole elution on to the second column, wash the column with 20 mM MHA-buffer, 30 mM NaCl, pH 7.0 until the baseline is reached (5 CV).
12. Use salt gradient from 30 to 750 mM NaCl (15 CV) for the cation-exchange chromatography. Collect the samples in 2 ml fractions and analyse each fraction by SDS-PAGE.

Note: Purification with the anion-exchange chromatography often does not need a salt gradient, because the scFv fragment may be in the flow-through.[b]

[a] The pH for the second washing step and the elution depends on the pI of the antibody fragment and on the type of the second column (i.e. if the antibody has a pI of 8.5 the pH should be adjusted to 7.0, a cation-exchange column will be the best choice).
[b] Most of the *E. coli* host proteins co-purified in IMAC have a pI less then 6.5, therefore they will bind to an anion-exchange column. A salt gradient for separation usually works very well.
[c] Imidazole as a storage buffer and as a sample component in SDS-PAGE is not desirable, because it will catalyse the hydrolysis of acid labile bonds (*The QIA-expressionist* (1992), 2nd edn, p. 43 (from Quiagen)). Therefore, the 2-step method might even be useful for those antibody fragments which are already pure enough after the IMAC step.
[d] This coupled method can also be carried out on standard chromatography system or on an FPLC set-up, using the general tubing diagram in *Figure 9*.

Acknowledgements

We thank Kristian Müller for designing prototypes of the benchtop cultivation flasks and all the group members for helpful discussions and constructive criticism throughout the method development. We are grateful to Sylke Fricke and Silke Steinbach for excellent technical assistance.

Appendix: Abbreviations for HCDC

d	doubling time of biomass, h
d_{set}	desired doubling time, h
FIA	flow-injection analysis
HCDC	high cell-density cultivation
m_c	weight of the culture, g
m_E	maintenance coefficient, for *E. coli*
$\dot{m}_{glc}$	substrate (glucose) mass flow rate, g h^{-1}
μ	specific growth rate ($\mu = (\ln 2) \cdot d^{-1}$), h^{-1}
μ_{max}	maximum specific growth, h^{-1}
μ_{set}	desired specific growth rate, h^{-1}
N	stirrer speed, r.p.m.
N_{max}	maximum stirrer speed
pO_2	dissolved oxygen concentration, % of saturation
$Q_{glc,F}$	volumetric substrate consumption rate at start of feeding, g L^{-1} h^{-1}
	$Q_{glc,F} = [\mu_{set} \cdot (Y_{X/glc}^{-1}) + m_E] \cdot X_F$
t	cultivation time, h
t_F	starting time of fed-batch mode, h
V_L	culture volume, $V_L(t) = m_c(t) \cdot s^{-1}$, L (s = density of culture, g litre^{-1})
X	cell dry-mass concentration (biomass), g L^{-1}
X_F	biomass at start of feeding, g L^{-1}
X_{max}	maximum biomass, g L^{-1}
$Y_{X/glc}$	yield coefficient for glucose (formed biomass per consumed glucose)

References

1. Plückthun, A. (1994). *Handbook of experimental pharmacology*, vol 3: *The pharmacology of monoclonal antibodies* (ed. M. Rosenberg and G. P. Moore), pp. 269–315. Springer-Verlag, Berlin.
2. Plückthun, A. (1994). In *Immunochemistry* (ed. C. J. van Oss and M. H. V. van Regenmortel), pp. 201–36. Marcel Dekker, New York.
3. Winter, G., Griffiths, A. D., Hawkins, R. E., and Hoogenboom, H. R. (1994). *Annu. Rev. Immunol.*, **12**, 433.
4. Kütemeier, G., Harloff, C., and Mocikat, R. (1992). *Hybridoma*, **11**, 23.
5. Adair, J. R. (1992). *Immunol. Rev.*, **130**, 5.
6. Prodromou, C. and Pearl, L. H. (1992). *Protein Eng.*, **5**, 827.
7. Desplancq, D., King, D. J., Lawson, A. D. G. and Mountain, A. (1994). *Protein Eng.*, **7**, 1027.
8. Whitlow, M., Filpula, D., Rollence, M. L., Feng, S. L., and Wood, J. F. (1994). *Protein Eng.*, **7**, 1017.
9. Griffiths, A. D., Malmquist, M., Marks, J. D., Bye, J. D., Embleton, M. J., McCafferty, J., Baier, M., Holliger, K. P., Gorick, B.D., Hughes-Jones, N. C., Hoogenboom, H. R., and Winter, G. (1993). *EMBO J.*, **12**, 725.

10. Holliger, K. P., Prospero, T., and Winter, G. (1993). *Proc. Natl Acad. Sci. USA*, **90**, 6444.
11. Zhou, H., Fisher, R. J., and Papas, T. S. (1993). *Nucl. Acids Res.*, **22**, 888.
12. Ørum, H., Andersen, P. S., Øster, A., Johansen, L. K., Riise, E., Bjønvad, M., Svendsen, I., and Engberg, J. (1993). *Nucl. Acids Res.*, **21**, 4491.
13. Kettleborough, C. A., Saldanha, J., Ansell, K. H., and Bendig, M. M. (1993). *Eur. J. Immunol.*, **23**, 206.
14. Knappik, A. and Plückthun, A. (1994). *Biotechniques*, **17**, 754.
15. Wentzell, L. M., Nobbs, T. J., and Halford, S. E. (1995). *J. Mol. Biol.*, **248**, 581.
16. Ge, L., Knappik, A., Pack, P., Freund, C., and Plückthun, A. (1995). In *Antibody engineering* (2nd edn) (ed. C. A. K. Borrebaeck), pp. 229–66. Oxford University Press.
17. M. J. Gait (ed.) (1984). *Oligonucleotide synthesis: a practical approach*. IRL Press, Oxford.
18. Nezlin, R. (1994). In *Immunochemistry* (ed. C. J. van Oss and M. H. V. van Regenmortel), pp. 3–45. Marcel Dekker, New York.
19. Huston, J. S., McCartney, J., Tai, M. S., Mottola, H. C., Jin, D., Warren, F., Keck, P., and Oppermann, H. (1993). *Int. Rev. Immunol.*, **10**, 195.
20. Raag, R. and Whitlow, M. (1995). *FASEB J.*, **9**, 73.
21. Jefferis, R. (1993). *Glycoconj. J.*, **10**, 358.
22. Lo, K. M., Roy, A., Foley, S. F., Coll, J. T., and Gillies, S. D. (1992). *Hum. Antibodies Hybridomas*, **3**, 123.
23. Bird, R. E., Hardman, K. D., Jacobson, J. W., Johnson, S., Kaufman, B. M., Lee, S., Lee, T., Pope, S. H., Riordan, G. S., and Whitlow, M. (1988). *Science*, **242**, 423.
24. Huston, J. S., Levinson, D., Mudgett-Hunter, M., Tai, M., Novotny, J., Margolies, M. N., Ridge, R. J., Bruccoleri, R. E., Haber, E., Crea, R., and Oppermann, H. (1988). *Proc. Natl Acad. Sci. USA*, **85**, 5879.
25. Glockshuber, R., Malia, M., Pfitzinger, I., and Plückthun, A. (1990). *Biochemistry*, **29**, 1362.
26. Reiter, Y., Brinkmann, U., Webber, K. O., Jung, S. H., Lee, B., and Pastan, I. (1994). *Protein Eng.*, **7**, 697.
27. Plückthun, A. (1993). In *Stability and stabilization of enzymes* (ed. W. J. J. van den Tweel, A. Harder, and R. M. Buitelaar), pp. 81–90. Elsevier Science Publishers, Amsterdam.
28. Knappik, A. and Plückthun, A. (1995). *Protein Eng.*, **8**, 81.
29. Huston, J. S., George, A. J. T., Tai, M. S., McCartney, J. E., Jin, D., Segal, D. M., Keck, P., and Oppermann, H. (1995). In *Antibody engineering* (ed. C. A. K. Borrebaeck) (2nd edn), pp. 185–227. Oxford University Press.
30. Huston, J. S., Mudgett-Hunter, M., Tai, M. S., McCartney, J., Warren, F., Haber, E., and Oppermann, H. (1991). In *Methods in enzymology* (ed. J. J. Langone, Vol. 203, pp. 46. Academic Press, San Diego.
31. Pack, P. and Plückthun, A. (1992). *Biochemistry*, **31**, 1579.
32. Pack, P., Kujau, M., Schroeckh, V., Knüpfer, U., Wenderoth, R., Riesenberg, D., and Plückthun, A. (1993). *Biotechnology*, **11**, 1271.
33. Pack, P., Müller, K., Zahn, R., and Plückthun, A. (1995). *J. Mol. Biol.*, **246**, 28.
34. Glockshuber, R., Schmidt, T., and Plückthun, A. (1992). *Biochemistry*, **31**, 1270.
35. Better, M., Chang, C. P., Robinson, R. R., and Horwitz, A. H. (1988). *Science*, **240**, 1041.

10: Producing antibodies in E. coli: from PCR to fermentation

36. Skerra, A. and Plückthun, A. (1988). *Science*, **240**, 1038.
37. Bardwell, J. (1994). *Mol. Microbiol.*, **14**, 199.
38. Colcher, D., Bird, R., Roselli, M., Hardman, K. D., Johnson, S., Pope, S., Dodd, S. W., Pantoliano, M. W., Milenic, D. E., and Schlom, J. (1990). *J. Natl Cancer Inst.*, **82**, 1191.
39. Gibbs, R. A., Posner, B. A., Filpula, D. R., Dodd, S. W., Finkelman, M. A. J., Lee, T. K., Wroble, M., Whitlow, M., and Benkovic, S. J. (1991). *Proc. Natl Acad. Sci. USA*, **88**, 4001.
40. Whitlow, M. and Filpula, D. (1991). *Methods: a companion to Methods in enzymology* **2**, 97.
41. Proba, K., Ge, L., and Plückthun, A. (1995). *Gene*, **159**, 203.
42. Derman, A. I., Prinz, W. A., Belin, D., and Beckwith, J. (1993). *Science*, **262**, 1744.
43. Freund, C., Ross, A., Plückthun, A., and Holak, T. A. (1994). *Biochemistry*, **33**, 3296.
44. Buchner, J. and Rudolph, R. (1991). *Biotechnology*, **9**, 157.
45. Rudolph, R. (1990). In *Modern methods in protein and nucleic acid research* (ed. H. Tschesche, H.), pp. 149–71. Walter de Gruyter, Berlin.
46. Pugsley, A. P. (1993). *Microbiol. Rev.*, **57**, 50.
47. Wülfing, C. and Plückthun, A. (1994). *Mol. Microbiol.*, **12**, 685.
48. Meerman, H. J. and Georgiou, G. (1994). *Biotechnology*, **12**, 1107.
49. Wülfing, C. and Plückthun, A. (1994). *J. Mol. Biol.*, **242**, 655.
50. Wülfing, C. and Plückthun, A. (1993). *Gene*, **136**, 199.
51. Nishihara, T., Iwabuchi, T., and Nohno, T. (1994). *Gene*, **145**, 145.
52. Reznikoff, W. S. (1992). *Mol. Microbiol.*, **6**, 2419.
53. Knaus, R. and Bujard, H. (1990). In *Nucleic acids and molecular biology* (ed. F. Eckstein and D. M. J. Lilley), Vol. 4, pp. 110. Springer Verlag, Berlin.
54. Lindner, P., Guth, B., Wülfing, C., Krebber, C., Steipe, B., Müller, F., and Plückthun, A. (1992). *Methods: a companion to Methods in enzymology* **4**, 41.
55. Munro, S. and Pelham, H. R. B. (1986). *Cell*, **46**, 291.
56. Thisted, T., Nielsen, A. K., and Gerdes, K. (1994). *EMBO J.*, **13**, 1950.
57. Dotto, G. P., Horiuchi, K., and Zinder, N. D. (1984). *Adv. Exp. Med. Biol.*, **179**, 185.
58. Lin-Chao, S., Chen, W. T., and Wong, T. T. (1992). *Mol. Microbiol.*, **6**, 3385.
59. Holmgren, A. (1989). *J. Biol. Chem.*, **264**, 13963.
60. Studier, F. W., Rosenberg, A. H., Dunn, J. J., and Dubendorff, J. W. (1990). In *Methods in Enzymology* (ed. D. V. Goeddel), Vol. 185, pp. 60. Academic Press, San Diego.
61. Carter, P., Kelley, R. F., Rodrigues, M. L., Snedecor, B., Covarrubias, M., Velligan, M. D., Wong, W. L. T., Rowland, A. M., Kotts, C. E., Carver, M. E., Yang, M., Bourell, J. H., Shepard, H. M., and Henner, D. (1992). *Biotechnology*, **10**, 163.
62. Ayala, M., Balint, R. F., Fernandezdecossio, M. E., Canaanhaden, L., Larrick, J. W., and Gavilondo, J. V. (1995). *Biotechniques*, **18**, 832.
63. Tormo, A., Almiron, M., and Kolter, R. (1990). *J. Bacteriol.*, **172**, 4339.
64. Liu, J. and Walsh, C. T. (1990). *Proc. Natl Acad. Sci. USA*, **87**, 4028.
65. Hockney, R. C. (1994). *Trends Biotechnol.*, **12**, 456.
66. Knappik, A., Krebber, C., and Plückthun, A. (1993). *Biotechnology*, **11**, 77.
67. Duenas, M., Vazquez, J., Ayala, M., Soderlind, E., Ohlin, M., Perez, L., Borrebaeck, C. A., and Gavilondo, J. V. (1994). *Biotechniques*, **16**, 476.
68. Bowden, G. A. and Georgiou, G. (1988). *Biotech. Prog.*, **4**, 97.
69. Sawyer, J. R., Schlom, J., and Kashmiri, S. V. S. (1994). *Prot. Eng.*, **7**, 1401.

70. Blackwell, J. R. and Horgan, R. (1991). *FEBS Lett.*, **295**, 10.
71. Bailey, J. E. and Ollis, D. F. (1986). *Biochemical engineering fundamentals.* McGraw-Hill Book Company, Singapore.
72. Riesenberg, D. (1991). *Curr. Opin. Biotechnol.*, **2**, 380.
73. Kleman, G. L. and Strohl, W. R. (1992). *Curr. Opin. Biotechnol.*, **3**, 93.
74. Kleman, G. L. and Strohl, W. R. (1994). *Curr. Opin. Biotechnol.*, **5**, 180.
75. Brown, T. A. (1991). In *Molecular biology LABFAX series* (ed. B. D. Hames and D. Rickwood), p. 6. Bios Scientific Publishers. Oxford.
76. Pfaff, M., Wagner, E., Wenderoth, R., Knüpfer, U., Guthke, R., and Riesenberg, D. (1995). In *Proceedings 6th International Conference on Computer Application in Biotechnology—CAB6* (ed. A. Munack and K. Schügerl), p. 6. IFAC Publications Elsevier Science Ltd., Oxford.
77. Riesenberg, D., Menzel, K., Schulz, V., Schumann, K., Veith, G., Zuber, G., and Knorre, W. A. (1990). *Appl. Microbiol. Biotechnol.*, **34**, 77.
78. Riesenberg, D., Schulz, V., Knorre, W. A., Pohl, H.-D., Korz, D., Sanders, E. A., Roβ, A., and Deckwer, W.-D. (1991). *J. Biotechnol.*, **20**, 17.
79. Hellmuth, K., Korz, D. J., Sanders, E. A., and Deckwer, W.-D. (1994). *J. Biotechnol.*, **32**, 289.
80. Strandberg, L., Andersson, L., and Enfors, S.-O. (1994). *FEMS Microbiol. Rev.*, **14**, 53.
81. Yee, L. and Blanch, H. W. (1993). *Biotechnol. Bioeng.*, **41**, 781.
82. Hahm, D. H., Pan, J., and Rhee, J. S. (1994). *Appl. Microbiol. Biotechnol.*, **42**, 100.
83. Kleman, G. L. and Strohl, W. R. (1994). *Appl. Environ. Microbiol.*, **60**, 3952.
84. Dedhia, N. N., Hottiger, T., and Bailey, J. E. (1994). *Biotechnol. Bioeng.*, **44**, 132.
85. San, K.-Y., Bennett, G. N., Aristidou, A. A., and Chou, C.-H. (1994). *Ann. NY Acad. Sci.*, **721**, 257.
86. Bauer, K. A., Ben-Bassat, A., Dawson, M., De La Puente, V. T., and Neway, J. O. (1990). *Appl. Environ. Microbiol.*, **56**, 1296.
87. Korz, D., Hellmuth, K., Sanders, E. A., Deckwer, W.-D., Knorre, W. A, and Riesenberg, D. (1991). *Proceedings of Strategies 2000—Fourth World Congress of Chemical Engineering.* p. 717, Brönners Breidenstein, Frankfurt (M).
88. Riesenberg, D., Pohl, H.-D., Schroeckh, V., and Knorre, W. A. (1994). In *ECB6: Proceedings of the 6th European Congress on Biotechnology* (ed. L. Alberghina, L. Frontali, and P. Sensi), p. 817. Elsevier Science BV, Amsterdam.
89. Elia, M. and Jennings, G. (1995). In *Physiological and clinical aspects of short-chain fatty acids* (ed. J. H. Cummings, J. L. Rombeau, and T. Sakata), p. 35. Cambridge University Press.
90. Akerstrom, B., Nilson, B. H., Hoogenboom, H. R., and Bjorck, L. (1994). *J. Immunol. Methods*, **177**, 151.
91. Nygren, P. A., Stahl, S., and Uhlen, M. (1994). *Trends Biotechnol.*, **12**, 184.
92. Porath, J. (1992). *Protein Expr. Purif.*, **3**, 263.
93. VanBogelen, R. A., Sankar, P., Clark, R. L., Bogan, J. A., and Neidhardt, F. C. (1992). *Electrophoresis*, **13**, 1014.
94. Afeyan, N. B., Fulton, S. P., Gordon, N. F., Maszaroff, I., Varady, L., and Regnier, F. E. (1990). *Biotechnology*, **8**, 203.

11

Production of single-chain Fv monomers and multimers

DAVID FILPULA, JEFFREY MCGUIRE, and MARC WHITLOW

1. Introduction

The Fv fragment (25 kDa) is a two-chain heterodimer of the antibody variable domains (V_L and V_H) and is regarded as the minimal structural component of an antibody required for antigen-binding activity. Fvs have been found to dissociate into V_L and V_H domains at low protein concentrations and under physiological conditions (1). This limited stability may be overcome by incorporation of a designed linker peptide to bridge isolated V_L and V_H domains into a single polypeptide (2). Basic research on recombinant single-chain Fv (scFv) proteins produced from *Escherichia coli* has recently provided some answers to at least two basic questions about this new antibody technology. First, may problems arise from linker obstruction of either antigen-binding sites or domain folding? Second, will scFv proteins display suitable *in vivo* stability, targeting, clearance, and tissue penetration?

Several antigen-binding studies reported for scFv proteins recognizing hapten antigens, polypeptide antigens, carbohydrate antigens, and tumour-associated antigens indicate that a similar monovalent binding activity and either equivalent or somewhat reduced K_a values, may be retained in diverse scFv proteins when compared to the monoclonal antibodies (mAbs) from which they were derived (3, 4). As further evidence for the identity of the binding site in mAbs and scFv proteins, an scFv version of a catalytic mAb was reported to produce nearly equivalent kinetic parameters (5). Finally, structural studies using NMR spectroscopy (6) or X-ray crystallography (7, 8) also support the fidelity of the scFv binding site.

scFv proteins lack the constant domains and bivalency of mAbs. mAbs (150 kDa) are six-fold larger than scFv proteins and display high avidity for repetitive epitopes. When compared to mAbs *in vivo*, anti-tumour scFv proteins clear more quickly from the blood and penetrate tumours with a rapid and even distribution (9, 10). A major application of scFv proteins will be in the production of recombinant single-chain immunoeffector proteins which link an scFv derived cell binding specificity to an effector protein domain, such as *Pseudomonas* exotoxin A. Single-chain immunotoxins have been shown to regress established tumours in rats and mice (11, 12). Furthermore, since scFv technology allows single transcript expression of antibody specificities *in vivo*, single-chain Fv constructions may be valuable in gene therapy applications of intracellular antibodies (13) or targeted viral vectors (14).

Commercialization of scFv technology will require large-scale protein production and purification. It would be valuable to develop a generic production process protocol, lacking affinity chromatography or proteolytic clipping steps, which might be applied to diverse scFv proteins including scFv expressed as insoluble proteins. In this chapter we describe our protocols for the design, construction, and purification of single-chain Fv proteins. We also describe the phenomenon of aggregation of scFv proteins to form multivalent Fv. Although the presence of scFv dimers and higher aggregates can be troublesome in the purification of scFv monomers, these stable Fv multimers have recently been identified as rearranged multivalent antibody fragments which may significantly extend scFv technology into bispecificity and cross-linking capabilities (15–17).

2. Linker designs

In natural antibodies, the V_L and V_H domains associate through non-covalent interactions. The designed linker polypeptide in scFv proteins creates a single subunit protein with two variable domains. In the description of these recombinant antibody sequences, we will follow the numbering system of Kabat *et al.* (18).

Since selected scFv linkers were initially designed to covalently connect the two variable chains without participation in the final Fv conformation, it could be predicted that active single-chain Fv proteins can be constructed in either orientation. Either V_L is the N-terminal domain followed by the linker and V_H (V_L–linker–V_H construction) or V_H is the N-terminal domain followed by the linker and V_L(V_H–linker–V_L construction). A retention of monovalent binding specificities and affinities in both types of purified recombinant scFv proteins has been reported (2–4). The choice of variable domain orientation may be relevant to optimal binding activity and this may be characteristic of individual Fv binding sites and selected linkers. For example, Desplancq *et al.* (17) report binding activity for B72.3 scFv to be dependent on variable chain order and upon linker length. Batra *et al.* (19) found that variable domain orientation did not affect the cytotoxic performance of an anti-Tac scFv-based immunotoxin.

The polypeptide linker sequences in scFv proteins were designed to span the ~3.5–4.0 nm between the C terminus of V_L and the N terminus of V_H, or between the C terminus of V_H and the N terminus of V_L. To provide flexibility, the linkers may include a motif of alternating G and S residues. This rationale is followed in the $(G_4S)_n$ linkers described by Huston *et al.* (4) and

Table 1. Linker design in V$_L$–linker–V$_H$ constructions

Linker name	Linker length[a]	Linker sequence[a]	Major oligomer forms of scFv	Reference
212	14	GSTSGSGKSSEGKG	Monomers, multimers	16, 21, 22
216	18	GSTSGSGKSSEGSGSTKG	Monomers	16, 21
217	12	GSTSGKPSEGKG	Multimers	16
218	18	GSTSGSGKPGSGEGSTKG	Monomers	21

[a] Linker peptide connects V$_L$ residue 107 to V$_H$ residue 1 (18)

serine-rich linkers described by Dorai et al. (20). However, our current linker designs include three charged residues (K, K, E) to enhance the solubility of the linker and its associated scFv. One of the K residues is placed close to the N terminus of V$_H$, to replace the positive charge lost when forming the peptide bond between the linker and the V$_H$. An additional consideration in linker design relates to proteolytic stability. Our studies with several scFv linkers that included the above selected characteristics have shown that protease susceptible sites may be identified and corrected. For example, the proteolytic clip in the 212 linker which occurred between K8 and S9 of the linker (see *Table 1*), may be protected by placing a proline residue at the position succeeding the K8 residue (21).

Single-chain Fv proteins are known to aggregate and form multimeric species. Recent reports (8, 15–17) have provided a better understanding of these aggregates. Multimeric scFv proteins result from intermolecular V$_L$/V$_H$ pairings between two or more scFv polypeptides that produce multivalent Fv molecules from monovalent scFvs (see *Figure 1*). The degree of multimer Fv formation is linker dependent. In general, longer linkers have a decreased aptitude to form multimeric Fv. An scFv protein with a short linker (0–10 residues), which hinders intramolecular V$_L$/V$_H$ domain pairing, forms predominately multimeric Fv (15). *Table 1* shows the linkers which we have recently utilized in *in vitro* and *in vivo* studies (5, 9, 10, 16, 21, 22). Since multivalent Fv formation is dependent upon individual variable domain sequences and buffer composition, as well as linker length, these are also important considerations in the interpretation of antigen-binding assays (16–17).

3. scFv gene construction from hybridoma cells

Large synthetic repertoires of variable genes which have been produced by bacteriophage surface display are increasingly the starting source for the isolation of an antibody specificity (Chapters 1 and 2, ref. 23). However, many researchers will wish to develop an scFv protein from a specific mAb produced by a hybridoma cell line. Sequence-specific DNA synthesis from mRNA isolated from mAb-producing hybridoma cells may be performed

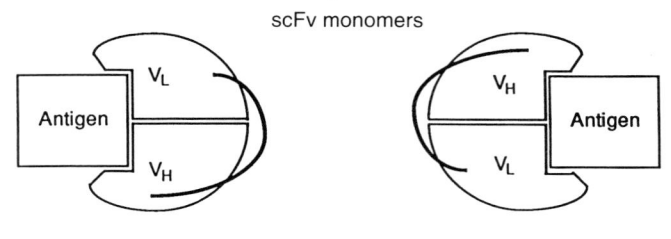

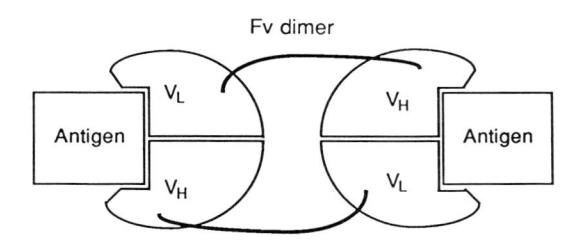

Figure 1. Schematic depiction of V_L/V_H rearrangement model. Top: Two single-chain Fv proteins bound to antigens. Linker peptides shown in bold print. Bottom: An scFv dimer with interchain V_L–V_H interfaces formed by V_L and V_H domains from two scFv monomer polypeptides.

using genetic cloning strategies which exploit the conserved sequences present in the signal and constant (C_H1 or C_L) domains which flank the variable domains, or alternatively, use the FR1 and FR4 framework regions which are on the boundary of either variable domain (see Chapters 1 and 6). Collections of oligonucleotide primers which are designed to prime synthesis into mouse or human antibody variable regions have been reported (e.g. ref. 23).

Whether PCR, oligonucleotide-directed mutagenesis, or wholly synthetic approaches are employed in scFv gene synthesis, it is convenient to design desired sequence features into the synthetic primers. The linker sequence itself and compatible restriction sites for insertion into the selected expression plasmid may be included in the oligonucleotides used. In the examples given below, the scFv gene has an *Aat* II site at the 5′ end; *Hin*dIII and *Pvu*II sites flank the linker; and a *Bam*HI site follows the translational stop codons. *Figure 2* shows the completed gene construction of an scFv protein derived from mAb CC49 which is currently in clinical trials for radioimmunodetection of colorectal cancer (9, 10). A PCR gene assembly method using 'splicing by overlap extension' (24–26) is one convenient approach to scFv gene synthesis. This general method can be used with previously isolated V_L and V_H cDNA clones or from total hybridoma first-strand cDNA. *Table 2* displays examples of oligonucleotide primers used in this protocol. Primers 1 and 2 are complementary to the FR1 and FR4 boundaries, respectively, of a specific

11: Single-chain Fv proteins

```
CC49 V_L                                                              20
         D   V   V   M   S   Q   S   P   S   S   L   P   V   S   V   G   E   K   V   T
        GAC GTC GTG ATG TCA CAG TCT CCA TCC TCC CTA CCT GTG TCA GTT GGC GAG AAG GTT ACT
        Aat II       CDR1                                                            34
         L   S   C   K   S   S   Q   S   L   L   Y   S   G   N   Q   K   N   Y   L   A
        TTG AGC TGC AAG TCC AGT CAG AGC CTT TTA TAT AGT GGT AAT CAA AAG AAC TAC TTG GCC
                                                                    CDR2              54
         W   Y   Q   Q   K   P   G   Q   S   P   K   L   L   I   Y   W   A   S   A   R
        TGG TAC CAG CAG AAA CCA GGG CAG TCT CCT AAA CTG CTG ATT TAC TGG GCA TCC GCT AGG
                                                                                      74
         E   S   G   V   P   D   R   F   T   G   S   G   S   G   T   D   F   T   L   S
        GAA TCT GGG GTC CCT GAT CGC TTC ACA GGC AGT GGA TCT GGG ACA GAT TTC ACT CTC TCC
                                                            CDR3                      94
         I   S   S   V   K   T   E   D   L   A   V   Y   Y   C   Q   Q   Y   Y   S   Y
        ATC AGC AGT GTG AAG ACT GAA GAC CTG GCA GTT TAT TAC TGT CAG CAG TAT TAT AGC TAT
                                                    107     218 Linker
         P   L   T   F   G   A   G   T   K   L   V   L   K   G   S   T   S   G   S   G
        CCC CTC ACG TTC GGT GCT GGG ACC AAG CTT GTG CTG AAA GGC TCT ACT TCC GGT AGC GGC
                                        Hind III
                                                CC49 V_H                              9
         K   P   G   S   G   E   G   S   T   K   G   Q   V   Q   L   Q   Q   S   D   A
        AAA CCC GGG AGT GGT GAA GGT AGC ACT AAA GGT CAG GTT CAG CTG CAG CAG TCT GAC GCT
            Sma I                                       Pvu II                        29
         E   L   V   K   P   G   A   S   V   K   I   S   C   K   A   S   G   Y   T   F
        GAG TTG GTG AAA CCT GGG GCT TCA GTG AAG ATT TCC TGC AAG GCT TCT GGC TAC ACC TTC
                CDR1                                                                  49
         T   D   H   A   I   H   W   V   K   Q   N   P   E   Q   G   L   E   W   I   G
        ACT GAC CAT GCA ATT CAC TGG GTG AAA CAG AAC CCT GAA CAG GGC CTG GAA TGG ATT GGA
        CDR2                                                                          68
         Y   F   S   P   G   N   D   D   F   K   Y   N   E   R   F   K   G   K   A   T
        TAT TTT TCT CCC GGA AAT GAT GAT TTT AAA TAC AAT GAG AGG TTC AAG GGC AAG GCC ACA
                                                                                      85
         L   T   A   D   K   S   S   S   T   A   Y   V   Q   L   N   S   L   T   S   E
        CTG ACT GCA GAC AAA TCC TCC AGC ACT GCC TAC GTG CAG CTC AAC AGC CTG ACA TCT GAG
                                            CDR3                                     107
         D   S   A   V   Y   F   C   T   R   S   L   N   M   A   Y   W   G   Q   G   T
        GAT TCT GCA GTG TAT TTC TGT ACA AGA TCC CTG AAT ATG GCC TAC TGG GGT CAA GGA ACC
                        112
         S   V   T   V   S   *   *
        TCA GTC ACC GTC TCC TAA TAG GAT CC
                                Bam H1
```

Figure 2. DNA sequence of CC49/218 scFv gene in V_L–218 linker–V_H orientation. The translated variable region sequences are numbered according to Kabat et al. (18).

mouse *kappa* V_L gene. Primers 3 and 4 are complementary to the FR1 and FR4 boundaries, respectively, of a specific mouse V_H gene. Primers 2 and 3 also have complementary 5' extensions which encode the scFv linker segment. The PCR amplification may be performed in a single tube first with all four primers, followed by a second PCR assembly using only the two outside primers 1 and 4 (see *Protocol 1*). Alternatively, the use of separate PCR tubes for the V_L and V_H gene amplifications may simplify the optimization of reaction conditions. The optional, engineered *Hin*dIII and *Pvu*II sites, which facilitate subsequent linker interchange, do not alter the protein sequence. The N terminal *Aat* II site can be fused to the *omp*A signal peptide for expression in *E. coli* (see Section 5). Selection of alternate or degenerate PCR primers for various mammalian variable region subgroups is

Table 2. Examples of PCR primers used in CC49/218 scFv assembly[a]

V_L forward *Primer 1*	5'AACACC<u>GACGTC</u>GTGATGTCACAGTCTCCATCCTC 3', *Aat* II
Linker/V_L back *Primer 2*	5'CACCACT<u>CCCGGG</u>TTTGCCGCTACCGGAAGTAGAGCCTTTC-AGCAC<u>AAGCTT</u>GGTCCCAGCACCGAACGTG 3', *Sma* I, *Hin*dIII
Linker/V_H forward *Primer 3*	5'AGCGGCAAA<u>CCCGGG</u>AGTGGTGAAGGTAGCACTAAAGGTC-AGGTT<u>CAGCTG</u>CAGCAGTCTGACGCTGAG 3', *Sma* I, *Pvu*II
V_H back *Primer 4*	5' GC<u>GGATCC</u>TATTAGGAGACGGTGACTGAGGTTCC 3', *Bam*HI

[a] Restriction sites underlined and listed after sequence.

facilitated by the availability of antibody variable gene sequence databases (see, e.g., Chapter 6 and ref. 18). Direct RNA sequencing of V_L and V_H regions from hybridoma mRNA (27) can rapidly provide nucleotide sequence information useful for PCR primer design in subsequent scFv gene assembly.

As an alternative to PCR, further engineering of the cloned scFv gene can be readily accomplished by multiple primed oligonucleotide-directed mutagenesis using the Sculptor™ system (Amersham Corporation) as described (5, 26). We have found that two independently primed mutations can be simultaneously introduced at > 95% frequencies using this convenient commercial system.

Protocol 1. Polymerase chain reaction assembly of scFv gene

Equipment and reagents

- Agarose gel electrophoresis equipment
- PCR thermocycler instrument (Perkin Elmer Cetus, GeneAmp 9600)
- 0.2 ml reaction tubes (e.g. Perkin Elmer MicroAmp Tubes, cat. no. N801–0540)
- MuLV reverse transcriptase (BRL or Perkin Elmer)
- *AmpliTaq* DNA polymerase (Perkin Elmer Cetus)
- Restriction enzymes (Boehringer Mannheim)
- Random hexanucleotides pd(N)$_6$ (Boehringer Mannheim)
- Oligonucleotide primers 1, 2, 3, and 4 (see *Table 2*)
- NuSieve GTG agarose (FMC Corporation)
- TAE buffer: 40 mM Tris–acetate, 1 mM EDTA, pH 7.5
- Ethidium bromide
- RNasin (Promega)
- RT buffer: 10 mM Tris–HCl (pH 8.3), 1 mM DTT, 50 mM KCl, 5 mM MgCl$_2$, 1 mM dATP, 1 mM dGTP, 1 mM dCTP, 1 mM TTP
- PCR buffer I: 10 mM Tris–HCl (pH 8.3), 50 mM KCl, 1.5 mM MgCl$_2$
- PCR buffer II: 10 mM Tris–HCl (pH 8.3), 50 mM KCl, 1.5 mM MgCl$_2$, 0.2 mM dATP, 0.2 mM dGTP, 0.2 mM dCTP, 0.2 mM TTP
- Ethidium bromide
- Phenol
- Chloroform
- Ethanol

Method

1. Combine, in a 0.2 ml tube, 1 µg of purified hybridoma mRNA, 50 pmol of pd(N)$_6$ primers, 50 U of *MuLV* reverse transcriptase, 20 U of RNasin in RT buffer to a total volume of 20 µl. Incubate for 10 min at 22°C.

2. Incubate for 1 h at 42°C, heat to 99°C for 5 min, and cool to 4°C.
3. Add 80 μl PCR buffer I containing 30 pmol of each of the four PCR primers (*Table 2*).
4. Heat the reaction mixture for 2 min at 95°C.
5. Add 2.5 U *AmpliTaq* DNA polymerase to the reaction mixture at 95°C.
6. Perform PCR synthesis in the thermal cycler. Each cycle consists of 30 sec denaturation at 95°C, 45 sec annealing at 62°C, and 30 sec extension at 72°C. Repeat each cycle 30 times and follow by a 5 min extension at 72°C. The optimal annealing temperature may be between 52°C and 72°C and can be determined empirically.
7. Remove a 1 μl aliquot from step 6 and add to 99 μl of PCR buffer II containing 30 pmol each of primer 1 and primer 4 (*Table 2*).
8. Perform the second PCR synthesis following steps 4, 5, and 6 in order.
9. Extract the PCR products with phenol and chloroform and precipitate the DNA with ethanol. Digest the resuspended DNA with *Aat*II plus *Bam*HI and purify the ~740 bp fragment by 4% NuSieve agarose electrophoresis in TAE buffer containing ethidium bromide (0.5 μg/ml).
10. Ligate, 'in gel' (3, 5), the fragment to the selected vector (see *Figure 3*).
11. Transform *E. coli* bacteria with the ligated DNA.

4. Expression of a single-chain Fv in *E. coli*

Since the first report of a low-yield secretion of functional scFv from *E. coli* (1), there have been continued efforts to achieve high yield (g/L) heterologous expression of scFv proteins which are soluble, active, and stable. Yeast (25), plants (28), baculovirus (29), bacillus (30), and mammalian cells (20) are all possible hosts. However, most work to date on scFv protein expression has employed *E. coli* systems. High-level cytoplasmic or periplasmic expression of scFv or scFv fusion proteins in *E. coli* may produce insoluble aggregates which can be renatured and purified. We will describe our *E. coli* production system which has provided us with 100 mg quantities of active clinical grade scFv protein following a protocol of dissolution, renaturation, and purification from an insoluble initial protein extract. Alternatively, research grade scFv proteins may be produced and rapidly purified using an engineered affinity tag (31) or fusion partner (32). For example, we have purified crude denatured or renatured scFv containing an engineered 5 histidine C terminal tail to about 90% purity on the nickel affinity resin Ni–NTA

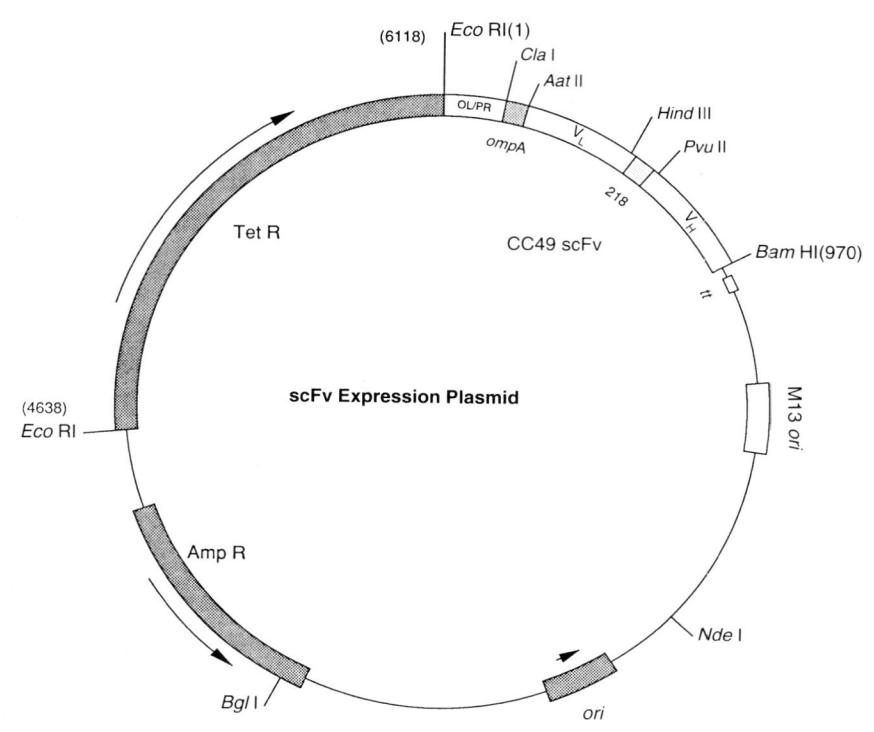

Figure 3. Enzon expression vector used for scFv production in *E. coli*. Contains the hybrid lambda phage promoter O_L/P_R, the *omp*A signal sequence, and a V_L–linker–V_H scFv gene.

(QIAGEN). However, we have thus far avoided affinity chromatography and fusion protein cleavage steps in the production of clinical grade samples which require stringent process validation.

The Enzon expression vector used for scFv expression in *E. coli* contains the hybrid lambda phage promoter O_L/P_R and the *omp*A signal sequence (see *Figure 3*). To produce the final expression strains, the completed scFv expression vectors are transformed into *E. coli* host strain GX6712 which contains the gene for the cI*ts*857 temperature-sensitive repressor. This provides a transcriptional regulation system in which induction of scFv synthesis occurs by raising the culture temperature from 32 °C to 42 °C. Fifteen independent murine-derived and one rat-derived scFv proteins have been expressed at 5–20% of total cell protein using this expression system. N terminal amino acid sequence analysis has confirmed the predicted signal sequence removal. High-level expression of secreted proteins in *E. coli* may result in the formation of protein aggregates in the periplasmic space. Since our expressed scFv proteins accumulate as insoluble aggregates, denaturation and refolding are required for purification (see Section 5).

11: Single-chain Fv proteins

Because expression of scFv proteins fused to the *omp*A signal sequence has resulted in mature scFv production at high levels for several distinct scFv proteins, we have included the signal sequence as part of our standard expression system.

5. Fermentation, renaturation, and purification of an scFv protein

In most reports, scFv and scFv fusion proteins have been produced in *E. coli* as insoluble aggregates that are subjected to denaturation and refolding prior to purification. Our heat-induction process (Section 4) results in partially lysed *E. coli* and the sedimented cellular debris also contains all of the expressed, insoluble scFv protein. When using alternate cytoplasmic expression systems, the insoluble scFv may accumulate in intracellular inclusion bodies. The purification process described here may be adapted to insoluble scFv derived from various crude extracts. In general, the insoluble pellet has been resuspended in a denaturant such as guanidine or urea and diluted 10- to 10000-fold with a renaturation buffer. We describe a generic production and purification protocol, which involves refolding and ion-exchange HPLC chromatographic steps, that we have used successfully on over ten different scFv proteins.

Protocol 2 presents a typical fermentation process for an *E. coli* GX6712 strain containing an scFv expression vector (*Figure 3*).

Protocol 2. Fermentation of scFv protein from *E. coli*

Equipment and reagents

- Chemap laboratory fermenter, 10 litre working volume (Chemapec Inc.)
- Inoculum (*E. coli* expression strain)
- M-63 (1 litre): 0.57 g H_3BO_3, 0.39 g $CuSO_4 \cdot 5H_2O$, 5 g $FeCl_3 \cdot 6H_2O$, 4 g $MnCl_2 \cdot 4H_2O$, 0.5 g $NaMoO_4 \cdot 2H_2O$, 5 g NaCl, 1 g $ZnSO_4 \cdot 7H_2O$, 2.9 ml H_2SO_4
- 5 M NaOH
- 2 M H_3PO_4
- Production medium (1 litre): dissolve in about 900 ml water, 3 g $(NH_4)_2SO_4$, 2.5 g K_2HPO_4. 30 g casein CE90MS (Deltown Specialties), 0.25 g $MgSO_4 \cdot 7H_2O$, 0.1 mg $CaCl_2 \cdot 2H_2O$, 10 ml M-63 salt concentrate, 0.2 ml MAZU 204 Antifoam (Mazer Chemicals). Adjust to pH 7.4 and bring to volume. Autoclave. Then add 30 g glucose, 0.1 mg biotin, 1 mg nicotinamide, 100 mg ampicillin (omit when inoculum is grown in LB medium plus tetracycline).

Method

1. Prior to fermentation, inoculate 0.5 litre of modified LB medium containing either 50 mg ampicillin or 25 mg tetracycline with one frozen 1 ml vial of *E. coli* expression strain. Shake flask at 32°C for 10 h.
2. Inoculate 9.5 litres of production medium in the Chemap fermenter with 0.5 L of inoculum.

Protocol 2. *Continued*

3. Adjust running parameters:
   (a) pH 7.2 ± 0.1 with titrants (5 M NaOH; 2 M H_3PO_4)
   (b) 1 volume air per volume medium per min aeration
   (c) 800 r.p.m. agitation
   (d) 32 °C temperature.

4. At an absorbance at 600 nm of 18–20, raise fermentation temperature to 42 °C. This temperature shift is achieved in 2 min in the Chemap unit.

5. At 1 h post shift-up, cool fermentation to 10 °C, then harvest cell paste at 7000 *g* for 10 min. The wet cell paste can be stored at −20 °C. Approximately 200–300 g of wet cell paste is normally recovered from a 10 litre fermentation.

Protocol 3 describes the resolubilization and renaturation of the scFv protein from the frozen cell pellet.

Protocol 3. Solubilization of scFv protein from *E. coli* cell paste

Equipment and reagents
- Cell homogenizer
- Chilling coil (Lauda/Brinkman)
- RC-5B centrifuge (Sorvall)
- Tissue homogenizer (Heat Systems Ultrasonics)
- UV spectrophotometer
- Cell lysis buffer: 50 mM Tris–HCl, 1.0 mM EDTA, 100 mM KCl, 0.1 mM PMSF (phenylmethylsulfonyl fluoride), pH 8.0
- Denaturing buffer: 6 M guanidine hydrochloride, 50 mM Tris–HCl, 10 mM $CaCl_2$, 50 mM KCl, pH 8.0
- Refolding buffer: 50 mM Tris–HCl, 10 mM $CaCl_2$, 50 mM KCl, 0.1 mM PMSF, pH 8.0
- 0.45 μm microporous membranes (Millipore, cat. no. HVLP0005)

A. *Cell lysis*

1. Thaw the cell paste from a 10 litre fermentation (200–300 *g*) overnight at 4 °C.

2. Gently resuspend the wet cell paste in 2.5 litres of the cell lysis buffer at 4 °C.

3. Pass the cell suspension through a Manton-Gaulin cell homogenizer three times. Because the cell homogenizer raises the temperature of the cell lysate to 25 ± 5 °C, the cell lysate is cooled to 5 ± 2 °C with a Lauda/Brinkman chilling coil after each pass.

B. *Washing the cell pellet*

1. Centrifuge the cell lysate at 24 300 *g* for 30 min at 6 °C. Discard the supernatant, for the pellet contains the insoluble scFv.

11: Single-chain Fv proteins

2. Wash the pellet by gently scraping it from the centrifuge bottles and resuspending it in 1.2 litres of cell lysis buffer.
3. Repeat steps 1 and 2 as many as five times. At any time during this washing procedure the material can be stored as a frozen pellet at −20°C.

C. *Solubilization and renaturation of the scFv protein*
1. Solubilize the washed cell pellet in freshly prepared denaturing buffer at 4°C, using 6 ml of denaturing buffer per gram of cell pellet. If necessary, a few quick pulses from a tissue homogenizer can be used to complete the solubilization.
2. Centrifuge the resulting suspension at 24 300 g for 45 min at 6°C and discard the pellet.
3. Determine the optical density at 280 nm of the supernatant. If the OD_{280} is above 30, additional denaturing buffer should be added to obtain an OD_{280} of approximately 25.
4. Slowly dilute the supernatant into cold (4–10°C) refolding buffer until a 1:10 dilution is reached. We have found that the best results are obtained when the supernatant is slowly added to the refolding buffer over a 2 h period, with gentle mixing.
5. Allow the solution to stand undisturbed for at least a 20 h period at 4°C.
6. Filter the solution through 0.45 µm microporous membranes at 4°C.
7. Concentrate the filtrate to about 500 ml at 4°C.

The fraction of multimer can be increased by treating the solubilized scFv solution from *Protocol 3* with 20% ethanol. The ethanol treatment results in some precipitation which must be filtered prior to purifying the scFv multimers. The exact ethanol concentration required can vary depending on the scFv.

Protocol 4. 20% ethanol treatment of the solubilized scFv, used to increase the fraction of multimers

Equipment and reagents
- Solubilized scFv solution from *Protocol 3*
- Ethanol
- 0.45 µm microporous membrane filter (Millipore, cat. no. HVLP0005)

Method
1. Under slow mixing add sufficient quantities of pure ethanol to bring the solubilized scFv solution to 20% ethanol. Sixteen litres of solubilized scFv would require 4 litres of ethanol.

Protocol 4. *Continued*

2. Allow the solution to stand undisturbed for at least 2 h. Longer periods of time (12–18 h) allow the precipitate to settle to the bottom of the tank, which makes the filtration easier.
3. Filter the solution through a 0.45 μm microporous membrane. Drawing the solution from the top of the tank after the precipitate has settled avoids premature clogging of the filter.
4. Discard the precipitate.

Since most scFv proteins have an isoelectric point between 8.0 and 9.4, cation-exchange chromatography is an appropriate purification approach. *Protocol 5* describes the purification methods by cation-exchange HPLC and the further characterization of the scFv by size-exclusion HPLC. The purification protocol yields scFv proteins that are greater than 95% pure as examined by SDS-PAGE. We have observed that the Amp^r gene product β-lactamase co-migrates with some scFv proteins on SDS-PAGE. This can have relevance in attempts to quantitate expression yields by gel analysis.

Protocol 5. HPLC purification of scFv protein

Equipment and reagents

- HPLC buffer A: 60 mM Mops, 0.5 mM Ca acetate, pH 6.4
- HPLC buffer B: 60 mM Mops, 10 mM Ca acetate, pH 7.5
- HPLC buffer C: 60 mM Mops, 100 mM CaCl$_2$, pH 7.5
- HPLC buffer D: 50 mM Mops, 100 mM NaCl, pH 7.5
- Conductivity meter
- UV spectrophotometer
- PolyCAT A column (Poly LC Inc., Columbia, MD)
- Pre-cast 4–20% acrylamide SDS-PAGE slab gels (Novex)
- TSK G3000SW column (Toso Haas)
- Waters HPLC system (Millipore)
- MacIntosh SE (Apple Computer)
- Dynamax software package (Rainin Instrument Co.)

A. *Cation-exchange HPLC purification*

1. Dialyse the renatured scFv solution against HPLC buffer A, until the conductivity is lowered to that of buffer A.
2. Equilibrate the 21.5 mm × 150 mm polyaspartic acid PolyCAT A column with HPLC buffer A for 20 min.
3. Load the dialysed sample on the PolyCAT A column. If more than 60 mg is loaded on this column the resolution begins to deteriorate, thus the sample must usually be divided into several PolyCAT A runs.
4. Determine optical density at 280 nm and calculate the protein concentration. Most scFv proteins have an extinction coefficient of about 2.0 mg ml^{-1} cm^{-1} at 280 nm and this can be used to determine the protein concentration.

11: Single-chain Fv proteins

5. Elute the sample from the PolyCAT A column with a 50 min linear gradient between HPLC buffers A and B. Most of the monomer scFv proteins that we have purified elute between 20 and 30 min using this gradient, while multimers elute later. We normally collect 3 min fractions.
6. Apply a final 6 min linear gradient at 15 ml/min (90 ml) to the PolyCAT A column to remove the remaining protein with HPLC buffer C. 50 ml fractions are analysed.
7. Analyse the collected fractions on 4–20% Tris–glycine SDS-PAGE gels.

B. *Size-exclusion (SE) HPLC characterization*

1. Equilibrate the 7.8 × 300 mm TSK G3000SW column with HPLC buffer D for 20 min.
2. Load 10–50 μl samples on the SE-HPLC column.
3. Elute the column with HPLC buffer D at a flow rate of 0.5 ml/min. Most of the scFv proteins that we have examined elute at 17–21 min.
4. Collect data using Dynamax software package.

Figure 4A shows an example of a PolyCAT A HPLC where the separation of monomeric and dimeric scFv molecules is apparent. *Figure 4B* displays a representative SE-HPLC chromatograph of a highly aggregated scFv where monomer, dimer, and trimer species are identified. Note that in both examples, the scFv proteins contain the 212 linker (see *Table 1*). scFv aggregates have recently been shown to contain intermolecular V_L/V_H pairs from rearranged scFv proteins (15, 16). Multivalent Fv stability is dependent on individual V_L and V_H sequences as well as linker length. Disassociating agents such as 0.5 M guanidine hydrochloride with 20% ethanol can catalyse an interconversion between monomeric and dimeric scFv species (16). Significantly, the multivalent Fv represents a novel antibody fragment which provides new avidity and bifunctionality possibilities to scFv technology.

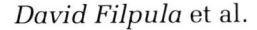

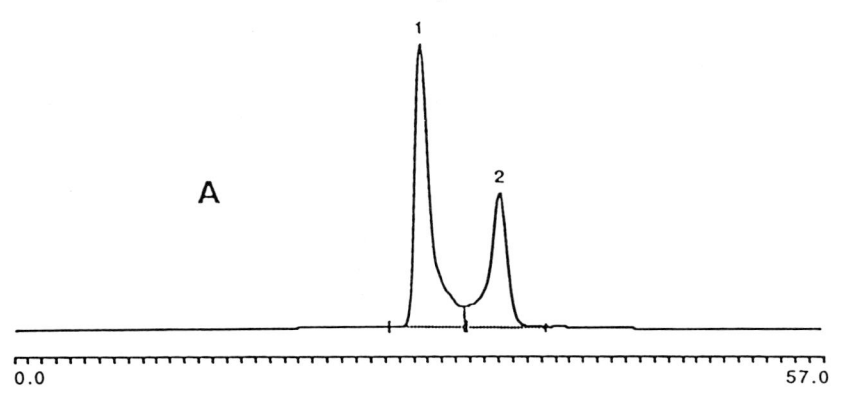

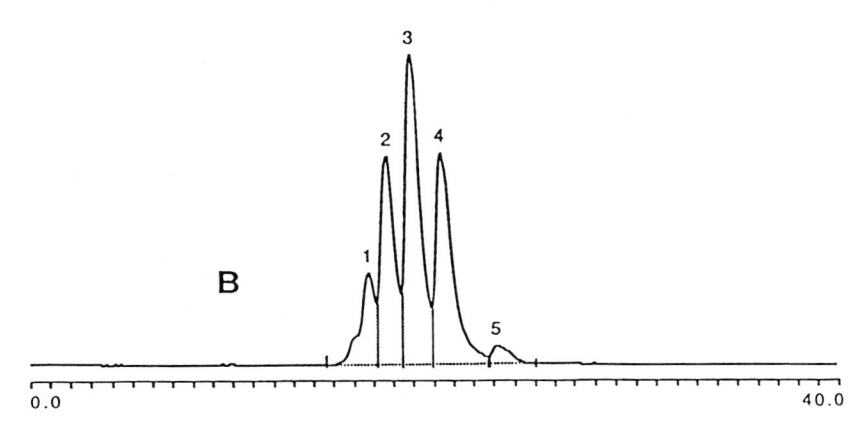

Figure 4. (A) PolyCAT A cation-exchange HPLC chromatographic separation of a 5 mg/ml sample of 4-4-20/212 scFv showing monomer (peak 1) and dimer (peak 2). Retention times of peaks 1 and 2 are 28.9 and 34.4 min, respectively. (B) HPLC size-exclusion chromatograph analysed on a TSK G3000SW column showing a highly aggregated CC49/212 scFv. The monomer (peak 4), dimer (peak 3), and trimer (peak 2) fractions elute at 20.4, 18.9, and 17.7 min, respectively.

References

1. Glockshuber, R., Malia, M., Pfitzinger, I., and Plückthun, A. (1990). *Biochemistry*, **29**, 1362.
2. Bird, R. E., Hardman, K. D., Jacobson, J. W., Johnson, S., Kaufman, B. M., Lee, S.-M., Lee, T., Pope, S. H., Riordan, G. S., and Whitlow, M. (1988). *Science*, **242**, 423.
3. Whitlow, M. and Filpula, D. (1991). *Methods, a companion to Methods in enzymology* **2**, 97.
4. Huston, J. S., Mudgett-Hunter, M., Tai, M.-S., McCartney, J., Warren, F., Haber,

E., and Oppermann, H. (1991). In *Methods in enzymology* (ed. J. J. Langone), Vol. 203, pp. 46–88. Academic Press, San Diego.
5. Gibbs, R. A., Posner, B. A., Filpula, D. R., Dodd, S. W., Finkelman, M. A. J., Lee, T. K., Wroble, M., Whitlow, M., and Benkovic, S. J. (1991). *Proc. Natl Acad. Sci. USA*, **88**, 4001.
6. Freund, C., Ross, A., Plückthun, A., and Holak, T. A. (1994). *Biochemistry*, **33**, 3296.
7. Zdanov, A., Li, Y., Bundle, D. R., Deng, S.-J., MacKenzie, C. R., Narang, S. A., Young, N. M., and Cygler, M. (1994). *Proc. Natl Acad. Sci. USA*, **91**, 6423.
8. Kortt, A. A., Malby, R. L., Caldwell, J. B., Gruen, L. C., Ivancic, N., Lawrence, M. C., Howlett, G. J., Webster, R. G., Hudson, P. J., and Colman, P. M. (1994). *Eur. J. Biochem.*, **221**, 151.
9. Milenic, D. E., Yokota, T., Filpula, D. R., Finkelman, M. A. J., Dodd, S. W., Wood, J. F., Whitlow, M., Snoy, P., and Schlom, J. (1991). *Cancer Res.*, **51**, 6363.
10. Yokota, T., Milenic, D. E., Whitlow, M., and Schlom, J. (1992). *Cancer Res.*, **52**, 3402.
11. Brinkmann, U., Pai, L. H., FitzGerald, D. J., Willingham, M. C., and Pastan, I. (1991). *Proc. Natl Acad. Sci. USA*, **88**, 8616.
12. Siegall, C. B., Chace, D., Mixan, B., Garrigues, U., Wan, H., Paul, L., Wolff, E., Hellström, I., and Hellström, K. E. (1994). *J. Immunol.*, **152**, 2377.
13. Marasco, W. A., Haseltine, W. A., and Chen, S. (1993). *Proc. Natl Acad. Sci. USA*, **90**, 7889.
14. Russell, S. J., Hawkins, R. E., and Winter, G. (1993). *Nucl. Acids Res.*, **21**, 1081.
15. Holliger, P., Prospero, T., and Winter, G. (1993). *Proc. Natl Acad. Sci. USA*, **90**, 6444.
16. Whitlow, M., Filpula, D., Rollence, M. L., Feng, S.-L., and Wood, J. F. (1994). *Protein Eng.*, **7**, 1017.
17. Desplancq, D., King, D. J., Lawson, A. D. G., and Mountain, A. (1994). *Protein Eng.*, **7**, 1027.
18. Kabat, E. A., Wu, T. T., Perry, H. M., Gottesman, K. S., and Foeller, C. (1991). *Sequences of proteins of immunological interest* (5th edn). US Department of Health and Human Services, Bethesda, MD.
19. Batra, J. K., FitzGerald, D., Gately, M., Chaudhary, V. K., and Pastan, I. (1990). *J. Biol. Chem.*, **265**, 15198.
20. Dorai, H., McCartney, J. E., Hudziak, R. M., Tai, M.-S., Laminet, A. A., Houston, L. L., Huston, J. S., and Oppermann, H. (1994). *Bio/Technology*, **12**, 890.
21. Whitlow, M., Bell, B. A., Feng, S.-L., Filpula, D., Hardman, K. D., Hubert, S. L., Rollence, M. L., Wood, J. F., Schott, M. E., Milenic, D. E., Yokota, T., and Schlom, J. (1993). *Protein Eng.*, **6**, 989.
22. Pantoliano, M. W., Bird, R. E., Johnson, L. S., Asel, E. D., Dodd, S. W., Wood, J. F., and Hardman, K. D. (1991). *Biochemistry*, **30**, 10117.
23. Griffiths, A. D., Williams, S. C., Hartley, O., Tomlinson, I. M., Waterhouse, P., Crosby, W. L., Kontermann, R. E., Jones, P. T., Low, N. M., Allison, T. J., Prospero, T. D., Hoogenboom, H. R., Nissim, A., Cox, J. P. L., Harrison, J. L., Zaccolo, M., Gherardi, E., and Winter, G. (1994). *EMBO J.*, **13**, 3245.
24. Johnson, S. and Bird, R. E. (1991). In *Methods in enzymology* (ed. J. J. Langone), Vol. 203, pp. 88–98. Academic Press, San Diego.
25. Davis, G. T., Bedzyk, W. D., Voss, E. W., and Jacobs, T. W. (1991). *Bio/Technology*, **9**, 165.

26. Whitlow, M. and Filpula, D. (1993). In *Tumor immunobiology: a practical approach* (ed. G. Gallagher, R. C. Rees, and C. W. Reynolds), pp. 279–91. IRL Press, Oxford.
27. Grant, F. J., Levin, S. D., Gilbert, T., and Kindsvogel, W. (1987). *Nucl. Acids Res.*, **15**, 5496.
28. Tavladoraki, P., Benvenuto, E., Trinca, S., DeMartinis, D., Cattaneo, A., and Galeffi, P. (1993). *Nature*, **366**, 469.
29. Laroche, Y., Demaeyer, M., Stassen, J-M., Gansemans, Y., Demarsin, E., Matthyssens, G., Collen, D., and Holvoet, P. (1991). *J. Biol. Chem.*, **266**, 16343.
30. Wu, X-C., Ng, S-C., Near, R. I., and Wong, S-L. (1993). *Bio/Technology*, **11**, 71.
31. Skerra, A., Pfitzinger, I., and Plückthun, A. (1991). *Bio/Technology*, **9**, 273.
32. LaVallie, E. R., DiBlasio, E. A., Kovacic, S., Grant, K. L., Schendel, P. F., and McCoy, J. M. (1993). *Bio/Technology*, **11**, 187.

12

Expression of immunoglobulin genes in mammalian cells

CHRISTOPHER BEBBINGTON

1. Introduction

Expression systems for antibodies in mammalian cells have recently been reviewed (1). This chapter describes detailed procedures for some of the most efficient, currently available expression systems for use in Chinese hamster ovary cells and mouse myeloma cell lines. These are the two cell types that have been most widely used for antibody production. Both can be grown readily in a variety of bioreactors including large-scale suspension fermentation systems, and have been used to produce antibodies for clinical use in humans.

The expression vectors for establishing permanent transfected cell lines described in this chapter are integrating vectors: i.e. integration of vector DNA into the host genome can be selected, thus permitting maintenance of vector sequences through multiple cell divisions. Strong viral promoter–enhancers are used to transcribe the immunoglobulin genes, and the vectors also contain a selectable marker gene which confers resistance to a toxic drug. Two selectable markers are described, encoding the enzymes dihydrofolate reductase (DHFR) and glutamine synthetase (GS). Both these enzymes can be inhibited by compounds which bind tightly to the enzyme and can be used to titrate enzyme activity. This permits the possibility of selecting rare variants of an antibody expressing cell line after transfection which have undergone an increase in copy number in the region of the chromosome containing the expression vector and hence produce higher levels of the selectable enzyme and incidentally higher levels of antibody. For this reason, DHFR and GS are generally described as 'amplifiable' selectable markers. (For a review of gene amplification, see ref. 2.) Such gene amplification can be important in achieving maximal levels of antibody gene expression.

Inhibitors of GS or DHFR enzyme activity can also be used in selecting initial transfectant clones to increase the stringency of selection. Current mammalian expression vectors integrate at apparently random sites within the host genome and are subject to significant position effects on the level of expression of genes present on the plasmid. Thus the use of a stringent selection regime can permit selection for transfectants in which the vector has integrated into relatively advantageous sites for expression such that the selectable marker is efficiently expressed, hence minimizing the screening necessary to identify efficient producers of antibody. Other vector systems for antibody production from mammalian cells using non-amplifiable selectable markers have been described elsewhere (3).

The factors to be considered in choosing an expression system are largely outside the scope of this chapter but may include, in addition to the requirement for efficient gene expression, the scale-up procedures available (NS0 cells adapt very readily to suspension fermenter growth whereas CHO cells perform well in small-scale roller bottle culture), minor glycosylation differences between the products of different cell types and fermentation conditions, and the effects of such differences on antibody properties (see ref. 1).

Establishing permanent transfected cell lines is time-consuming and it is often important to produce small amounts of antibody more rapidly. For this purpose, the efficient COS-cell transient expression system is described which permits the production of microgram amounts of antibody within 1–3 days from transfection. This can permit the screening of a number of different recombinant antibodies or combinations of heavy and light chains, for example, in a variety of *in vitro* functional assays before the construction of permanent cell lines is undertaken.

2. Transient expression of antibodies in COS cells

COS cells are derivatives of an African green monkey cell line, permissive for the replication of many bacterial plasmids, into which has been inserted a short sequence (about 340 bp) including the replication origin of the monkey virus SV40. This has been made possible by the transformation of the monkey kidney line, CV1, with an origin-defective SV40 virus to provide all the *trans*-acting components for replication (4). Thus expression vectors with the SV40 origin can replicate to many tens of thousands of copies/cell within 1–2 days of transfection with concomitant efficient expression. The high copy number eventually kills the host cell so such a system can only be used for transient expression over about 3 days, but this can be very useful for producing quantities of antibodies in the microgram range relatively rapidly (5).

2.1 Choice of cell line and medium

Either of two widely available COS cell lines, COS-1 and COS-7 work well. COS-1 is perhaps more generally used. COS cells grow attached to plastic and have few special growth requirements. Therefore they can be grown in various tissue culture media. The method described here uses Dulbecco modified Eagle medium (DMEM).

12: Expression of immunoglobulin genes in mammalian cells

2.2 Choice of expression vector

The vector must contain the SV40 origin of replication and have a plasmid backbone which does not have the sequences near the origin of replication of pBR322 which inhibit replication in mammalian cells (the so-called 'poison' sequences (6)). Suitable plasmid backbones include pML, pBR328, and pEE6 (7).

Expression vectors which combine these features and also contain the powerful hCMV-MIE promoter and enhancer to direct efficient transcription in COS cells are pEE6hCMV.gpt, pEE14, pEE12, and pEE13 (see *Figures 1, 2, 3,* and *4* respectively). Separate expression plasmids containing the light and heavy chains of immunoglobulins can be co-transfected into COS cells. Alternatively, both chains can be inserted into a single vector as described in Section 3.2.

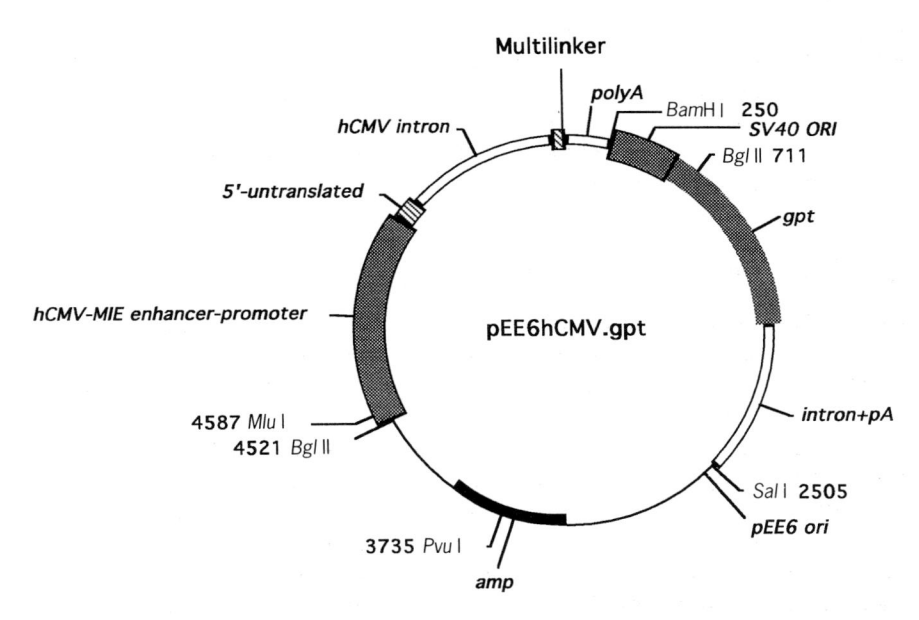

Figure 1. pEE6hCMVgpt (6.7 kb) (5) consists of a 2.1 kb fragment from human cytomegalovirus containing the major immediate early (hCMV-MIE) enhancer–promoter and 5' untranslated region, including a 830 bp hCMV intron; a synthetic multilinker cloning site; the SV40 Early region polyadenylation signal (poly A) (*Bcl*I–*Bam*HI); an SV40 origin of replication (SV40 ORI) including the SV40 Early promoter (the 344 bp *Hin*dIII–*Pvu*II fragment of SV40); a gpt coding sequence; and plasmid vector pEE6 sequences (7) (thin line) containing a β-lactamase gene (amp) and origin of replication (pEE6 ori). The unique restriction sites in the multilinker, reading clockwise, are *Hin*dIII, *Xba*I, *Xma*I, *Sma*I, *Eco*RI, and *Bcl*I. Figure reproduced courtesy of Celltech Therapeutics Ltd.

Christopher Bebbington

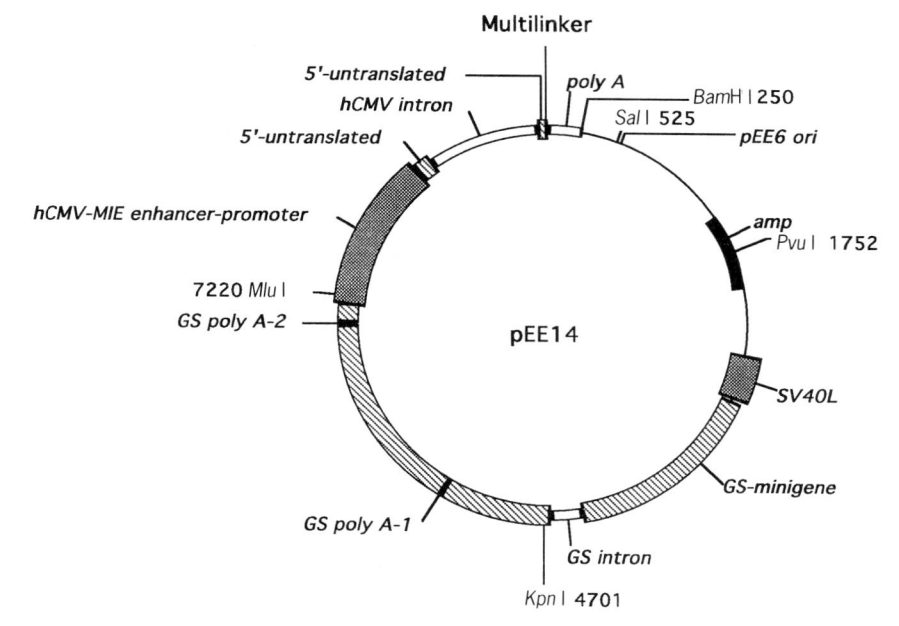

Figure 2. pEE14 (9.4 kb), an expression vector for CHO cells containing a GS selectable marker. A hamster GS minigene (containing a single GS intron and polyadenylation signals from the GS gene (GS polyA-1 and polyA-2), expressed from the SV40 Late promoter (SV40L; the 344 bp *HindIII–Pvu*II fragment spanning the origin of replication) has been inserted into the *Bgl*II site of pEE6hCMV (7). The GS cassette is from pSVLGS1 (8), but a number of restriction sites have been removed to facilitate cloning into the multilinker. Other features are as *Figure 1*. Figure reproduced courtesy of Celltech Therapeutics Ltd.

Protocol 1. Transfection of COS-1 cells

Equipment and reagents

- COS-1 cells (ATCC, cat. no. CRL 1650)
- DMEM (Gibco-BRL, cat. no. 41965)
- Non-essential amino acids (NEAA) (Gibco-BRL, cat. no. 11140)
- Fetal calf serum (FCS): heat inactivate at 56°C for 30 min and store at –20°C
- Penicillin–streptomycin (Gibco-BRL, cat. no. 15070)
- Supplemented DMEM: 500 ml DMEM, 5 ml NEAA, 5 ml penicillin–streptomycin, 50 ml FCS
- Serum-free supplemented DMEM: as above but without FCS
- DMEM–Tris: mix 200 ml serum-free supplemented DMEM with 50 ml 0.25 M Tris–HCl (pH 7.5)
- Tris buffered saline (TBS): mix 25 ml 1 M Tris-HCl (pH 7.4), 27.5 ml 5 M NaCl, 5 ml 1 M KCl, 0.7 ml 1 M CaCl$_2$, 0.5 ml 1 M MgCl$_2$, and 2 ml 0.3 M Na$_2$HPO$_4$. Make up to 1 L and filter-sterilize. Store at 4°C.
- DEAE–Dextran solution. Dissolve at 1 mg/ml in TBS. Store at –20°C.
- 90 mm Petri dishes

12: Expression of immunoglobulin genes in mammalian cells

- 10% DMSO/HBS: dissolve in just less than 1 L, 8 g NaCl, 0.375 g KCl, 0.1 g Na$_2$HPO$_4$, 1.1 g glucose, 5 g Hepes. Adjust the pH to 7.1 with 1 M NaOH and make up to 1 L. Add 110 ml DMSO, filter-sterilize and store at 4°C.

- Circular recombinant plasmid DNA. Purify the DNA, e.g. by CsCl centrifugation, then precipitate from ethanol, wash the DNA pellet with 70% ethanol and leave the DNA pellet to dry in a tissue culture laminar flow cabinet to ensure sterility. Dissolve in sterile distilled water at 0.5 mg/ml.

Method

1. Plate out cells the day before transfection at 2 × 10^6 cells per 90 mm Petri dish.

2. On the day of transfection, rinse the cells twice by adding 10 ml serum-free supplemented DMEM and aspirating.

3. Set up tubes containing (quantities are per dish to be transfected):
   - 15–30 µg plasmid DNA
   - 4 ml DMEM–Tris
   - 1 ml DEAE–Dextran solution

4. Aspirate the serum-free supplemented DMEM from the cells and add the DNA/DEAE–Dextran complex (step 3) to each dish.

5. Return the dishes to the tissue culture incubator for 6–8 h.

6. Remove the complex and add 5 ml 10% DMSO/HBS. Leave for 2–5 min.

7. Remove the DMSO/HBS by aspirating, wash once with serum-free supplemented DMEM, and replace with supplemented DMEM containing 10% serum.

8. Culture the cells for up to 72 h and harvest samples of the medium for analysis of antibody production.

3. Dihydrofolate reductase selection in Chinese hamster ovary (CHO) cells

3.1 Choice of cell line and medium

Typically, a *dhfr*$^-$ variant CHO cell line is used, such as DUKX-B11 (also called DXB11) (10), isolated from the proline auxotroph CHO-K1, or DG44 (11), isolated from the CHO (Toronto) cell line. In DXB11, only one of the alleles of the *dhfr* gene is defective, whereas in DG44 both alleles are mutant. Consequently, there is a low but appreciable reversion frequency of DXB11 to a *dhfr*$^+$ phenotype. The reversion frequency of DG44 is essentially undetectable.

The *dhfr*$^-$ cell lines require exogenous nucleosides and glycine for survival. DXB11 also has a requirement for proline because the CHO-KI parent is incidentally *pro*$^-$. All of these nutritional requirements are provided by either

Christopher Bebbington

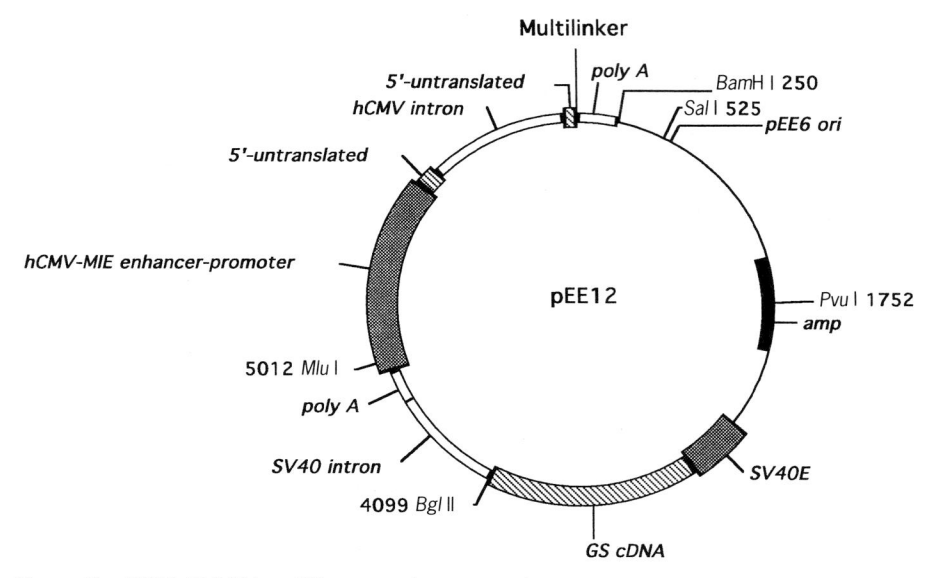

Figure 3. pEE12 (7.1 kb), a GS expression vector for myeloma cells. It consists of a GS expression cassette inserted into pEE6hCMV (7). The GS cassette contains an SV40 Early promoter (SV40E; a 323 bp *Pvu*II–*Hin*dIII fragment spanning the SV40 origin of replication); a hamster GS cDNA; an SV40 intron (SV40 nucleotides 4099–4710); and a *Bcl*I–*Bam*HI fragment of SV40, containing the Early region polyadenylation signal (poly A). The SV40-GS cassette is derived from pSV2.GS (8), but a number of restriction sites have been removed to facilitate cloning into the multilinker. Other features are as *Figure 1*. For a review of the genetic manipulation of antibodies and the reconstruction of antibody fragments to produce immunoglobulins, see ref. 17. Figure reproduced courtesy of Celltech Therapeutics Ltd.

of the non-selective media given in *Protocol 2*. Two alternative sets of media are described based upon either MEM-Alpha medium or Dulbecco modified Eagle medium (DMEM).

3.2 DHFR vector design

A variety of different vectors containing DHFR coding sequences can transform the cells to a *dhfr*⁺ phenotype, thus allowing them to grow without added nucleosides and glycine (see, for example, refs 12 and 13). It is preferable to use a vector with a weakly expressed DHFR gene, so that the levels of MTX required for amplification are minimized, e.g. pSVM.*dhfr* (see *Figure 5*). A very efficient expression cassette for immunoglobulin coding sequences can be provided by the plasmid pEE6hCMV.gpt (*Figure 1*) used in conjunction with pSVM*dhfr*. A single plasmid can be constructed containing both an antibody light-chain gene and the DHFR gene by using the single *Bam*HI restriction site downstream of the amplifiable gene in pSVM.*dhfr*. This forms

274

12: Expression of immunoglobulin genes in mammalian cells

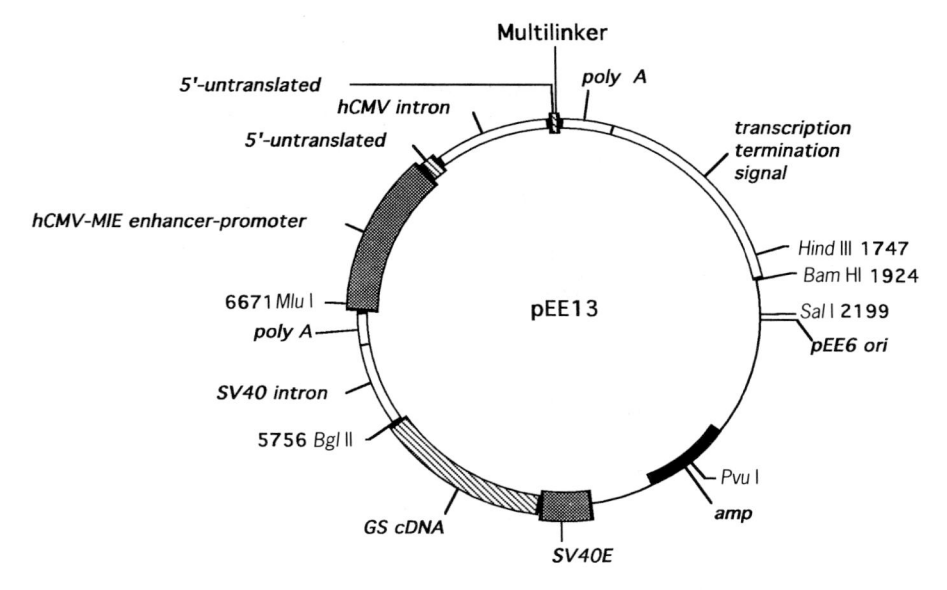

Figure 4. pEE13, a GS expression vector for expression of dimeric proteins in myeloma cells. It is identical to pEE12, but has a 1.7 kb BclI–BamHI sequence from the 3′-flanking region of the mouse immunoglobulin μ-chain gene inserted at the BamHI site which acts as a transcription termination signal (9). Unique sites in the multilinker are XmaI, SmaI, EcoRI, and BclI. Figure reproduced courtesy of Celltech Therapeutics Ltd.

a suitable site for introducing a complete transcription unit consisting of the BglII–BamHI fragment of pEE6hCMV-B containing the light-chain gene. The process can then be repeated to introduce the heavy-chain transcription unit once this has been cloned, for instance into pEE6hCMV.gpt.

Vector DNA introduced into a host cell by calcium phosphate-mediated transfection can frequently become ligated into high molecular weight species so that multiple plasmid molecules ultimately integrate at the same chromosomal location. For this reason, it is not always necessary for the amplifiable gene and the non-selected gene to be on the same vector. Thus, although it is normally preferable to construct a single vector containing all the required genes, if this is not practicable, the plasmids can be co-transfected simultaneously.

3.3 Transfection of CHO cells

Transfection can be carried out using a calcium phosphate co-precipitation procedure as described in *Protocol 2*. Alternatively, DHFR vectors can be transfected using electroporation precisely as described for NS0 cells in *Protocol 6*.

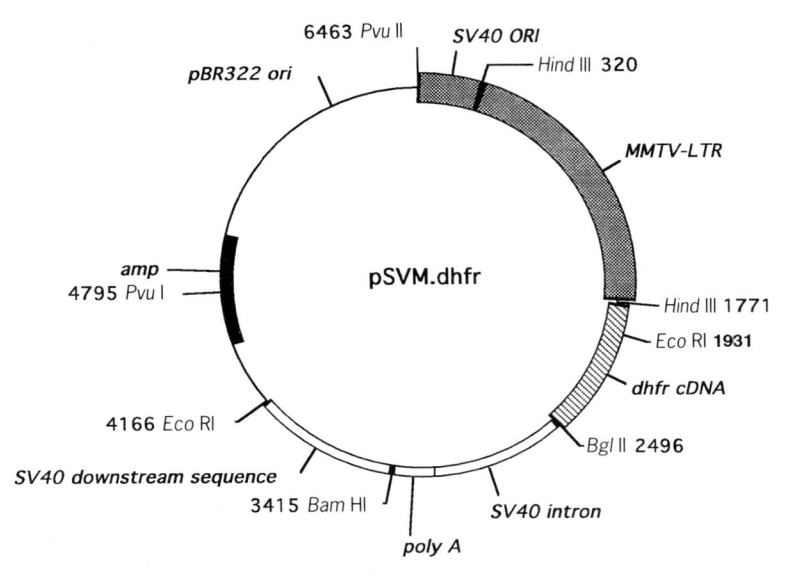

Figure 5. pSVM.dhfr (6.45 kb), an amplifiable vector for use in dhfr⁻ CHO cells (12) contains the 323 bp PvuII–HindIII fragment of SV40 spanning the SV40 origin of replication (SV40 ORI; a 1.4 kb long terminal repeat of mouse mammary tumour virus (MMTV-LTR) containing a glucocorticoid regulatable promoter; a 735 bp mouse *dhfr* cDNA; an SV40 intron (SV40 nucleotides 4099–4710); a *Bcl*I–*Eco*RI fragment of SV40 (nucleotides 2770–1782) containing the Early region polyadenylation signal (poly A) and additional SV40 downstream sequence; and the PvuII–EcoRI fragment of pBR322 (2295 bp) containing the origin of replication and β-lactamase gene (*amp*) which confers resistance to ampicillin. Figure reproduced courtesy of Celltech Therapeutics Ltd.

Protocol 2. Transfection and selection of DHFR vectors in *dhfr*⁻ CHO cells

Equipment and reagents

- *dhfr*⁻ CHO cell line DXB11 or DG44 (e.g. from Dr L. Chasin, Columbia University, New York) growing exponentially in non-selective medium
- Circular recombinant plasmid DNA, including plasmid without the gene to be expressed, e.g. pSVM.*dhfr*, as a control. Purify the plasmid DNA, for example by CsCl centrifugation, then precipitate from ethanol, wash the DNA pellet with 70% ethanol and leave the DNA pellet to dry in a tissue culture laminar-flow cabinet to ensure sterility. Dissolve in sterile distilled water at 0.5 mg/ml
- 200 mM L-glutamine stock solution (Gibco-BRL, cat. no. 15032). Store at −20°C in 5 ml aliquots.
- Non-essential amino acids (NEAA) (Gibco-BRL, cat. no. 043-01140)
- Fetal calf serum (FCS) : heat inactivate at 56°C for 30 min and store at −20°C. Use for DMEM-based medium only.
- Dialysed FCS (Gibco-BRL, cat. no. 063-6300): heat inactivate at 56°C for 30 min and store at −20°C
- Non-selective medium (Ham's-based): 500 ml Ham's F12 nutrient mix (Flow, cat. no. 12-432), 50 ml FCS, 5 ml L-glutamine
   OR
   Non-selective medium (MEM Alpha medium-based): MEM Alpha medium with nucleosides (Gibco-BRL, cat. no. 22571), 50 ml FCS

12: Expression of immunoglobulin genes in mammalian cells

- Selective medium (DMEM-based): 500 ml DMEM (Gibco-BRL, cat. no. 41965), 50 ml dialysed FCS, 5 ml NEAA
  OR
  Selective medium (MEM Alpha medium-based): 500 ml MEM Alpha medium *without* nucleosides (Gibco-BRL, cat. no. 22561), 50 ml dialysed FCS
- Tissue culture Petri dishes (90 mm diameter).
- Sterile plastic tubes (e.g. PQ Universal tubes)
- Non-selective medium without serum
- 2 × HBS: dissolve 8.18 g NaCl, 5.95 g Hepes, 2 g Na_2HPO_4 (anhydrous) in 400 ml distilled water, adjust to pH 7.1 using 1 M NaOH and make up to 500 ml. Filter-sterilize and store at 4°C
- 2 M $CaCl_2$. Filter-sterilize and store at 4°C
- Glycerol/HBS: 15 ml glycerol, 50 ml 2 × HBS, and 53 ml distilled water. Filter-sterilize and store at 4°C
- Vortex mixer

Method

1. Plate the cells at a density of approximately 10^6 cells per 90 mm Petri dish, the day before transfection, in non-selective medium (either alternative works well).
2. On the day of transfection, bring the 2 M $CaCl_2$ and 2 × HBS to room temperature.
3. In a sterile plastic tube, mix 62 µl $CaCl_2$ 10 µg DNA, and distilled water to make the volume up to 0.5 ml (quantities are for each dish of cells). In a second tube, add an equal volume of 2 × HBS (i.e. 0.5 ml per dish of cells to be transfected).
4. Add the DNA/$CaCl_2$ solution to the 2 × HBS, with continuous agitation to form a pale milky co-precipitate of calcium phosphate with DNA.
5. Immediately vortex the mixture for 10 seconds and add 2.5 ml of non-selective serum-free medium.
6. Rinse each dish of cells once in 10 ml non-selective serum-free medium by aspirating, adding medium, and re-aspirating.
7. Add the DNA/calcium phosphate mixture to the cells and return the dishes to the 37°C incubator for 3.5–4 h. Also treat some dishes of cells using the identical transfection procedure, but without adding DNA, to act as negative controls. During the incubation in the presence of DNA, ensure that the cells on the dishes do not dry out, this may necessitate rocking the dishes occasionally.
8. Aspirate the dishes to remove the calcium phosphate/DNA mixture, and add 3 ml glycerol/HBS to each dish for 1.5–2 min.
9. Add 10 ml non-selective serum-free medium to rinse off the glycerol, re-aspirate, and replace with 10 ml non-selective medium. Allow the cells to recover in non-selective medium in the tissue culture incubator overnight.
10. The next day, remove the medium and replace with the selective medium (either alternative works well).
11. Change the medium again after 3–4 days when substantial cell death should be apparent. (The dead cells float off the bottom of the Petri dish).

Protocol 2. *Continued*

12. Change the medium again every 3-4 days. After 7-8 days, many small adherent colonies of cells may appear but wait until 10-12 days after transfection before scoring the number of stably transformed colonies since many colonies survive for only a few days due to transient expression of the introduced DHFR genes.
13. Calculate the transfection frequency for the stable transformants. The transfection frequency with pSVM.*dhfr* should be about $1-2 \times 10^{-4}$ transfected cells/10 µg DNA, but the exact frequency will depend on the plasmid and the particular experiment.
14. Cell lines can be established from individual colonies by trypsinization and assaying for expression of the gene of interest.

3.4 Selection for amplification of DHFR vectors with methotrexate (MTX)

The concentration of MTX needed for amplification of a *dhfr* gene depends on the efficiency with which the gene is expressed, which is, in part, determined by the vector chosen (see Section 3.2). There is, however, also variability between individual transfected clones. Therefore cells from each clone to be amplified should be selected using a range of MTX concentrations, using *Protocol 3*.

Protocol 3. Amplification of DHFR vectors with methotrexate

Reagents

- 100 µM MTX. Prepare this as a sterile solution in water and store at -20°C in 1 ml aliquots. **CAUTION** MTX is toxic; wear gloves when handling and observe full safety precautions including the use of a face mask. Weigh out the solid MTX in a fume hood. Dispose of waste solutions safely.
- CHO cell transfectants containing a DHFR vector grown in selective medium (see *Protocol 2*)

Method

1. Add MTX to tissue culture dishes to final concentrations of 10, 20, 50, 100, 200, and 500 nM. Return to the tissue culture incubator.
2. After 10 days, examine the plates, by eye, for surviving colonies.
3. Isolate colonies from the dishes with the highest concentration of MTX yielding resistant colonies and analyse these individually for levels of product expression. Alternatively, pool the MTX-resistant colonies and subject them to a second round of selection at, for instance, 500 nM and higher concentrations of MTX.

Gene amplification can be up to 10-fold in the first round of amplification using *Protocol 3*, as measured by gene copy number or expression of a linked gene. In subsequent rounds of amplification, gene copy number is often roughly proportional to MTX-resistance and up to 2000 copies have been reported after three or more rounds of selection. However, expression levels may be saturated at lower copy numbers.

4. Glutamine synthetase selection in Chinese hamster ovary (CHO) cells

Glutamine is a key metabolite in a number of biosynthetic and catabolic pathways and must either be provided as a medium component or must be synthesized from glutamate and ammonia by means of GS. Some mammalian cell lines, such as CHO cell lines, express sufficient GS enzyme to survive in an appropriate glutamine-free medium. Under these conditions, GS is an essential enzyme and inhibition of GS by the specific inhibitor methionine sulphoximine (MSX) is lethal (14). A transfected GS gene in a mammalian expression vector can then act as a dominant selectable marker in CHO cells if it confers resistance to concentrations of MSX which are just sufficient to kill both wild-type cells and the natural MSX-resistant variants which result from amplification of the endogenous GS genes. Amplification of vector copy-number can subsequently be achieved using elevated levels of MSX.

4.1 Choice of cell type and medium

The preferred cell line is CHO-K1. For routine growth of this cell line, the preferred medium is GMEM-S with 10% dialysed FCS. CHO-K1 cells grow well in this medium and MSX is toxic at concentrations above 3 μM. Dialysed FCS is required since normal FCS can contain significant amounts of glutamine. It can be purchased already dialysed. Each batch should be tested for the ability to support growth of CHO-K1 and for the absence of glutamine (batches containing less than 20 μg/ml glutamine are adequate).

4.2 Vector design

The expression vector used, pEE14 (*Figure 2*), contains a GS minigene, with a single intron of the GS gene and about 1 Kb of 3' flanking DNA, transcribed from the SV40 Late promoter. This transcription unit is derived from pSVLGS.1 (8) and has been chosen because it typically yields transfectant cell lines with higher vector copy-numbers than some other GS expression vectors (unpublished results).

A number of restriction sites have been removed from the GS gene in pEE14, resulting in several convenient sites in the multilinker which now occur only once in the plasmid and these can be used for insertion of the gene sequences to be expressed. The rest of the vector sequences in pEE14 are also present in pEE6hCMV.gpt (*Figure 1*).

For expression of antibodies, the plasmid pEE6hCMV.gpt is needed in addition to pEE14. The preferred scheme for vector construction for antibody-expression is generally as follows. Insert a light-chain coding sequence (which must contain a translation–initiation signal) into the polylinker of pEE14. Similarly, insert the heavy-chain coding sequence into pEE6hCMV.gpt. Isolate a complete transcription unit from the heavy-chain vector as a *Bgl*II–*Bam*HI fragment and insert this at the *Bam*HI site of the light-chain plasmid such that both genes are in the same orientation. (For manipulation of human antibody sequences see Bendig *et al.*, Chapter 7, Walter and Tomlinson, Chapter 6, Pope *et al.*, Chapter 1).

4.3 Cell transfection

Cells should be kept growing exponentially in GMEM-S medium (*Protocol 4*) for at least a few days prior to transfection and should not be allowed to reach confluence prior to transfection. The DNA used for transfection is circular plasmid DNA purified by CsCl density centrifugation and checked by agarose gel electrophoresis to ensure that it is essentially free of chromosomal DNA and RNA. The transfection procedure is described in *Protocol 4* and selection for amplification in *Protocol 5*.

Protocol 4. Transfection and selection of GS vectors in CHO-K1 cells[a]

Equipment and reagents

- Dialysed FCS (Gibco-BRL, cat. no. 014–06300): heat inactivate at 56°C for 30 min and store at –20°C. Each batch should be tested for the presence of glutamine.
- 100 mM L-MSX stock solution (Sigma): prepare 18 mg/ml solution of L-MSX in PBS. Filter-sterilize and store at –20°C. **CAUTION**: MSX is toxic. Wear gloves when handling. Weigh out the solid in a fume hood and wear a face mask.
- 100 × non-essential amino acids (NEAA) (Gibco-BRL, cat. no. 11140): store at 4°C. For Gibco-BRL GMEM-S medium only.
- 100 × glutamate + asparagine (G + A): add 600 mg glutamic acid (Sigma) to 600 mg asparagine (Sigma). Make up to 100 ml with distilled water, sterilize through a 2 μm filter (Nalgene), and store at 4°C. For Gibco-BRL GMEM-S medium only.
- 100 mM sodium pyruvate (Gibco-BRL, cat. no. 11360). For Gibco-BRL GMEM-S medium only.
- 50 × nucleosides: add 35 mg adenosine, 35 mg guanosine, 35 mg cytidine, 35 mg uridine, and 12 mg thymidine (each from Sigma). Make up to 100 ml with distilled water, filter-sterilize and store at –20°C in 10 ml aliquots. For Gibco-BRL GMEM-S medium only.
- GMEM-S medium (Gibco-BRL formulation)[a,b]: add the following, in order, and aseptically, 400 ml autoclaved distilled water, 50 ml 10 × GMEM basal medium *without* glutamine (Gibco-BRL, cat. no. 042–2541), 18 ml 7.5% sodium bicarbonate (Gibco-BRL, cat. no. 043–05080), 5 ml NEAA, 5 ml G + A, 5 ml 100 mM sodium pyruvate, 10 ml 50 × nucleosides, and 50 ml dialysed FCS.
OR
- GMEM-S medium (JRH Biosciences formulation): add the following, in order, and aseptically, 500 ml GMEM-S medium (cat. no. 51492), 10 ml 50 × GS supplement (58672), and 50 ml dialysed FCS.

12: Expression of immunoglobulin genes in mammalian cells

- Circular recombinant plasmid DNA, including pEE14 as a control. Purify the plasmid DNA by CsCl centrifugation, then precipitate from ethanol, wash the DNA pellet with 70% ethanol and leave the DNA pellet to dry in a tissue culture laminar-flow cabinet to ensure sterility. Dissolve in sterile distilled water at 0.5 mg/ml.
- CHO-K1 cells (available from ATCC, cat. no. CCL61) growing exponentially in GMEM-S medium
- Tissue culture Petri dishes (90 mm diameter)
- Equipment and reagents for calcium phosphate transfection procedure (*Protocol 2*)

Method

1. On the day prior to transfection, trypsinize the CHO cells and seed 9 cm Petri dishes at 10^6 cells per dish in 10 ml GMEM-S medium. Incubate in the tissue culture incubator overnight.
2. Transfect the cells with plasmid DNA using the calcium phosphate co-precipitation procedure (*Protocol 2*). Also treat several plates by the calcium phosphate procedure but without DNA to act as controls. Return the dishes to the tissue culture incubator.
3. One day after transfection, add MSX to a final concentration of 25 μM in GMEM-S medium.
4. Count the number of surviving colonies 2–3 weeks after transfection.
5. Isolate individual colonies and analyse expression of the desired product. When expanding cultures, maintain the MSX concentration but, after trypsinization, wait one day before adding the MSX to allow the cells to recover.

[a] The products of two different medium-manufacturers are given. The JRH Biosciences medium has recently been made available and is the simpler to prepare. Gibco-BRL reagents have been more widely used, but different sets of components are available in the US[b] from those in the UK.
[b] Gibco-BRL US formulation: 500 ml GMEM *without* glutamine (cat. no. 320–1710, this is not a stock item and will need to be made to order), 5 ml NEAA (cat. no. 11140), 5 ml G + A (as UK), 5 ml 100 mM sodium pyruvate (cat. no. 11360), 10 ml 50 × nucleosides (as UK), and 50 ml dialysed FCS (cat. no. 220–6300).

Protocol 5. Selection for GS vector amplification in CHO cells

Equipment and reagents

- Individual cell lines derived from colonies transfected with GS vector (from *Protocol 4*)
- GMEM-S medium + 10% dialysed FCS (*Protocol 4*)
- 90 mm tissue culture dishes or 75 cm tissue culture flasks
- 100 mM MSX stock solution (*Protocol 4*)

Method

1. Take several individual transfected cell lines[a] producing significant amounts of the desired product (from *Protocol 4*) and plate out each cell line on several dishes or in flasks at a density of approximately

Protocol 5. *Continued*

10^6 cells per dish or flask in GMEM-S + 10% dialysed FCS. Incubate for 24 hours in the tissue culture incubator.[b]

2. Replace the medium with fresh GMEM-S + 10% dialysed FCS containing various concentrations of MSX, ranging from 100 µM to 1 mM.

3. Incubate the Petri dishes for 10–14 days, changing the medium once during this time (replace it with fresh GMEM-S + 10% dialysed FCS containing the same concentration of MSX). After this time, considerable cell death should have occurred and colonies resistant to the higher levels of MSX should have appeared. The maximum concentration of MSX at which colonies survive will depend on the particular initial transfectant but is typically between 250 µM and 500 µM.

4. Isolate several discrete colonies which have survived at the highest MSX concentration. Either pick the colonies and assay them individually or alternatively pool all colonies from one initial cell inoculum (step 1) and assay these together.[c]

5. Clone those amplified individual cell lines or pools which exhibit high production rates of product using limiting dilution cloning. Re-analyse the expression of product in the cloned cell lines.

[a] In our experience, independent transfectants amplify more efficiently than pools of transfectants.
[b] Whenever trypsinizing GS-selected cells, leave the cells for 24 hours to recover before re-applying MSX selection.
[c] The increased production rate of the desired product can be up to 10-fold in this first round of amplification. It is not normally appropriate to select for subsequent rounds of amplification because the production rate does not usually increase significantly at higher levels of MSX.

5. Glutamine synthetase selection in NS0 myeloma cells

Some mammalian cell lines, such as rodent myelomas and hybridomas, do not express sufficient GS to survive without added glutamine. In this case, a transfected GS gene can function as a selectable marker by permitting growth in a glutamine-free medium. Amplification of vector copy-number can subsequently be achieved using elevated levels of MSX. The basic system (8) has been updated using modified vectors to simplify plasmid construction.

5.1 Choice of cells and medium

The preferred cell line is NS0 (15). It is highly transfectable and yields essentially no glutamine-independent variants (less than 1 in 10^8 cells). Other rodent myelomas may also be used provided that they can be transfected by

12: Expression of immunoglobulin genes in mammalian cells

electroporation and do not grow or yield variants which can grow in the glutamine-free selection medium.

Two alternative selective medium formulations are described in *Protocol 6*. G-DMEM (based on DMEM) or G-IMDM (based on the Iscove modification of DMEM).

5.2 GS-vector design

The basic expression vector used here is pEE12 (*Figure 3*). This is identical to pEE14 (*Figure 2*), used for CHO cells, except that the GS minigene in pEE14 has been replaced with a GS cDNA expressed from the SV40 Early promoter, derived from the transcription unit from pSV2.GS (8). This GS gene gives a higher transfection efficiency in NS0 cells than the GS gene in vector pEE14. In constructing pEE12, several restriction sites have been removed from the GS gene so that sites in the multilinker remain unique.

For expression of antibodies, the coding sequences for the two polypeptides should be separately inserted into vectors downstream of an hCMV promoter and a single vector subsequently constructed containing transcription units for both chains and the GS selectable marker. The vector pEE6hCMV.gpt (*Figure 1*) is useful for this purpose. The preferred scheme is generally as follows. Insert a light-chain coding sequence (which should contain a translation initiation signal) at the multilinker of pEE12 and the heavy-chain coding sequence with translation–initiation signal at the multilinker of pEE6hCMV.gpt. Then isolate the complete hCMV-heavy chain-SV40 transcription unit as a *Bgl*II–*Bam*HI cassette and insert this at the *Bam*HI site of the pEE12-light-chain plasmid, such that the light- and heavy-chain genes will be transcribed in the same orientation. Light-chain coding sequences can be present as cDNAs, but genomic constant region segments have generally been used for the heavy chain. It is not known whether removal of the introns present in the heavy chain constant region would affect expression.

This type of plasmid has worked well for the synthesis of a number of antibodies. A single round of selection for gene amplification, or sometimes no amplification at all, should suffice to provide high-yielding cell lines. In some cases, the productivity of such lines has been stably maintained even in the absence of continued MSX selection (16).

It is not necessarily the case that a vector such as that described above will be optimal for every antibody and so, if resources permit, it may be useful to try additional vector constructs. The plasmid construction described above has the theoretical disadvantage that light-chain transcription may be more efficient than heavy-chain transcription due to promoter occlusion of the downstream hCMV promoter by the upstream hCMV promoter. Heavy-chain expression may therefore limit productivity. Whether this is in fact the case will also depend on the efficiency of translation of the mRNAs for the two chains and this may be antibody dependent. It may therefore be advantageous to prepare and test the alternative plasmid, produced by cloning the heavy-chain gene into pEE12 and the light-chain gene into pEE6hCMV.B and combining the two so that the heavy chain is now upstream of the light chain. Another approach makes use of a modified pEE12 plasmid, pEE13 (*Figure 4*), which has a transcription–termination signal from a mouse immunoglobulin μ-chain gene downstream of the SV40 polyadenylation signal. This serves to isolate the two immunoglobulin genes in the final plasmid and minimizes the potential promoter occlusion, so leading to higher overall expression of assembled immunoglobulin in several cases (unpublished results).

5.3 Cell transfection

The NS0 cells should be kept growing exponentially in non-selective medium and have a cell viability greater than 90% prior to transfection. Cells are grown at 37°C in an atmosphere of 5% CO_2. Linearized DNA is introduced into the cells by electroporation. Transfected cells are then selected for the ability to grow in glutamine-free medium by first gradually reducing the glutamine concentration (*Protocol 6*). The frequency with which glutamine-independent transfectants are isolated should be at least 1 in 10^5 cells plated, using this procedure for NS0 cells.

Protocol 6. Transfection of GS-vectors into NS0 cells[a]

Equipment and reagents

- NS0 cells (ECACC, cat. no. 85110503)
- Electroporation apparatus (e.g. Bio-Rad 'Gene Pulser'; note that a capacitance extender is not needed for this procedure)
- CsCl density centrifugation equipment and reagents
- 0.4 cm electroporation cuvettes (Bio-Rad, cat. no. 165-2088)
- Restriction enzyme (e.g. *Bam*HI or *Sal*I) and appropriate buffers
- Ethanol
- Suitable tissue culture equipment including humidified CO_2 incubators, set at 37°C and 5% CO_2
- 96-well and 24-well tissue culture trays
- PBS (e.g. Flow Laboratories Ltd, cat. no. 28-203-05), made up according to the manufacturer's instructions
- Benchtop centrifuge
- Recombinant plasmid DNA (a GS-vector based on pEE12 or pEE13)
- 40 μg pEE12 DNA as a control for transfection
- Dialysed FCS (*Protocol 4*)
- 200 mM L-glutamine (e.g. Gibco-BRL, cat. no. 043-05030[c]): store in 5 ml aliquots at −20°C

- Non-dialysed FCS (treat as described in *Protocol 4*)
- G + A (*Protocol 4*)
- NEAA (*Protocol 4*)
- 50 × nucleosides (*Protocol 4*)
- GS supplement (*Protocol 4*)
Use either DMEM or IDMM medium[b]
- Celltech DME (JRH Biosciences, cat. no. 51435)
  OR
  DMEM base *without* glutamine and ferric nitrate but *with* 1 mM sodium pyruvate (specially prepared by Gibco-BRL)
- IMDM base (JRH Biosciences, cat. no. 51472)
  OR
  IMDM base (Sigma, cat. no. I-2762 or I-4136)
- Non-selective medium: 500 ml DMEM or IMDM base medium, 5 ml 200 mM L-glutamine, and 50 ml FCS
- Selective medium (G-DMEM, JRH Biosciences formulation): 500 ml Celltech DME, 10 ml GS supplement, and 50 ml dialysed FCS

12: Expression of immunoglobulin genes in mammalian cells

OR
 Selective medium (G-DMEM, Gibco-BRL formulation): 500 ml DMEM, 5 ml NEAA, 5 ml G + A, 10 ml 50 × nucleosides, and 50 ml dialysed FCS

OR
 Selective medium (G-IMDM, JRH Biosciences formulation): 500 ml IMDM, 10 ml GS supplement, and 50 ml dialysed FCS

OR
 Selective medium (G-IMDM, Sigma formulation): 500 ml IMDM, 10 ml G + A, 10 ml 50 × nucleosides, and 50 ml dialysed FCS

Method

1. Purify the plasmid DNA by CsCl density centrifugation, ethanol precipitate the DNA and then linearize it by digestion with a restriction enzyme which cuts once in the bacterial plasmid sequence (e.g. *Bam*HI or *Sal*I) using an appropriate buffer. Re-precipitate the DNA using ethanol, wash the pellet with 70% ethanol and allow to dry in a tissue culture laminar-flow hood to ensure sterility. Resuspend 40 µg of the DNA in 40 µl sterile distilled water.

2. On the day of transfection, count the cells. 10^7 cells will be needed per plasmid transfection, plus 10^7 cells to be 'mock' transfected without DNA and 10^7 cells to be transfected with pEE12 as a positive control.

3. Centrifuge the cells (1200 r.p.m. for 5 min in a benchtop centrifuge) and wash the cells once by centrifugation in 50 ml cold PBS. From this stage, maintain the cells on ice or at 4°C.

4. Resuspend the cells in PBS at 10^7 cells/ml.

5. Add 10^7 cells to each electroporation cuvette on ice. Add 40 µl plasmid DNA to one cuvette, the pEE12 DNA (positive control) to another, and 40 µl water to another cuvette containing cells to act as the mock-transfected sample. Mix each sample gently with a pipette but avoid bubbles and excess liquid up the sides of the cuvette.

6. Leave the cuvettes containing the cells on ice for 5 min.

Note: *Before using the electroporation apparatus, read the manufacturer's instructions and observe all safety precautions.*

7. Wipe the outside of each cuvette dry and use the 'Gene Pulser' to deliver two consecutive pulses at 1500 V, 3 µF according to the manufacturer's instructions.

8. Return the cuvettes to ice for 2–5 min and then add each sample of cells to 30 ml of non-selective culture medium, pre-warmed to 37°C, and mix well.

9. Plate out each of the samples of transfected cells (including the pEE12 positive control transfection and the mock transfection) as follows. It is suggested that three dilutions of the original cell

Protocol 6. *Continued*

suspension in step 8 are made to ensure that colonies can be picked from plates containing, on average, less than one transfected cell/well. This minimizes the risk of picking two or more colonies together.

  (a) Distribute 20 ml of cell suspension from step 6 into four 96-well tissue culture plates (i.e. approximately 50 µl per well).

  (b) Dilute 10 ml of cell supension with a further 30 ml of non-selective medium (i.e. 1 in 4 dilution) and distribute over another five 96-well plates (50 µl per well).

  (b) Dilute 10 ml of diluted cell suspension from (b) with 40 ml of non-selective medium (i.e. 1 in 16 dilution) and distribute over a further five 96-well plates (50 µl per well).

10. Return all the plates to a 37 °C tissue culture incubator and incubate overnight.

11. The next day, add 150 µl of selective medium to each well of the transfected plates without removing the medium already there. Return the plates to the incubator and incubate until substantial cell death has occurred and discrete surviving colonies appear.[a]

12. Three weeks post-transfection, viable colonies of glutamine-independent transfectants should be visible amongst the background of dead cells. Because the selection procedure depends on the cells depleting the medium of glutamine gradually, cells in plates which have been diluted before plating out will take longer to die and may even grow to form colonies for a few days after transfection. However, by three weeks any such growth should have died off. Viable colonies can be distinguished microscopically by the occurrence of bright round cells. There should be no surviving colonies on the mock-transfected plates. The frequency of viable colonies should be 2–5 per 10^5 cells plated for the pEE12 positive control plasmid. Certain recombinant plasmids may show a somewhat reduced transfection efficiency. Identify wells containing single healthy colonies and mark these on the lid of the plate with a pen.

13. Collect spent culture medium from the chosen wells, once the medium has begun to turn orange–yellow, and use this to assay for product secretion (see Section 5).

14. Do not expand the transfectant clones too rapidly. (Direct inoculation of a flask from the 96-well plate may lead to reduced growth rate or significant cell death.) Transfer the cells from one well of a confluent 96-well plate to one well of a 24-well plate. Then, after several days, take the contents of the well to inoculate a small flask. At each stage, include in the transfer of cells as much as possible of the medium in which the cells were growing. As a precaution, re-feed the empty

12: Expression of immunoglobulin genes in mammalian cells

wells after transfer of cells. In this way, if the transferred cells fail to survive, residual cells in the original well can be used for a second attempt.

15. Once in flasks, maintain the cultures at between 10^5 and 10^6 cells/ml. The doubling time will vary between transfectants but should be 20–40 h. Cell lines growing more slowly than this are unlikely to be useful.

[a] This procedure allows the cell to deplete the medium of residual glutamine so the glutamine concentration declines gradually.

[b] Two different base media are described and each can be obtained from two different manufacturers. DMEM is the preferred base, but IMDM is adequate if DMEM is not available. The JRH Biosciences media formulations have recently been made available and are simpler to prepare. The alternative formulations based on Gibco-BRL and Sigma reagents have been more widely used, but require more components to be added to the base medium.

5.4 Selection for GS-gene amplification in NS0 cells using MSX

Amplification of the GS-vector in the NS0 cells can be selected using MSX as described in *Protocol 7*.

Protocol 7. Selection for GS-vector amplification in NS0 cells

Equipment and reagents

- Several (at least 5) independent transfected cell lines producing significant amounts of product from *Protocol 6*[a]
- 24-well tissue culture plates
- Selective medium (*Protocol 6*)
- 100 mM L-MSX (*Protocol 4*)

Method

1. Distribute the transfected cells in 24-well plates at a density of 2–5 × 10^5 cells/0.5 ml/well in selective medium.

2. Add 0.5 ml of selective medium containing MSX to each well to *final* concentrations ranging between 5 μM and 500 μM (use several wells for each concentration of MSX).

3. Incubate the plates until discrete MSX-resistant colonies appear (typically 3–4 weeks).

4. Isolate pools or individual colonies from each independent transfectant at the highest MSX concentration at which MSX resistance occurs. This varies widely between different transfectants.

5. Assay the MSX-resistant cell lines or pools for protein product.[b]

Protocol 7. *Continued*

6. Clone the amplified pools or cell lines secreting product at the highest rate by limiting dilution cloning, maintaining MSX selection throughout. Screen the clones obtained for the highest levels of product secretion.
7. Store stocks of these frozen in liquid nitrogen.
8. Once frozen cell stocks have been secured, the stability of the production rate can be tested by growth of the cells in the presence and absence of MSX for extended periods (e.g. 2 months).

[a] Pools of transfectants do not amplify as efficiently as independent cell lines.
[b] The secretion rate (in μg/10⁶ cells/day) should be 2–10-fold higher than the original transfectant if vector amplification has occurred. Not all primary transfectants will show significant amplification.

In some cases highly amplified arrays of vector sequences can be retained in the absence of the selective agent (at least for several weeks in culture) while in other cases, the amplified sequences are lost very rapidly if the selection is removed. For many purposes it may be acceptable to maintain cells in the presence of the selective agent, but if stability in the absence of selective agent is considered important, it would be advisable to screen a number of independently derived lines to identify any which show stable productivity under such conditions.

6. Concluding remarks

The methods described in this chapter, making use of highly efficient transcription cassettes and stringent selection methods, have made it possible to produce a number of immunoglobulins and related molecules rapidly for *in vitro* analysis. Both CHO and NS0 cell lines are also compatible with growth on a large scale in a variety of different fermentation systems, thus permitting economic production of many of these important proteins.

References

1. Bebbington, C. R. (1995). In *Monoclonal antibodies: the second generation* (ed. H. Zola), pp. 165–81. Bios Scientific, Oxford.
2. Stark, G. (1986). *Cancer Surv.*, **5**, 1.
3. Coloma, M. *et al.* (1992). *J. Immunol. Methods*, **152**, 89–104.
4. Gluzman, Y. (1981). *Cell*, **23**, 175–82.
5. Whittle, N., Adair, J., Lloyd, C., Jenkins, L., Devine, J., Schlom, J., Raubitschek, A., Colcher, D., and Bodmer, M. (1987). *Protein Eng.*, **1**, 499–505.
6. Seidman, M. (1989). *Mutat. Res.*, **220**, 55–60.
7. Stephens, P. and Cockett, M. (1989). *Nucl. Acids Res.*, **17**, 7110.
8. Bebbington, C. R., Renner, G., Thomson, S., King, D., Abrams, D., and Yarranton, G. T. (1992). *Bio/Technology*, **10**, 169–75.
9. Law, R., Kuwabara, M. D., Briskin, M., Fasel, N., Hermanson, G., Sigman, D. S., and Wall, R. (1987). *Proc. Natl Acad. Sci. USA*, **84**, 9160–4.

12: Expression of immunoglobulin genes in mammalian cells

10. Urlaub, G. and Chasin, L. A. (1980). *Proc. Natl Acad. Sci. USA*, **77**, 4216–20.
11. Urlaub, G., Kas, E., Carothers, A. M., and Chasin, L. A. (1983). *Cell*, **33**, 405–12.
12. Lee, F., Mulligan, R., Berg, P., and Ringold, G. R. (1981). *Nature*, **294**, 228.
13. Crouse, G. F., McEwan, R. N., and Pearson, M. L. (1983). *Mol. Cell. Biol.*, **3**, 257–66.
14. Wilson, R. (1993). In *Gene amplification in mammalian cells* (ed. R. Kellems), pp. 301–11, Marcel Dekker Inc., N.Y.
15. Galfre, G. and Milstein, C. (1981). In *Methods in enzymology* (ed. J. J. Langone and H. Van Vunakis), Vol. 73(B), pp. 3–46. Academic Press, London.
16. Brown, M., Renner, G., Field, R. P., and Hassell, T. (1992). *Cytotechnology*, **9**, 231–6.
17. Ward, E. S. and Bebbington, C. R. (1995). *Monoclonal antibodies* (ed. J. R. Birch and E. S. Lennox), p. 137. Wiley–Liss, NY.

13

Preparation and uses of Fab' fragments from *Escherichia coli*

PAUL CARTER, MARIA L. RODRIGUES, JOHN W. PARK, and GERARDO ZAPATA

1. Introduction

1.1 Overview

Antibody (Ab) engineering has been revolutionized by the advent of methods to express functional Fv (1) and Fab (2) from *Escherichia coli* by co-secretion of corresponding light and heavy chain fragments (see also Chapter 10). The low Ab fragment titres obtained using shake-flask cultures (typically < 5 mg/litre) are useful for many research applications. However, such titres are inconveniently low for biophysical or X-ray crystallographic studies and unsuitable for biotechnology or human therapy where gram or even kilogram quantities of material may be required. This prompted us to develop a system for very high-level expression (up to 2 g/litre) of Ab fragments in *E. coli* grown to high cell-density in the fermenter (3). This high-level production of soluble and functional Ab fragments (reviewed in ref. 4) has facilitated biophysical (5, 6), X-ray crystallographic (7, 8), and pharmacokinetic (9) studies as well as preclinical efficacy studies in Rhesus monkeys.

Other systems for high-level expression of Ab fragments in *E. coli* have subsequently been described by others (refs 10, 11, Chapter 10). These studies, taken together with our own, demonstrate that high-level expression of various Ab fragments is possible utilizing a variety of different promoters (*pho*A, *lac*Z, arabinose) and signal sequences (*st*II, *pel*B, *pho*A, *omp*A) (reviewed in ref. 12).

Here we focus on Fab' fragments which differ from Fab fragments by the addition of a few (typically less than ten) hinge region residues at the carboxy terminus of the heavy chain C_H1 domain, in this case a single cysteine followed by two alanine residues. The starting point for the protocols presented here are *E. coli* fermentation pastes following Fab' expression using a plasmid such as pAK19 (ref. 3, *Figure 1*). Fab' fragments are first recovered from *E. coli* fermentation pastes with the single unpaired hinge cysteine in the free thiol form (Fab'–SH) (*Protocols 1* and *2*). Applications of such fragments are presented which take advantage of directed chemical coupling using the free cysteine residue: construction of monospecific (Ms) F(ab')$_2$ (*Protocol 4*), bispecific (Bs) F(ab')$_2$ (*Protocol 5*), Fab'–PEG (*Protocol 6*), immunoliposomes for targeted drug delivery (*Protocol 7*), and immobilization of Fab' on a solid support for affinity purification (*Protocol 8*).

1.2 High-level production of Ab fragments in *E. coli*

Plasmid pAK19 (3) directs the co-expression of light chain and heavy chain Fd' fragment of the humanized variant of the anti-p185^{HER2} Ab, HuMAb4D5-8 (13), from a synthetic dicistronic operon (*Figure 1*). Functional HuMAb4D5-8 Fab' is expressed at titres of up to 2 g/litre from *E. coli* strain 25F2 containing plasmid pAK19 grown to high cell-density (120–150 OD_{550}) in the fermenter (3). The Fab' fragment of HuMAb4D5-8 may be a broadly useful template for high-level *E. coli* expression of humanized Fab' molecules since anti-CD3 and anti-CD18 Abs created by CDR grafting were expressed at titres of up to 700 mg/litre (14) and 2 g/litre, respectively (B. Snedecor, personal communication). In contrast, corresponding chimeric Fab' fragments containing murine variable domains are expressed at 10- to 50-fold lower levels than their humanized counterparts. Thus high titres of Ab fragments in the fermenter are dependent upon the primary sequence. In addition, it is crucial to use tightly regulated promoters to avoid plasmid loss during culturing (reviewed in ref. 12).

1.3 'Targeting' of Fab' fragments to the periplasmic space of *E. coli*

Fab' fragments accumulate predominantly in the periplasmic space following fermentation growth at slow agitation rates (3), whereas high agitation rates leads to preferential accumulation in the media (5), following passive leakage of Fab' from cells. Functional Fab' fragments are readily and efficiently

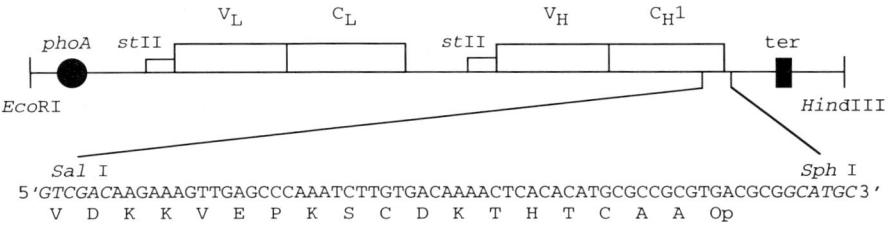

```
            V_L       C_L              V_H      C_H1
phoA  stII  ┌────┐────┐      stII  ┌────┐────┐          ter
 ●─────────┤    │    ├───────────┤    │    ├──────────■
EcoRI                                                      HindIII

Sal I                                                   Sph I
5'GTCGACAAGAAAGTTGAGCCCAAATCTTGTGACAAAACTCACACATGCGCCGCGTGACGCGGCATGC3'
  V  D  K  K  V  E  P  K  S  C  D  K  T  H  T  C  A  A  Op
```

Figure 1. Schematic representation of the dicistronic operon for Fab' expression in plasmid pAK19 (3). Expression is under the transcriptional control of the *E. coli* alkaline phosphatase promoter (*phoA*) which is inducible by phosphate starvation. Heavy and light chains are preceded by the *E. coli* heat-stable enterotoxin II (*stII*) signal sequence to direct secretion to the periplasmic space of *E. coli*. The 3' end of the C$_H$1 gene is flanked by *Sal*I and *Sph*I (sequence in italics) to facilitate modification of the hinge region, in this case encoding for the hinge CAA (single letter code). The C$_H$1 gene is flanked on its 3' side by the bacteriophage λ t$_o$ transcriptional terminator (ter).

recovered from fermentation pastes with the unpaired hinge cysteine predominantly (65–90%) in the free thiol form (Fab'–SH) (3, 14, 15). In contrast, Fab' in the media are present as covalent adducts, and regeneration of free thiol requires mild reduction (10). We prefer to prepare Fab' fragments from fermentation pastes rather than media, as this avoids the reduction step which is not completely selective for the hinge cysteine and may result in unwanted reduction of the interchain disulfide bond between heavy and light chains (16).

2. Recovery of Fab' fragments from E. coli

2.1 Release of soluble Fab'–SH fragments from E. coli

Fab' fragments are first released from *E. coli* fermentation pastes in near quantitative yield (as judged by antigen-binding ELISA) by partial digestion of the bacterial cell wall using hen egg-white lysozyme (step 1 of *Protocols 1* and *2*). Similarly efficient release can also be achieved by osmotic shock in the presence of 20% (w/v) sucrose 200 mM NaCl (PC, unpublished data). The most complete release of Fab' is achieved by fully disrupting the cells by sonication or by using a French press or microfluidizer (GZ, unpublished data). However, the improved release is offset by a much more challenging purification from both protein and nucleic acid. Recovery of Fab' is performed at pH ~5.5, to maintain the free thiol in the less reactive protonated form, and in the presence of EDTA, to chelate metal ions capable of catalysing disulfide bond formation. Alternatively, the free thiol can be protected, e.g. with pyridyl disulfide, followed by deprotection prior to any directed coupling step (GZ, unpublished data).

2.2 Recovery of Fab'–SH fragments from E. coli

Streptococcal Protein G affinity chromatography is a reliable method for the purification of Fab' fragments of chimeric and humanized Abs (ref. 3, step 4 of *Protocols 1* and *2*) that takes advantage of the Protein G binding site located on the C_H1 domain of human IgG1 (17). Staphylococcal Protein A has proved useful for affinity purification of some humanized Fab fragments (5) but is not as broadly useful as Protein G. In addition, Fab fragments purified using Protein A are commonly more heterogeneous than those purified using Protein G.

Ab fragments have been engineered with affinity-purification tails to permit affinity purification by other methods (reviewed in ref. 12), e.g. a poly His tail at the C-terminus of V_L has been used for affinity purification of single-chain Fv (scFv) fragments by immobilized metal affinity chromatography (18). More recently a peptide tag for affinity purification of scFv fragments using streptavidin has also been developed (19). Bakerbond ABx resin (J. T. Baker) offers an economical and broadly useful alternative to Protein A and

Protein G for the affinity purification of Ab that is compatible with Fab′ and other fragments obtained from fully disrupted *E. coli* cells (GZ, unpublished data, see also *Protocol 6*). ABx behaves like an ion exchanger in that Ab and Ab fragments may be resolved by manipulating the ionic strength and/or the pH. In addition, ABx has affinity matrix-like properties in that it binds Ab but virtually none of the contaminating proteins from *E. coli* or from mammalian culture fluids. ABx utilizes mixed modes in binding to Ab: weak cation-exchange, mild anion-exchange, and mild hydrophobic interaction. *Protocol 1* has been successfully scaled down for the purification of small (microgram) quantities of Fab fragments obtained from low cell-density shake-flask cultures (20). It is anticipated that Fab′ fragments might be prepared from shake-flask cultures in a similar manner.

2.3 Endotoxin removal

Removal of endotoxin from *E. coli*-derived Ab fragments is a prerequisite for most *in vivo* applications. We routinely obtain MsF(ab′)$_2$, BsF(ab′)$_2$, and Fab′–PEG fragments with ≤ 0.5 endotoxin units per mg (*Protocol 1* followed by *Protocols 4, 5*, and *6*, respectively) as judged by the limulus amoebocyte lysate test (Associates of Cape Cod). The following simple precautions are recommended in the purification of Ab fragments free of endotoxin:

- Use endotoxin-free sterile water to prepare all buffers.
- Avoid the use of all glassware which may have been washed with water contaminated with endotoxin.
- Purge gel-filtration column of endotoxin prior to use with 0.1 M NaOH.

PyroBind ST filters are effective in purging Fab′–PEG or F(ab′)$_2$ preparations of low levels of contaminating endotoxin if necessary. These precautions can be ignored for the sake of simplicity and cost for applications of Fab′ fragments where residual endotoxin is not important. The endotoxin removal methods described here are convenient for research-scale preparations of Ab fragments. Methods such as ion-exchange chromatography and ultrafiltration are more readily amenable to large-scale production as described by Gavit *et al.* (21).

Protocol 1. Preparation of Fab′–SH fragments from *E. coli* fermentation pastes

Equipment and reagents

- Endotoxin-free sterile water (McGaw) is used to prepare all buffers
- 0.1 M acetic acid, pH ~ 2.8 (unbuffered)
- 0.1 M PMSF (Sigma, cat. no. P. 7626) in anhydrous ethanol stored at –20 °C
- 0.25 M EDTA, pH 8.0
- Hen egg-white lysozyme (Canadian Lysozyme)
- DEAE fast-flow (Pharmacia, cat. no. 17–0709–01)

13: Fab' fragments from E. coli

- 500 ml capacity tissue culture filter (Corning, cat. no. 25992–500)
- Protein G fast-flow (Pharmacia, cat. no. 17-0618-02). Pack a column using ≥ 1 ml resin per 4 mg of Fab' to be purified. Regenerate the resin after use by washing with 2 column volumes (cv) 3 M potassium thiocyanate in 30 mM Tris–HCl, pH 8.0. Store the column in PBS containing 20% (v/v) ethanol at 4°C.
- PBS: 137 mM NaCl, 2.68 mM KCl, 7.96 mM Na_2HPO_4, 1.47 mM KH_2PO_4, pH 7.2
- ME buffer: 10 mM MES, pH 5.5, 1 mM EDTA
- AE buffer: 100 mM acetic acid, pH 2.8, 1 mM EDTA
- 0.5 M MES, pH 5.5
- Millex-GV 0.22 μm filter (Millipore, cat. no. SLGVO25LS)
- Cryogenic vials (Corning, cat. no. 25078)
- Orbital shaker (Bellco Biotechnology, cat. no. 7744-02020)
- 350 ml sintered-glass funnel (Corning, cat. no. ASTM 40–60 C)
- FPLC or low-pressure chromatography system

Method

1. *Preparation of shockate.* To 100 g frozen fermentation paste add:

    - 0.1 M acetic acid, pH ~ 2.8   180 ml
    - 0.25 M EDTA   20 ml
    - 0.1 M PMSF   2 ml

    Mix vigorously for 30 min on an orbital mixer at room temperature (RT) to thaw paste. Add 100 mg hen egg-white lysozyme and stir for 30 min at RT. Pellet debris by centrifugation at 27 500 *g* for 10 min at 4°C.

2. *DEAE fast-flow batch chromatography*[a]. Equilibrate ~100 ml DEAE fast-flow resin with ME buffer on a 350 ml sintered-glass funnel using a water (or other low pressure vacuum) pump. Pass the sample over the resin and then suck resin dry.

3. *Filtration.* Remove the residual debris from the sample by passing through a 0.22 μm tissue culture filter.

4. *Protein G Sepharose 4 fast-flow chromatography.* Equilibrate the Protein G column with ME buffer. Load the sample and then wash the column with ≥ 4 column volumes (cv) of ME buffer (60 cm h^{-1}). Elute the Fab' with 1.5–2.5 cv AE buffer (12 cm h^{-1}). Pool the Fab'-containing fractions and add 0.2 volumes (vol) 0.5 M MES, pH 5.5.[b]

5. *Sterile filtration.* Pass the Fab' preparation through a 0.22 μm filter.

6. *Yield of Fab'–SH.* Determine the Fab' concentration from the absorbance at 280 nm and the free thiol content by analysis with DTNB (5,5'-dithiobis(2-nitrobenzoic acid) as described in *Protocol 3*. The concentration of Fab' is commonly >1 mg/ml with a recovery of 100–900 μg Fab' per gram of paste depending upon the Ab and the fermentation run. A 10 litre fermentation run with plasmid pAK 19 will typically yield 1.1–1.4 kg paste (3). The free thiol content is usually in the range 65–90%.

Protocol 1. *Continued*

7. *Storage.* Aliquot the purified Fab' into cryogenic vials, flash-freeze in liquid nitrogen, and then store at –70 °C until required.

[a] The DEAE fast-flow step serves to reduce contaminating cell wall debris, nucleic acid, and endotoxin in the sample and improves the efficiency of the subsequent Protein G chromatography as well as helping to preserve the life of the protein G column.
[b] The slow elution rate prevents undue dilution of Fab' eluted from Protein G. This pool is then raised in pH to minimize acid denaturation of the Fab' fragment.

Protocol 2. Preparation of Fab'–SH fragments from *E. coli* fermentation pastes with very low residual endotoxin

Reagents

As for *Protocol 1* with the following additions and modifications:

- Phenyl Sepharose 6 fast-flow (Pharmacia, cat. no. 17-0973-05). Pack a column using ≥ 1 ml resin per 5 mg of Fab' to be purified. Regenerate the column by washing in turn with: 4 cv 1 M NaOH, 3 cv endotoxin-free sterile water, and 5 cv 70% (v/v) ethanol. Store the column in PBS containing 20% (v/v) ethanol at RT.
- SPEAS buffer: 25 mM NaH_2PO_4, 1 mM EDTA, 25 mM $(NH_4)_2SO_4$, adjusted to pH 3.0 using concentrated HCl
- PyroBind ST (Sepracor, cat. no. MPAPST)
- 0.5 M MES buffer, pH 5.5

Method

1. Proceed as for *Protocol 1*, steps 1–4, except that the Fab' is eluted from the Protein G column with SPEAS buffer.

2. *Phenyl Sepharose 6 fast-flow chromatography.* Equilibrate the phenyl Sepharose column with 4 cv SPEAS buffer and load the Protein G-purified Fab'. Pool the break-through fractions and add 0.2 vol 0.5 M MES, pH 5.5.

3. *Endotoxin removal.* Purge the Fab' preparation of residual endotoxin by multiple passage (≥ 10) over PyroBind ST using two 60 ml syringes and a 3-way stopcock as a female–male Luer adapter.

4. Proceed as *Protocol 1*, steps 5–7. The yield of Fab' usually approaches that from *Protocol 1* and the free thiol content is improved and may approach 100%.

2.4 Determination of free thiol content

The yield of Fab'–SH determined by analysis with 5,5'-dithiobis(2-nitrobenzoic acid, DTNB (*Protocol 3*), is routinely 65–90%. Mass spectrometric analysis is consistent with the remaining Fab' including a covalent adduct with glutathione (J. Bourell, personal communication).

13: Fab' fragments from E. coli

> **Protocol 3.** DTNB analysis to determine free thiol content of Fab' preparations[a]
>
> *Equipment and reagents*
> - 100 mM sodium phosphate, pH 7.0
> - 3 mM DTNB in 100 mM sodium phosphate, pH 7.0. Store at −70°C
> - Dual-beam spectrophotometer
> - Optically matched pair of 1 cm path-length quartz cuvettes
>
> *Method*
> 1. Add 950 μl 100 mM sodium phosphate, pH 7.0 to quartz reference cuvette and 750 μl to sample cuvette. Auto-zero the spectrophotometer at a wavelength of 280 nm.
> 2. Add 200 μl Fab' preparation to the sample cuvette and measure the absorbance at both 280 nm and 412 nm.
> 3. Add 50 μl 3 mM DTNB stock to first the reference and then to the sample cuvettes. Follow the absorbance at 412 nm until a plateau is reached (1–5 min).
> 4. Estimate the concentration of Fab' from the measured absorbance at 280 nm.[b]
> 5. Calculate the concentration of free thiol from the increase in absorbance at 412 nm between steps 3 and 4 using $\varepsilon_{412} = 14\,150$ cm^{-1} M^{-1} for the thionitrobenzoate anion.
> 6. Calculate the percentage of Fab' present as the free thiol form using estimates of total Fab' concentration and free thiol concentration from steps 4 and 5, respectively.
>
> [a] Adapted from Creighton (22).
> [b] For huMAb4D5-8 Fab, $\varepsilon^{1\%} = 1.56$ and $M_r = 47\,738$ (3).

3. Applications of Fab' fragments from E. coli

3.1 Construction of monospecific F(ab')$_2$ fragments

Thioether-linked and disulfide-linked monospecific (Ms) F(ab')$_2$ are readily constructed by coupling Fab' fragments using a bifunctional maleimide (e.g. o-PDM, ref. 3, or bismaleimidohexane) or DTNB, respectively (9). We recommend disulfide-linked F(ab')$_2$ for *in vitro* studies as they are simpler to prepare and are functionally indistinguishable from thioether-linked F(ab')$_2$ (9). On the other hand, thioether-linked MsF(ab')$_2$ are preferred for *in vivo* applications as they have pharmacokinetic advantages to disulfide-linked F(ab')$_2$. In particular, a thioether-linked F(ab')$_2$ was found to have a 3-fold longer permanence time, *T*, in normal mice than the corresponding disulfide-linked F(ab')₂:

$$T = AUC/C_0;$$

where *AUC* is the area under the plasma concentration *versus* time-curve and C_0 is the extrapolated initial concentration. Potential drawbacks of thioether-linked F(ab')₂ for clinical applications are immunogenicity or even mutagenicity of the linker.

Research grade MsF(ab')₂ has been prepared according to *Protocol 4* in yields > 70% (e.g. *Figure 2*) and purity > 90% as judged by scanning laser densitometry (9). Coupling reactions are performed at pH 5.3 and 4°C to limit the extent of hydrolysis of the maleimide rings which would otherwise diminish the efficiency of coupling as described by Glennie *et al.* (23). Small amounts of disulfide-linked MsF(ab')₂ may be present in the Fab'–SH preparation or formed under the coupling conditions used. For *in vivo* applications, at least, we recommend eliminating this unwanted disulfide-linked MsF(ab')₂ by mild reduction of the hinge disulfide (*Protocol 4*, step 4) and alkylation of any free thiol generated (*Protocol 4*, step 5) which leaves the thioether-linked MsF(ab')₂ intact. This *E. coli* F(ab')₂ technology has been further refined to prepare multiple gram quantities of clinical grade MsF(ab')₂ (GZ, unpublished data).

Protocol 4. Preparation of thioether-linked monospecific F(ab')₂ fragments

Equipment and reagents

- Endotoxin-free sterile water (McGaw) is used to prepare all buffers
- 1 M Tris–HCl, pH 8.0
- 40 mM *N,N*-1,2-phenylenedimaleimide (*o*-PDM, Sigma, cat. no. P-7518). Prepare in prechilled *N,N*-dimethyl formamide (Aldrich cat no. D15, 855–0) and store at −70°C
- Centriprep-30 concentrators (Amicon, cat. no. 4306)
- 40 mM L-cysteine. Prepare in pre-chilled water and store at −70°C
- 0.5 M iodoacetamide in 100 mM Tris–HCl, pH 7.5. Prepare using pre-chilled buffer and store at −70°C
- Millex-GV 0.22 µm filter (Millipore, cat. no. SLGVO25LS)
- Sephacryl S100-HR (Pharmacia, cat. no. 17-0612-01). Pack a 2.5–5 cm × 100–150 cm column. The recommended sample loading is ≤ 10 mg protein per cm² of column cross-section in a volume of ≤ 0.02 cv. Purge the column of contaminating endotoxin with 2 cv 0.1 M NaOH prior to equilibrating and running in PBS. Sephacryl S200-HR (Pharmacia, cat. no. 17-0584-01) is an acceptable alternative to Sephacryl S100-HR. Store the column in PBS containing 20% (v/v) ethanol.
- Cryogenic vials (Corning, cat. no. 25078).
- FPLC or low-pressure chromatography system

Method

1. Adjust the Fab'–SH preparation from *Protocol 1* to pH 5.3 using 1 M Tris–HCl, pH 8.0 (~ 0.1 vol) and chill to 4°C. We recommend coupling the Fab' at a concentration of 2–5 mg/ml. If necessary the Fab' preparation may be concentrated using Centricon-30 concentrators.

13: Fab' fragments from E. coli

2. *Coupling reaction.* Couple the Fab'–SH to itself by dropwise addition of pre-chilled 40 mM *o*-PDM corresponding to 0.5 molar equivalents of the free thiol. Incubate overnight at 4°C with gentle stirring. Pellet any precipitate by centrifuging at 45 000 *g* for 10 min at 4°C.
3. *Ultrafiltration.* Concentrate the coupling reaction to 5–7 ml using Centriprep-30 concentrators.
4. Add 0.2 vol 1 M Tris–HCl, pH 8.0 to the coupling reaction followed by 0.1 vol 40 mM L-cysteine and incubate for 15 min at 4°C.[a]
5. Add 0.1 vol 0.5 M iodoacetamide to the coupling reaction and incubate for 15–30 min at 4°C.[b] Pass the sample through a 0.22 µm filter to remove particulates.
6. *Size-exclusion chromatography.* Purify the F(ab')$_2$ away from unreacted Fab' by S100-HR size-exclusion chromatography using PBS as a running buffer (12 cm h^{-1}).
7. *Yield.* Estimate the concentration of F(ab')$_2$ from the measured absorbance at 280 nm.[c]
8. *Sterile filtration.* Pass the purified F(ab')$_2$ through a 0.22 µm filter.
9. *Storage.* Aliquot the F(ab')$_2$ into cryogenic vials, flash freeze in liquid nitrogen, and store at –70°C.

[a] Reduces any unwanted disulfide-linked F(ab')$_2$.
[b] Blocks any free thiol generated.
[c] For huMAb4D5–8 F(ab')$_2$, $\varepsilon^{1\%}$ = 1.56 and M_r = 95 965 (3).

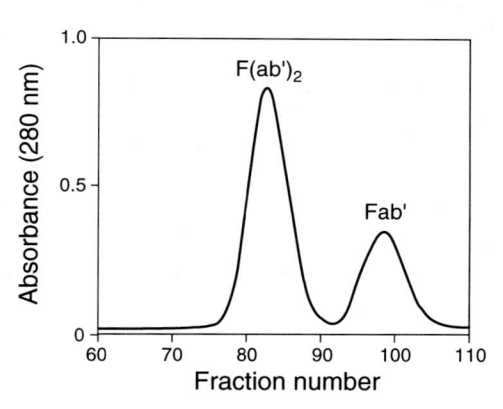

Figure 2. Purification of MsF(ab')$_2$ by size-exclusion chromatography. Shown is a representative chromatograph (S200-HR, 5 cm × 100 cm) after coupling 240 mg chimeric MAbH52 (anti-CD18) Fab' (93% Fab'–SH) with 0.5 molar equivalents of *o*-PDM as in *Protocol 4*. The yield of MsF(ab')$_2$ after concentration was 147 mg at 0.3 endotoxin units per mg.

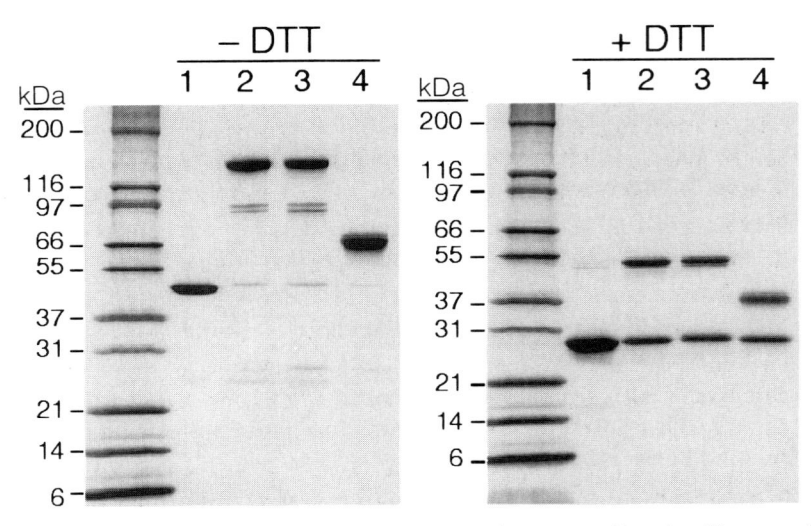

Figure 3. SDS-PAGE analysis of *E. coli*-derived Ab fragments. Proteins (2 μg per lane) were electrophoresed on a 4–20% gel (Novex) under non-reducing conditions after reaction with 50 mM iodoacetamide or and after reduction with 10 mM DTT and then stained with Coomassie Brilliant Blue R250. Lane 1, HuMAb4D5-8 Fab' (*Protocol 1*); Lane 2, HuMAb4D5-8 MsF(ab')$_2$ (*Protocol 4*); Lane 3, HuMAb4D5-8 × HuMAbUCHT1-9 BsF(ab')$_2$ (*Protocol 5*); Lane 4, HuMAb4D5-8 Fab'–PEG (*Protocol 6*). Also shown are molecular weight standards.

3.2 Construction of bispecific F(ab')$_2$ fragments

Research-grade thioether-linked bispecific (Bs) F(ab')$_2$ has been prepared in yields of > 50% and purity > 90% (ref. 14, *Protocol 4*). The level of contamination of BsF(ab')$_2$ with MsF(ab')$_2$ is very low as judged by the absence of detectable F(ab')$_2$ in mock-coupling reactions with either Fab'–mal or Fab'–SH alone (ref. 14). Furthermore, the coupling reaction is subjected to a mild reduction step followed by alkylation to remove trace amounts of disulfide-linked F(ab')$_2$ that might be present. SDS-PAGE of the purified BsF(ab')$_2$ under non-reducing conditions gives one major band of the expected size (~ 96 kDa) (*Figure 3*). In contrast, SDS-PAGE under reducing conditions gives two major bands with electrophoretic mobility and amino terminal sequence anticipated for free light chain and thioether-linked heavy chain dimers (*Figure 3*).

Protocol 5. Preparation of bispecific F(ab')$_2$ fragments[a]

Reagents

As for *Protocol 4* with the following additions:
- SAE buffer: 20 mM sodium acetate, 5 mM EDTA, pH 5.3

Method

1. *Preparation of Fab'–mal*. Adjust one Fab'–SH Protein G pool (from *Protocol 1*) to pH 5.3 with 1 M Tris–HCl (pH 8.0). Add 0.1 vol 40 mM o-PDM and incubate for 30 min at 4°C. Pellet any precipitate by centrifugating at 45 000 g for 10 min at 4°C. We recommend using 5–200 mg Fab'–SH depending upon the amount of BsF(ab')$_2$ required.

2. Remove excess o-PDM by Protein G purification (as *Protocol 1*, step 4).

3. *Ultrafiltration and diafiltration[b]*. Buffer exchange and concentrate Fab'–mal into SAE buffer, using Centriprep-30 concentrators (3 or 4 changes). Estimate the concentration of Fab'–mal from the measured absorbance at 280 nm.[c]

4. *Coupling reaction*. Mix the Fab'–mal with an equimolar amount of the Fab'–SH of second specificity, based upon the amounts of the maleimide derivative and free thiol. Concentrate the coupling reaction to 2–5 mg/ml and incubate for 14–48 h at 4°C.

5. Proceed as *Protocol 4*, steps 3–9.

[a] Protocol modified after Glennie *et al.* (23).
[b] Also removes traces of o-PDM.
[c] For huMAb4D5-8 Fab, $\varepsilon^{1\%}$ = 1.56 and M_r = 47 738.

3.3 Construction of Fab'–PEG

Modification of proteins with PEG (PEGylation) has been used to increase their permanence time and reduce their immunogenicity *in vivo* (24–26). PEGylation is routinely performed through lysine residues resulting in heterogeneous populations of molecules modified at different sites and with varying stoichiometry. Alternatively, cysteine mutations have been introduced to permit site-specific modification with the advantages of more uniform populations of molecules with defined stoichiometry and site of attachment of PEG (27). Similarly, PEGylation of Fab'–SH fragments using the single free thiol gives a rather homogeneous population of molecules with defined stoichiometry (1:1) and site (hinge cysteine) of modification (*Protocol 6*).

Fab'–PEG can often be fully resolved from unreacted Fab' and small amounts of disulfide-linked MsF(ab')$_2$ formed by ABx chromatography (*Figure 4A*). If necessary, residual traces of unreacted Fab' can be removed by hydrophobic interaction chromatography using phenyl Toyopearl (*Figure 4B*). Modification of HuMAb4D5-8 Fab' with PEG (M_r 5000) increases the serum permanence time in normal mice by approximately 3.9-fold (J. Mordenti, G. Osaka, PC, and GZ, unpublished data).

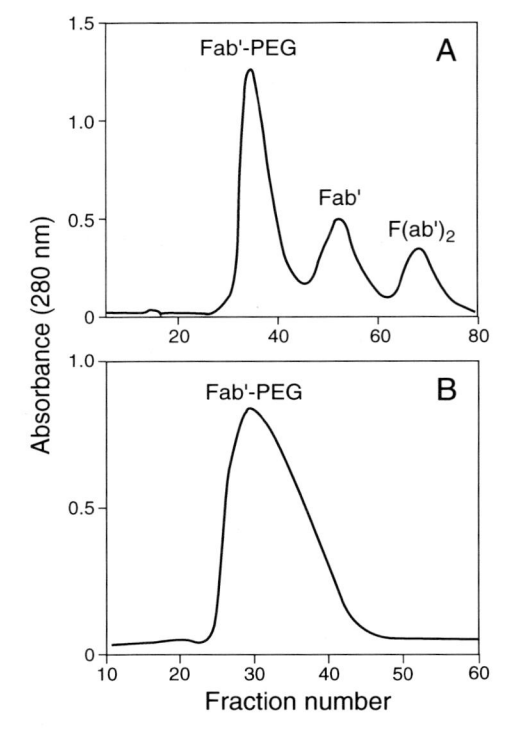

Figure 4. Purification of Fab'–PEG. 400 mg HuMAbH52 OZ Fab' (8) (72% Fab'–SH) were reacted in the presence of 5 molar equivalents of MeO–PEG–mal prior to purification by (A) ABx chromatography followed by (B) phenyl Toyopearl chromatography.

Protocol 6. Preparation of Fab'–PEG

Equipment and reagents

- Endotoxin-free sterile water (McGaw) is used to prepare all buffers
- Methoxy–PEG–maleimide (MeO–PEG–mal), M_r 5000, or specified molecular weight (Shearwater Polymers). MeO–PEG–mal is prepared in 20 mM Tris–HCl, pH 6.5, and either used immediately or stored at –70 °C until required
- 1 M Tris–HCl, pH 8.0
- Bakerbond ABx 40 μm 275A (J. T. Baker, cat. no. 7269–02). Pack a column using ≥ 1 ml ABx resin per 25 mg of PEG-modified protein. Regenerate the column after use by washing with 3 M guanidine–HCl, 20 mM Hepes, pH 7.8, followed by 1% (v/v) acetic acid containing 20 mM phosphoric acid and then distilled water. Store the column in 20% (v/v) ethanol at RT

- Phenyl-650M Toyopearl (TosoHaas, cat. no. 14477). Pack a column using ≥ 1 ml resin per 20 mg of PEG-modified protein. Regenerate and store the column in 0.1 M NaOH
- A buffer: 20 mM MES, pH 6.0
- B buffer: 20 mM MES, 100 mM $(NH_4)_2SO_4$, pH 6.0
- C buffer: 50 mM sodium acetate, 1.5 M $(NH_4)_2SO_4$, pH 5.4
- D buffer: 50 mM sodium acetate, 0.15 M $(NH_4)_2SO_4$, pH 5.4
- E buffer: 50 mM sodium acetate, 3.0 M $(NH_4)_2SO_4$, pH 5.4
- FPLC or low-pressure gradient chromatography system
- Conductivity meter

Method

1. Adjust the Fab'–SH preparation from *Protocol 1* to pH 6.5 using 1 M Tris–HCl, pH 8.0. We recommend using 10–400 mg Fab'–SH at 2–5 mg/ml depending upon the amount of Fab'–PEG required. If necessary the Fab'–SH preparation may be concentrated using Centricon-30 concentrators.

2. *Coupling reaction.* Couple the Fab'–SH (thiol content determined using *Protocol 3*) in the presence of 5 molar equivalents of MeO–PEG–mal.

3. *ABx chromatography.* Wash the ABx column with A buffer until both the pH and conductivity of the flow-through match that of A buffer. Dilute the coupling reaction with sterile endotoxin-free water until the conductivity equals that of A buffer (approximately 2.0 mS). Load the coupling reaction on to the column and wash the unreacted MeO–PEG–mal from the column with 2 cv A buffer (200 cm h^{-1}). Elute the Fab'–PEG with a linear gradient from 100% A buffer to 100% B buffer over 15 cv.

4. *Phenyl Toyopearl chromatography.* Wash the phenyl Toyopearl column with C buffer until both the pH and conductivity of the flow-through match that of buffer C. Dilute the ABx purified Fab'–PEG pool with 1 vol E buffer. Load the diluted pool on to the column and wash with 2 cv C buffer (200 cm h^{-1}). Elute the Fab'–PEG with a linear gradient from 100% C buffer to 100% D buffer over 15 cv.

5. Proceed as *Protocol 4*, steps 6–9.

3.4 Construction of Fab'–immunoliposomes

Site-specific attachment of phospholipid to Fab'–SH for immunospecific targeting of liposomes to cells was first demonstrated over a decade ago using Fab' fragments obtained by limited proteolysis and mild reduction of intact Abs (28). The ready availability of humanized Fab'–SH fragments together with recent advances in liposome technology, such as sterically stabilized liposomes (29), significantly enhance the clinical potential of immunoliposomes.

Anti-p185^{HER2} immunoliposomes prepared according to *Protocol 7* bind specifically to p185^{HER2}-overexpressing breast cancer cells and inhibit their proliferation (30). Furthermore, the cytotoxicity of doxorubicin encapsulated in anti-p185^{HER2} immunoliposomes is specific for p185^{HER2}-overexpressing tumour cells (30). Immunoliposomes constructed using Fab' fragments prepared according to *Protocol 1* were found to be prone to leakage of the encapsulated drug. This problem was overcome by additional purification of the Fab' using phenyl Sepharose plus the use of PyroBind ST filters (*Protocol 2*) which reduce contaminating endotoxin and perhaps *E. coli*-derived lipid (JP, PC, unpublished data).

Protocol 7. Preparation of Fab'-immunoliposomes[a]

Equipment and reagents

- Liposomes, preferably small unilamellar liposomes containing neutral phospholipids such as hydrogenated soy phosphatidylcholine (HSPC), cholesterol (Chol), and polyethylene glycol (M_r = 1900)-derivatized phosphatidylethanolamine (PEG-PE). PEG-PE may be synthesized as described by Allen *et al.* (32)
- *N*-[4-(*p*-maleimidophenyl)butyryl] phosphatidylethanolamine (M-PE) (Molecular Probes, cat. no. M-1617)
- 200 mM stock β-mercaptoethanol
- Sephacryl S-400 HR (Pharmacia, cat. no. 17–0609–01). Pack a 1.5 cm × 12 cm column for preparations containing up to 1.5 mg Fab'-SH. Purge column of contaminating endotoxin with 2 cv 0.1 M NaOH prior to equilibrating and running in MEN buffer. Store column in PBS containing 20% (v/v) ethanol
- MEN buffer: 100 mM MES, 2 mM EDTA, 180 mM NaCl, pH 5.5
- HN buffer: 20 mM Hepes, 150 mM NaCl, pH 7.2
- Centriprep-10 concentrators (Amicon, cat. no. 4305)
- Bradford protein assay (Bio-Rad, cat. no. 500–0001)
- Desferrioxamine mesylate (Sigma, cat. no. D-9533)
- Argon gas supply
- membrane filters (25 mm diameter, (Poretics); PCTE membrane, 50 nm pore size (cat. no. 11005) 100 nm pore size; (cat. no. 11010))
- 250 mM $(NH_4)_2SO_4$
- Sephadex G-75 (Pharmacia, cat. no. 17–0050–01)
- Doxorubicin-HCl (Aldrich cat. no. 86, 036–0)
- Conc. NaOH
- β-mercaptoethanol

Method

1. *Preparation of sterically stabilized liposomes.* Prepare unilamellar liposomes by repeated freeze-thawing, using hydrogenated soy phosphatidylcholine (HSPC), cholesterol (Chol), and polyethylene glycol-derivatized phosphatidylethanolamine (29). The final liposome composition should consist of HSPC:Chol at a ratio of 3:2 and PEG-PE at 0–6 mol (of total phospholipid). In addition, 2 mol % M-PE (of total phospholipid) is included in the lipid mixture in chloroform prior to the formation of liposomes to provide the attachment site for Fab'. Extrude the liposomes repeatedly under positive pressure with argon gas through polycarbonate membrane filters of defined pore size sequentially from 100 nm to 50 nm, to yield liposomes of 60–120 nm diameter. The size distribution may be confirmed by dynamic light scattering (33). Determine the liposome concentration by phosphate assay (34).

2. *Encapsulation of doxorubicin in sterically stabilized liposomes.* Prepare the liposomes as in step 1 except for the addition of 250 mM $(NH_4)_2SO_4$ and 1 mM desferrioxamine mesylate at pH 5.5, to allow subsequent efficient trapping and encapsulation of doxorubicin (35). Unencapsulated $(NH_4)SO_4$ is removed by gel filtration using Sephadex G-75. Dissolve the doxorubicin-HCl directly in the liposome suspension at 0.1 mg doxorubicin-HCl per µmol phospholipid. Dox-

13: Fab' fragments from E. coli

orubicin becomes sequestered in the interior of the liposome with a loading efficiency of > 99% using 1 mg doxorubicin-HCl per 10 μmol phospholipid.

3. *Conjugation reaction.* Prepare Fab'–SH in MEN buffer as described in *Protocol 2.* Add 0.8 mg Fab'–SH to a liposome suspension containing 20 μmol liposomal phospholipid in a total volume of 2 ml. This should provide sufficient immunoliposomes for cell-culture studies and small-scale *in vivo* studies in mice. Adjust the pH of the resulting mixture to pH 7 using concentrated NaOH to facilitate the nucleophilic addition of Fab'–SH to the maleimide moiety of M-PE. Allow the reaction to proceed at 4°C with gentle shaking for 4–12 h.

4. *Deactivation of maleimide groups.* Add β-mercaptoethanol to the reaction mixture at a 2-fold molar excess to M-PE.

5. *Recovery of immunoliposomes.* Remove any unreacted Fab'–SH and β-mercaptoethanol from the immunoliposomes by gel filtration using Sephacryl S-400 HR. Collect the immunoliposomes immediately after the void volume. Pool the immunoliposome-containing fractions and concentrate using Centriprep-10 concentrators.

6. *Analysis of immunoliposomes.* Determine the Fab' loading in the resulting immunoliposomes using the Bradford protein assay. Determine the liposome concentration by the phosphate assay (34). Typically > 50% of added Fab'–SH can be conjugated to liposomes, resulting in approximately 50–100 Fab' molecules per liposome.

[a] Protocol modified after Martin and Papahadjopoulos (31).

3.5 Immobilization of Fab'–SH for use in affinity purification

The possibility of using Abs for the one-step purification of proteins has long been recognized. Unfortunately immunoaffinity purification has only rarely been useful to date because it is arduous, time-consuming, and expensive to obtain large amounts of suitable Abs and immobilize them on appropriate matrices (reviewed in ref. 36). Rapid isolation of Abs using very large Ab-phage libraries (37) together with a system for high-level *E. coli* expression and recovery of Fab'–SH (3) provides a solution to most of these problems.

Here the utility of immobilized Fab' fragments for immunoaffinity purification is demonstrated. In particular, a chimeric version of the anti-CD3 Ab UCHT1 (15) was first immobilized on an activated thiol support (*Protocol 8*). The immobilized Fab' was then used to purify the extracellular domain (ECD) of the ε chain of CD3 to near homogeneity in a single step starting from a crude preparation (*Figure 5*).

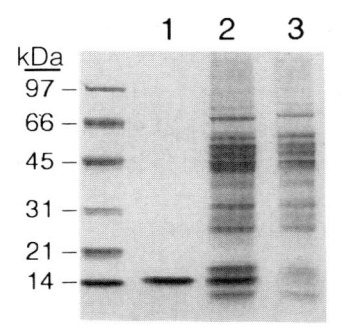

Figure 5. SDS-PAGE analysis of affinity-purified extracellular domain (ECD) of the CD3 ε chain. Proteins were electrophoresed on a 4–20% gel (Bio-Rad) under non-reducing conditions and then stained with Coomassie Brilliant Blue R250. Lane 1, CD3 ε ECD affinity-purified using immobilized chimeric anti-CD3 Fab'; Lane 2, E. coli shockate containing unpurified CD3 ε ECD; Lane 3, E. coli shockate following affinity purification of CD3 ε ECD using a chimeric Fab' fragment of the anti-CD3 Ab, UCHT1 (15) immobilized according to Protocol 8.

Protocol 8. Immobilization of Fab'–SH on a solid support

Equipment and reagents

- Thiopropyl-Sepharose 6B (Pharmacia, cat. no. 17–0420–01). 1 g of dried resin reaches a final volume of ~ 4 ml after swelling. Use 0.5 g resin per mg of Fab'–SH to be immobilized.
- 100 ml sintered-glass funnel (Corning, ASTM 10-15 M)
- 0.02% (w/v) sodium azide in PBS
- Deprotection buffer: 10 mM Tris–HCl, pH 8.0, 1 mM EDTA, 50 mM β-mercaptoethanol
- NME buffer: 10 mM MES, pH 5.5, 1 mM EDTA, 200 mM NaCl
- AE buffer: 100 mM acetic acid, pH 2.8, 1 mM EDTA

Method

1. *Preparation of Fab'–mal.* Prepare o-PDM derivatized Fab'–SH according to steps 1 and 2 of *Protocol 5*.

2. *Preparation of thiopropyl-Sepharose 6B resin.* Swell the freeze-dried resin in distilled water for 15 min at RT. Wash the resin with 20 vol of distilled water on a 100 ml sintered-glass funnel using a water (or other low-pressure vacuum) pump.

3. *Deprotection of thiopropyl-Sepharose 6B resin.* Remove the 2-thiopyridone protecting group from the resin by washing with 4 vol deprotection buffer followed by ~ 50 vol ME buffer to remove the reducing agent.

4. *Coupling reaction.* Mix the Fab'–mal with the deprotected resin and 2 vol NME buffer. Incubate for 12–18 h at 4°C with very gentle mixing.

13: Fab' fragments from E. coli

5. Pour the slurry into an appropriate-sized column and wash the resin with, in turn, 10 cv NME buffer followed by 4 cv AE buffer. Wash the resin with 4 cv 0.02% (w/v) sodium azide in PBS and store at 4°C.
6. Estimate the coupling efficiency from the initial OD_{280} of the Fab'–mal compared to the OD_{280} of the flow-through at step 5.

Acknowledegements

We thank Brad Snedecor and Jim Bourell for sharing unpublished fermentation titre and mass spectrometry data, respectively, Wayne Anstine for preparing *Figures 2–5*, and Paul Godowski for support.

References

1. Skerra, A. and Plückthun, A. (1988). *Science*, **240**, 1038.
2. Better, M., Chang, C. P., Robinson, R. R., and Horwitz, A. H. (1988). *Science*, **240**, 1041.
3. Carter, P., Kelley, R. F., Rodrigues, M. L., Snedecor, B., Covarrubias, M., Velligan, M. D., Wong, W. L. T., Rowland, A. M., Kotts, C. E., Carver, M. E., Yang, M., Bourell, J. H., Shepard, H. M., and Henner, D. (1992). *Bio/Technology*, **10**, 163.
4. Carter, P., Rodrigues, M. L., and Shalaby, M. R. (1994). In *Handbook of experimental pharmacology*, Vol. 113, *The pharmacology of monoclonal antibodies* (ed. M. Rosenberg and G. P. Moore), pp. 135–45, Springer–Verlag, Heidelberg.
5. Kelley, R. F., O'Connell, M. P., Carter, P., Presta, L., Eigenbrot, C., Covarrubias, M., Snedecor, B., Bourell, J. H., and Vetterlein, D. (1992). *Biochemistry*, **31**, 5434.
6. Kelley, R. F. and O'Connell, M. P. (1993). *Biochemistry*, **32**, 6828.
7. Eigenbrot, C., Randal, M., Presta, L., Carter, P., and Kossiakoff, A. A. (1993). *J. Mol. Biol.*, **229**, 969.
8. Eigenbrot, C., Gonzales, T., Mayeda, J., Carter, P., Werther, W., Hotaling, T., Fox, J., and Kessler, J. (1994). *Proteins: Structure, Function and Genetics*, **18**, 49.
9. Rodrigues, M. L., Snedecor, B., Chen, C., Wong, W. L. T., Garg, S., Blank, G. S., Maneval, D., and Carter, P. (1993). *J. Immunol.*, **151**, 6954.
10. Better, M., Bernhard, S. L., Lei, S.-P., Fishwild, D. M., Lane, J. A., Carroll, S. F., and Horwitz, A. H. (1993). *Proc. Natl Acad. Sci. USA*, **90**, 457.
11. Pack, P., Kujau, M., Schroeckh, V., Knüpfer, U., Wenderoth, R., Riesenberg, D., and Plückthun, A. (1993). *Bio/Technology*, **11**, 1271.
12. Plückthun, A. (1994). In *Handbook of experimental pharmacology*, Vol. 113, *The pharmacology of monoclonal antibodies* (ed. M. Rosenberg and G. P. Moore), pp. 269–315. Springer–Verlag, Heidelberg.
13. Carter, P., Presta, L., Gorman, C. M., Ridgway, J. B. B., Henner, D., Wong, W. L. T., Rowland, A. M., Kotts, C., Carver, M. E., and Shepard, H. M. (1992). *Proc. Natl Acad. Sci. USA*, **89**, 4285.

14. Rodrigues, M. L., Shalaby, M. R., Werther, W., Presta, L., and Carter, P. (1992). *Int. J. Cancer Suppl.*, **7**, 45.
15. Shalaby, M. R., Shepard, H. M., Presta, L., Rodrigues, M. L., Beverley, P. C. L., Feldmann, M., and Carter, P. (1992). *J. Exp. Med.*, **175**, 217.
16. Parham, P. (1983). In *Cellular immunology* (ed. E. M. Weir) (4th ed.), Vol. 1, pp. 14.1–14.23. Blackwell Scientific Press, CA.
17. Derrick, J. P. and Wigley, D. B. (1992). *Nature*, **359**, 752.
18. Skerra, A., Pfitzinger, I., and Plückthun, A. (1991). *Bio/Technology*, **9**, 273.
19. Schmidt, T. G. M. and Skerra, A. (1993). *Protein Eng.*, **6**, 109.
20. Zhu, Z. and Carter, P. *J. Immunol.* **155**, 1903
21. Gavit, P., Walker, M., Wheeler, T., Bui, P., Lei, S.-P., and Weickmann, J. (1992). *Bio/Pharmacology*, **5**, 28.
22. Creighton, T. E. (1990). In *Protein structure: a practical approach* (ed. T. E. Creighton), pp. 155–67. IRL Press, Oxford.
23. Glennie, M. J., McBride, H. M., Worth, A. T., and Stevenson, G. T. (1987). *J. Immunol.*, **139**, 2367.
24. Lisi, P. J., Van Es, T., Abuchowski, A., Palczuk, N. C., and Davis, F. F. (1984). *J. Appl. Biochem.*, **4**, 19.
25. Abuchowski, A., Kazo, G. M., Verhoest Jr., C. R., Van Es, T., Kafkewitz, D., Nucci, M. L., Viau, A. T., and Davis, F. F. (1982). *Cancer Biochem. Biophys.*, **7**, 175.
26. Katre, N. V., Knauf, M. J., and Laird, W. J. (1987). *Proc. Natl Acad. Sci. USA*, **84**, 1487.
27. Goodson, R. J. and Katre, N. V. (1992). *Bio/Technology*, **8**, 343.
28. Martin, F. J., Hubbell, W. L., and Papahadjopoulos, D. (1981). *Biochemistry*, **20**, 4229.
29. Papahadjopoulos, D., Allen, T. M., Gabizon, A., Mayhew, E., Matthay, K., Huang, S. K., Lee, K.-D., Woodle, M. C., Lasic, D. D., Redemann, C., and Martin, F. (1991). *Proc. Natl Acad. Sci. USA*, **88**, 11460.
30. Park, J. W., Hong, K., Carter, P., Asgari, H., Guo, L. Y., Keller, G. A., Wirth, C., Shalaby, R., Kotts, C., Wood, W. I., Papahadjopoulos, D., and Benz, C. C. (1995). *Proc. Natl Acad. Sci. USA*, **92**, 1327.
31. Martin, F. J. and Papahadjopoulos, D. (1982). *J. Biol. Chem.*, **257**, 286.
32. Allen, T. M., Hansen, C., Martin, R., Redemann, C., and Yau-Young, A. (1991). *Biochim. Biophys. Acta*, **1066**, 29.
33. Gabizon, A. and Papahadjopoulos, D. (1988). *Proc. Natl Acad. Sci. USA*, **85**, 6949.
34. Bartlett, G. R. (1959). *J. Biol. Chem.*, **234**, 466.
35. Lasic, D. D., Frederik, P. M., Stuart, M. C. A., Barenholz, Y., and McIntosh T. J. (1992). *FEBS Lett.*, **312**, 255.
36. Bailon, P. and Roy, S. K. (1990). In *Protein purification: from molecular mechanisms to large-scale processes* (ed. M. R. Ladisch, R. C. Willson, C-d. C. Painton, and S. E. Builder), pp. 150–67, American Chemical Society Symposium Series no. 427, Washington, DC.
37. Griffiths, A. D., Williams, S. C., Hartley, O., Tomlinson, I. M., Waterhouse, P., Crosby, W. L., Kontermann, R. E., Jones, P. T., Low, N. M., Allison, T. J., Prospero, T. D., Hoogenboom, H. R., Nissim, A., Cox, J. P. L., Harrison, J. L., Zaccolo, M., Gherardi, E., and Winter, G. (1994). *EMBO J.*, **13**, 3245.

A1

Addresses of suppliers

ABI
ABI, Kelvin Close, Birchwood Science Park North, Risley, Warrington WA3 7PB, UK.
ABI, Foster City, USA.
Amersham International
Amersham International, White Lion Road, Amersham, Bucks HP9 9LL, UK.
Amersham International, Arlington Heights, FL, USA.
Amicon, 72 Cherry Hill Drive, Beverley, MA 01915, USA.
Aminco (distributed by Sopra)
Anachem Ltd, Anachem House, 20 Charles St., Luton, Bedfordshire LU2 0EB, UK.
Apple Computers, Cupertino, CA, USA.
Associates of Cape Cod, 704 Main St, Falmouth, MA 02540, USA.
ATCC, Parklawn Drive, Rockville, MD 20852, USA.
Azlon, Silobeud Industrial Park, 205–1 Kelsey Lane, Tampa, FL 33619, USA.
B. Braun Biotech International, Melsungen, Germany.
J. T. Baker, 222 Red School Lane, Phillipsburg, NJ 08865, USA.
BDH
BDH, distributed in UK by various suppliers, e.g. LIG Supplies Ltd.
BDH, distributed in USA by Gallard Schlesinger Industries.
Becton Dickinson
Becton Dickinson, Between Towns Rd, Cowley, Oxford OX4 3LY, UK.
Becton Dickinson, 2350 Qume Drive, San Jose, CA 95131–1807, USA.
Becton Dickinson, Denderstraat 24, B–9440, Erembodegem, Belgium.
Bellco Biotechnology, 340 Edrudo Rd, Vineland, NJ 08360, USA.
BIAcore, Pharmacia Biosensor, 23 Grosvenor Rd, St. Albans, Herts, AL1 3AW, UK.
J. Bibby Science Products Ltd, Stone, Staffordshire ST15 05A, UK.
Biogenesis Ltd, 12 Yeomans' Park, Bournemouth BH8 OBJ, UK.
Biometra Ltd, P.O. Box 167, Maidstone, Kent ME14, UK.
Bio-Rad Laboratories
Bio-Rad Laboratories, Caxton Way, Watford, Hertfordshire, UK.
Bio-Rad Laboratories, Life Science Group, 2000 Alfred Nobel Drive, Hercules, CA 94547, USA.

Addresses of suppliers

BIO-101, PO Box 2284, La Jolla, CA 92038-2284, USA.
Bio-stat Diagnostics Ltd, Bio-stat House, Pepper Road, Hazel Grove, Stockport, Cheshire SK7 5BW, UK.
Biosys, 21 Quai du Clos des Roses, 60200 Compiegne, France.
Biotecx Laboratories Inc.
Biotecx Laboratories Inc., distributed in UK by Biogenesis.
Biotecx Laboratories Inc., 6023 South Loop East, Houston, Texas 77033, USA.
Boehringer Mannheim
Boehringer Mannheim, Bell Lane, Lewes, East Sussex BN7 1LG, UK.
Boehringer Mannheim, PO Box 504114, Indianapolis, IN 46250-0414, USA.
Boehringer Mannheim, PO Box 310120, D-6800 Mannheim 31, Germany.
Brinkmann Instruments, Westbury, NY, USA.
BRL, see Life Technologies Ltd.
Cambridge Bioscience, 24-25 Signet Court, Newmarket Rd, Cambs CB5 8LA, UK.
Cambridge Research Biochemicals
Cambridge Research Biochemicals, Gadbrook Park, Northwich, Cheshire CW9 7RA, UK.
Cambridge Research Biochemicals, Wilmington, DE 19897, USA.
Canadian Lysozyme, 31212 Peardonville Rd, Abbotsford, British Columbia V2S 5W6, Canada.
Chemapec Inc., Woodbury, NJ 14831, USA.
Cinna/Biotecx from Biogenesis
Clontech Laboratories Inc., 1020 East Meadows Circle, Palo Alto, CA 94303-4230, USA.
Corning, HP-AB-)03, Corning, NY, USA.
Costar
Costar, 10 The Valley Centre, Gordon Rd, High Wycombe, Bucks HP13 6EQ, UK.
Costar, 205 Broadway, Cambridge, MA, USA.
DAKO Corporation
DAKO Corporation, 6392 Via Real, Carpinteria, CA 93013, USA.
DAKO Corporation, Produktionsvej 42, PO Box 1359 DK 2600, Glostrup, Denmark.
Deltown Specialties, Fraser, NY, USA.
Difco
Difco, Detroit, MI 48232-7058, USA.
Difco, P.O. Box 14B, Central Avenue, West Molesey, Surrey KT8 2SE, UK.
Dupont Medical Products, Biotechnology Systems Division, 31 Pecks Lane, PO Box 5509, Newton, CT 06470-5509, USA.
Dupont UK Ltd, Wedgwood Way, Stevenage, Herts SG1 4QN, UK.

Addresses of suppliers

Dynal
Dynal, P.O. Box 158, Skøyen N–0212, Oslo 2, Norway.
Dynal, 45 North Station Plaza, Great Neck, NY 11021, USA.
Dynal, 24–26 Grove St., New Ferry, Wirral L62 5AZ, UK.
Dynatech
ECACC (European Collection of Animal Cell Cultures), Department of Cell Resources, Porton Down, Salisbury, Wilts SP4 0JG, UK,
Eppendorf, Netherler-Hinz GmbH, 22331 Hamburg, Germany.
Falcon (see Beckton Dickinson).
Fisons Applied Sensor Technology, Saxon Way, Bar Hill, Cambridge, UK.
Flow Laboratories
Flow Laboratories, see ICN Biomedicals Ltd for UK distributor.
Flow Laboratories, PO Box 12621, Research Triangle Park, NC 27709, USA.
Flowgen Instruments Ltd, Broad Oak Enterprise Village, Broad Oak Rd, Sittingbourne, Kent ME9 8AQ, UK.
Fluka Chemical Corporation
Fluka Chemical Corporation, 980 South Second Street, Ronkonkoma, NY 11779–7238, USA.
Fluka Chemical Corporation, Industriestrasse 25, 9470 Buchs, Switzerland.
FMC
FMC, 191 Thomaston St, Rockland, ME 04814–2994, USA.
FMC, Testarellogasse 11, 1130 Wein, Austria.
Fragol Industrieschmierstoff, D–4330, Mühlheim/Ruhr, Germany.
Fuji X-Ray Products, 125 Finchley Rd, London NW3 6JH, UK
Gallard Schlesinger Industries, 584 Mineola Ave, Carl Place, NY 11514–1731, USA.
Gaulin Corporation, Everett, MA, USA.
Gibco, see Life Technologies Ltd.
Greiner Labortechnik Ltd, 13 Station Rd, Cam, Dursley, Gloucestershire GL11 5NS, UK.
Hartmann and Braun, Frankfurt, Germany.
Heat Systems-Ultrasonics Inc., Farmingdale, NY, USA.
IBI Kodak
IBI Kodak, PO Box 9588, New Haven, CT 06535, USA.
IBI Kodak, 1 rule Jacquard, BP 53, 69684 Chassieu Cedex, France.
ICN Biomedicals Ltd, Unit 18, Thame Park Business Centre, Wenman Rd, Thame, Oxon OX9 3XA, UK.
Invitrogen Corporation
Invitrogen Corporation, see R and D Systems for UK supplier.
Invitrogen Corporation, 3985-B, Sorrento Valley Boulevard, San Diego, CA, USA.
Jackson ImmunoResearch Laboratories Inc.
Jackson ImmunoResearch Laboratories Inc., for UK distributor see Stratech Scientific.

Addresses of suppliers

Jackson ImmunoResearch Laboratories Inc., PO Box 9, West Grove, PA 19390, USA.
Jackson ImmunoResearch Laboratories Inc., Kronenburgstraat 45, 9th Floor, B-200 Antwerp, Belgium.
JRH Biosciences, PO Box 14848, Lenexa, KS 66215–0848, USA.
Kodak, see IBI Kodak
Labsystems, Pulttitie 8, SF–00881 Helsinki, Finland.
Life Technologies Ltd
Life Technologies Ltd, 3 Fountain Drive, Inchinnan Business Park, Paisley, Renfrewshire PA4 9RF, Scotland.
Life Technologies Ltd, 8400 Helgerman Court, PO Box 6009, Gaithersburg, MD, USA.
LIG Supplies Ltd, Westry Ave, March Trading Park, March, Cambridgeshire PE15 0BN, UK.
LKB-Pharmacia, S–75182 Uppsala, Sweden.
Lüdi, H. & Co. AG, Buhnrain 30, CH-8052, Zurich, Switzerland.
Mazer Chemicals, Gurnee, IL, USA.
McGaw, PO Box 19791, Irvine, CA 92714, USA.
Medingen, Freital, Germany.
Merck & Co. Inc., Building 80y-la93, 126 East Lincoln Ave, Rahway, NJ 07065, USA.
Millipore
Millipore, The Boulevard, Blackmoor Lane, Watford WD1 8YW, UK.
Millipore, 397 Williams St, Marlborough, MA 01752, USA.
Molecular Probes, 4849 Pitchford Ave, Eugene, OR 97402, USA.
New Brunswick Scientific (UK) Ltd, 163 Dixons Rd, North Mymms, Hatfield, Hertfordshire AL9 7JE, UK.
New England Biolabs Ltd
New England Biolabs Ltd, 67 Knowl Piece, Wilbury Way, Hitchin, Hertfordshire SG4 0TY, UK.
New England Biolabs Ltd, 32 Tozer Rd, Beverly, MA 01915–5599, USA.
Novex, 4202 Sorrento Valley Boulevard, San Diego, CA 92121, USA.
Nunc, see Life Technologies Ltd.
Nycomed Pharma, A/S, Postboks 5012, Major stua, N–0301, Oslo, Norway (distributed by Life Technologies).
Oxoid UK, see Unipath Ltd.
Perkin Elmer, see ABI
PerSeptive Biosystems
PerSeptive Biosystems, Riegelerstrasse 2, D–79111 Freiburg, Germany.
PerSeptive Biosystems Inc., University Park at MIT, 38 Sidney St., Cambridge, MA 02139, USA.
Pierce Chemical Co.
Pierce Chemical Co., PO Box 117, Rockford, IL 61105, USA.

Addresses of suppliers

Pierce Chemical Co., PO Box 1512, 3260 BA oud-Beijerland, The Netherlands.
Peirce and Warriner, 44 Upper Northgate St, Chester CH1 4EF, UK.
Pharmacia
  Pharmacia, 800 Centennial Ave, Piscataway, NJ 08854, USA.
  Pharmacia, 751 82 Uppsala, Sweden.
Pharmacia Biotech, 23 Grosvenor Rd, St Albans, Hertfordshire AL1 3AW, UK.
Polygen Corp., 200 Fifth Avenue, Waltham, MA 02254, USA.
Poly LC Inc., Columbia, MD, USA.
Poretics, 111 Lindbergh Avenue, Livermore CA 94550, USA.
Promega
  Promega, Epsilon House, Enterprise Rd, Chilworth Research Centre, Southampton SO1 7NS, UK.
  Promega, 2800 Woods Hollow Rd, Madison, WI 53711-5399, USA.
Qiagen
  Qiagen, Chatsworth, CA, USA
  Qiagen GmbH, Hilden, Germany.
Rainin Instruments Co., Woburn, MA, USA.
R and D Systems Europe, 4-10 The Quadrant, Barton Lane, Abingdon OX14 3YS, UK.
Sartorius, Göttingen, Germany.
Schleicher and Schuell Gmbh, P.O. Box 4, D-37582 Dassel, Germany;
  Schleicher and Schuell Inc., Optical Avenue, Keene, NH 03431, USA.
Schott Glasswerk, Postfach 2480, D-55014 Mainz, Germany.
Scotlab Ltd, Kirkshaws Rd, Coatbridge, Strathclyde, UK.
Sepracor, 33 Locke Drive, Marlborough, MA 01752, USA.
Severn Biotech, Unit 23 Hoobrook Enterprise Centre, Worcester Rd. Kidderminster DY10 1HY, UK.
Shearwater Polymers, 2130 Memorial Parkway SW, Huntsville, AL 35801, USA.
Sigma Chemical Co.
  Sigma Chemical Co., Fancy Rd, Poole, Dorset BH17 7NH, UK.
  Sigma Chemical Co., PO Box 14508, St Louis, MO 63178, USA.
Silicon Graphics Ltd, 1530 Arlington Business Park, Theale, Reading, Berks RG7 4SB
Skatron, Lierbyen, Norway.
SLT Labinstruments, Gesellschaft mbH, Untersbergstraße 1A, A-5082, Grödig, Austria.
SOPRA, GmbH, Schubertstrasse 9-11, D-6087 Büttelborn, Germany.
SOPRA Inc., 33 Nagog Park, P.O. Box Z619, Acton MA 01720-6619, USA.
Sorvall, see Dupont Medical Products
Souther Biotechnology Association Inc., 160A Oxmoor Blvd, Birmingham AL, 35209, USA.

Addresses of suppliers

Sterilin
Sterilin, for UK distributor see J. Bibby Science Products Ltd
Sterilin, for USA distributor see Azlon
Stratagene Cloning Systems
Stratagene Cloning Systems, 11099 N. Torrey Pines Rd, La Jolla, CA 92037, USA.
Stratagene Cloning Systems, Wipplingerstrasse 19, 1010 Wien, Austria.
Stratagene Ltd, 140 Cambridge Innovation Centre, Cambridge Science Park, Milton Rd, Cambridge CB4 4GF, UK.
Stratech Scientific, 61–63 Dudley St, Luton, Bedfordshire LU2 0NP, UK.
Sun Microsystems Inc., Mountain View, CA, USA.
Tecator, Sweden.
Techne, Duxford, Cambridgeshire CB2 4PZ, UK.
Titertek, see Flow
TosoHaas, 156 Keystone Drive, Montgomeryville, PA 18036, USA.
Toyo Roshi International, 6785-A, Sierra Court, Dublin, CA 94568, USA.
Treff AG
Treff AG, for UK distributor see Scotlab
Treff AG, CH-9113, Degersheim, Schweig, Switzerland
Unipath Ltd, Wade Rd, Basingstoke, Hampshire RG24 8PW, UK.
USB
USB, for UK distributor see Amersham International
USB, Cleveland, USA.
Walker Safety Cabinets Ltd, 3 Pyegrove, Glossop, Derbyshire SK13 8RA, UK.
Watson Marlow, Falmouth, UK.
Whatman plc, Springfield Mill, Maidstone, Kent ME14 2LE, UK.
Wisag, Oelikonerstrasse 88, CH-8057, Zurich, Switzerland.

A2

Sequencing primers for antibody V genes

(Sequences of PCR primers for antibody genes are also given in Chapters 1, 6, and 7)

Primer	5'/3'	Location	5' Sequence 3'
(A) for scFV:			
Gene3 lead	5'	fd gene3 leader	TTA TTA TTC GCA ATT CCT TTA GTT GTT CCT
FDTSEQ1	3'	fd gene3	GTC GTC TTT CCA GAC GTT AGT
HIS6SEQ	3'	scFv His tag	ATG GTG ATG ATG ATG TGC GG
LMB2	3'	3' M13/pUC (−20)	GTA AAA CGA CGG CCA GT
LMB3	5'	5' M13/pUC (−25)	CAG GAA ACA GCT ATG AC
myc seq 10	3'	scFv myc tag	CTC TTC TGA GAT GAG TTT TTG
PCR-H-LINK	3'	scFv linker	ACC GCC AGA GCC ACC TCC GCC
PCR-L-LINK	5'	scFv linker	GGC GGA GGT GGC TCT GGC GGT
pelB BACK	5'	pelB leader	GAA ATA CCT ATT GCC TAC GG
pUC19reverse	5'	pUC backbone	AGC GGA TAA CAA TTT CAC ACA GG
(B) for diabodies:			
SEQ RBS FOR	3'	diabody central rbs	CAC TGA CTG TCT CCT TGT C
SEQ RBS BACK	5'	diabody central rbs	CTC TTT GCC ATA CCA CTC
(C) for Fabs:			
CH1(γ1)-SEQ	3'	Human γ1 CH1	GGT GCT CTT GGA GGA GGG TGC
Cκ-SEQ	3'	Human Cκ	CAA CTG CTC ATC AGA TGG CG
Cλ-SEQ	3'	Human Cλ	GTG GCC TTG TTG GCT TGA AGC
(D) for eukaryotic vectors:			
PEC-SEQ1	5'	Human κ leader intron	GCA GGC TTG AGG TCT GGA C
pG1D1 Back (γ1, λ)	5'	HCMV promoter	CGC CTG GAG ACG CCA TCC ACG
pG1D1 For (γ1)	3'	Human γ1 intron	GTG CCA TGT GAC CGC GGT GTG
pG4D100 For (γ4)	3'	Human γ4 intron	CGC CAC CTG CGT CAC CTT AGC
pKN100 Back (γ4, κ)	5'	HCMV promoter intron	GCT GAC AGA CTA ACA GAC TG
pKN100 For (κ, λ)	3'	Human κ intron	CAG GGC ATG TTA GGG ACA GAC

A3

Sequences of the human germline V_H, V_κ, J_H, and J_κ segments

Amino acid sequences of the functional segments are given in single letter code. Sequence alignments, numbering, and CDRs are according to Kabat *et al.* (1), except for the CDR1 of both the V_H and V_κ regions where alignments and numbering are according to Chothia *et al.* (2) and Tomlinson *et al.* (3) respectively. Labelling of H1, H2, H3, L1, L2, and L3 is according to Chothia *et al.* (2) and Tomlinson *et al.* (3) Sequences shown are from January 1996. The 'V BASE' directory is on the Worldwide Web, and will be updated to include the sequences of the human germline V_λ, D, and J_λ segments, (see Chapter 6).

References

1. Kabat, E. A., Wu, T. T., Perry, H. M., Gottesman, K. S., and Foeller, C. (1991). *Sequences of proteins of immunological interest* (5th edn). US Department of Health and Human Services, Bethesda, Maryland.
2. Chothia, C., Lesk, A. M., Gherardi, E., Tomlinson, I. M., Walter, G., Marks, J. D., Llewelyn, M. B., and Winter, G. (1992). *J. Mol. Biol.* **227**, 799.
3. Tomlinson, I. M., Cox, J. P. L., Gherardi, E., Lesk, A. M., and Chothia, C. (1995). *EMBO J.*, **14**, 4628.

Sequences of human germline V_H, V_κ, J_H, and J_κ segments

The Human Germline V_κ repertoire

```
                                         L1                            L2                                              L3
                          FR1           CDR1            FR2           CDR2            FR3                              CDR3
                    1    10    20       30           40      50      60    70    80                                90
```

		FR1	CDR1 (L1)	FR2	CDR2 (L2)	FR3	CDR3 (L3)
$V_\kappa I$	O12/O2 (DPK9)	DIQMTQSPSSLSASVGDRVTITC	RASQSISS------YLN	WYQQKPGKAPKLLIY	AASSLQS	GVPSRFSGSGSGTDFTLTISSLQPEDFATYYC	QQSYSTP
	O18/O8 (DPK1)	DIQMTQSPSSLSASVGDRVTITC	QASQDISN------YLN	WYQQKPGKAPKLLIY	DASNLET	GVPSRFSGSGSGTDFTFTISSLQPEDIATYYC	QQYDNLP
	A20 (DPK4)	DIQMTQSPSSLSASVGDRVTITC	RASQSISN------YLA	WYQQKPGKVPKLLIY	AASTLQS	GVPSRFSGSGSGTDFTLTISSLQPEDVATYYC	QKYNSAP
	A30	DIQMTQSPSSLSASVGDRVTITC	RASQGIRN------DLG	WYQQKPGKAPKRLLIY	AASSLQS	GVPSRFSGSGSGTEFTLTISSLQPEDFATYYC	LQHNSYP
	L14 (DPK2)	NIQMTQSPSAMSASVGDRVTITC	RARQGISN------YLA	WFQQKPGKVPKHLIY	AASSLQS	GVPSRFSGSGSGTEFTLTISSLQPEDFATYYC	LQHNSYP
	L1	DIQMTQSPSSLSASVGDRVTITC	RASQGISN------YLA	WFQQKPGKAPKSLIY	AASSLQS	GVPSRFSGSGSGTDFTLTISSLQPEDFATYYC	QQYNSYP
	L15 (DPK7)	DIQMTQSPSSLSASVGDRVTITC	RASQSISS------WLA	WYQQKPEKAPKSLIY	AASSLQS	GVPSRFSGSGSGTDFTLTISSLQPEDFATYYC	QQNSYP
	L4/L18	AIQLTQSPSSLSASVGDRVTITC	RASQSISS------ALA	WYQQKPGKAPKLLIY	DASSLES	GVPSRFSGSGSGTDFTLTISSLQPEDFATYYC	QQFNSYP
	L5/L19 (DPK5/DPK6)	DIQMTQSPSSVSASVGDRVTITC	RASQSISS------WLA	WYQQKPGKAPKLLIY	AASSLQS	GVPSRFSGSGSGTDFTLTISSLQPEDFATYYC	QQANSFP
	L8 (DPK8)	DIQLTQSPSFLSASVGDRVTITC	RASQSISS------YLN	WYQQKPGKAPKLLIY	AASTLQS	GVPSRFSGSGSGTEFTLTISSLQPEDFATYYC	QQLNSYP
	L23	AIRMTQSPFSLSASVGDRVTITC	WASQGISS------YLA	WYQQKPAKAPKLFIY	YASSLQS	GVPSRFSGSGSGTDYILTISSLQPEDFATYYC	QQYYSTP
	L9	AIRMTQSPSSFSASTGDRVTITC	RASQSISS------YLA	WYQQKPGKAPKLLIY	AASTLQS	GVPSRFSGSGSGTDFTLTISCLQSEDFATYYC	QQYYSYP
	L24 (DPK10)	VIWMTQSPSLLSASTGDRVTISC	RMSQSISS------YLA	WYQQKPGKAPELLIY	AASTLQS	GVPSRFSGSGSGTDFTLTISSLQPEDFATYYC	QQYYSFP
	L11 (DPK3)	AIQMTQSPSSLSASVGDRVTITC	RASQGIRN------DLG	WYQQKPGKAPKLLIY	AASSLQS	GVPSRFSGSGSGTDFTLTISSLQPEDFATYYC	LQDYNYP
	L12	DIQMTQSPSTLSASVGDRVTITC	RASQSISS------WLA	WYQQKPGKAPKLLIY	DASSLES	GVPSRFSGSGSGTEFTLTISSLQPDDFATYYC	QQYNSYS
$V_\kappa II$	O11/O1 (DPK13)	DIVMTQTPLSLPVTPGEPASISC	RSSQSLLDSDDGNTYLD	WYLQKPGQSPQLLIY	TLSYRAS	GVPDRFSGSGSGTDFTLKISRVEAEDVGVYYC	MQRIEFP
	A17 (DPK18)	DVVMTQSPLSLPVTLGQPASISC	RSSQSLVYS-DGNTYLN	WFQQRPGQSPRRLIY	KVSNWDS	GVPDRFSGSGSGTDFTLKISRVEAEDVGVYYC	MQGTHWP
	A1 (DPK19)	DVVMTQSPLSLPVTLGQPASISC	RSSQSLVYS-DGNTYLN	WFQQRPGQSPRRLIY	KVSNWDS	GVPDRFSGSGSGTDFTLKISRVEAEDVGVYYC	MQGTHWP
	A18 (DPK28)	DIVMTQTPLSLSVTPGQPASISC	KSSQSLLHS-DGKTYLY	WYLQKPGQSPQLLIY	EVSNRFS	GVPDRFSGSGSGTDFTLKISRVEAEDVGVYYC	MQGIHLP
	A2 (DPK12)	DIVMTQTPLSLSVTPGQPASISC	KSSQSLLHS-DGKTYLY	WYLQKPGQPPQLLIY	EVSNRFS	GVPDRFSGSGSGTDFTLKISRVEAEDVGVYYC	MQSIQLP
	A19/A3 (DPK15)	DIVMTQSPLSLPVTPGEPASISC	RSSQSLLHS-NGYNYLD	WYLQKPGQSPQLLIY	LGSNRAS	GVPDRFSGSGSGTDFTLKISRVEAEDVGVYYC	MQALQTP
	A23 (DPK16)	DIVMTQTPLSSPVTLGQPASISC	RSSQSLVHS-DGNTYLS	WLQQRPGQPPRLLIY	KISNRFS	GVPDRFSGSGAGTDFTLKISRVEAEDVGVYYC	MQATQFP
$V_\kappa III$	A27 (DPK22)	EIVLTQSPGTLSLSPGERATLSC	RASQSVSSS-----YLA	WYQQKPGQAPRLLIY	GASSRAT	GIPDRFSGSGSGTDFTLTISRLEPEDFAVYYC	QQYGSSP
	A11 (DPK20)	EIVLTQSPATLSLSPGERATLSC	GASQSVSSS-----YLA	WYQQKPGLAPRLLIY	DASSRAT	GIPDRFSGSGSGTDFTLTISRLEPEDFAVYYC	QQYGSSP
	L2/L16 (DPK21)	EIVLTQSPATLSVSPGERATLSC	RASQSVSS------NLA	WYQQKPGQAPRLLIY	GASTRAT	GIPARFSGSGSGTEFTLTISSLQSEDFAVYYC	QQYNNWP
	L6	EIVLTQSPATLSLSPGERATLSC	RASQSVSS------YLA	WYQQKPGQAPRLLIY	DASNRAT	GIPARFSGSGSGTDFTLTISSLEPEDFAVYYC	QQRSNWP
	L20	EIVLTQSPATLSLSPGERATLSC	RASQSVSS------YLA	WYQQKPGQAPRLLIY	DASNRAT	GIPARFSGSGSGTDFTLTISSLEPEDFAVYYC	QQRSNWH
	L25 (DPK23)	EIVLTQSPATLSLSPGERATLSC	RASQSVSSS-----YLS	WYQQKPGQAPRLLIY	GASTRAT	GIPARFSGSGSGTDFTLTISSLQPEDFAVYYC	QQDYNLP
$V_\kappa IV$	B3 (DPK24)	DIVMTQSPDSLAVSLGERATINC	KSSQSVLYSSNNQNYLA	WYQQKPGQPPKLLIY	WASTRES	GVPDRFSGSGSGTDFTLTISSLQAEDVAVYYC	QQYYSTP
$V_\kappa V$	B2	ETTLTQSPAFMSATPGDKVNISC	KASQDIDD------DMN	WYQQKPGEAAIFIIQ	EATTLVP	GIPPRFSGSGYGTDFTLTINNIESEDAAYYC	LQHDNFP
$V_\kappa VI$	A26/A10 (DPK26)	EIVLTQSPDFQSVTPKEKVTITC	RASQSIGS------SLH	WYQQKPDQSPKLLIK	YASQSFS	GVPSRFSGSGSGTDFTLTINSLEAEDAATYYC	HQSSSLP
	A14 (DPK25)	DVVMTQSPAFLSVTPGEKVTITC	QASEGIGN------YLY	WYQQKPDQAPKLLIK	YASQSIS	GVPSRFSGSGSGTDFTFTISSLEAEDAATYYC	QQGNKHP

The Human Germline J_κ repertoire

```
                          L3
                         CDR3
                         100
```

$J_\kappa 1$	WTFGQGTKVEIKR
$J_\kappa 2$	YTFGQGTKLEIKR
$J_\kappa 3$	FTFGPGTKVDIKR
$J_\kappa 4$	LTFGGGTKVEIKR
$J_\kappa 5$	ITFGQGTRLEIKR

Sequences of human germline V_H, V_κ, J_H, and J_κ segments

The Human Germline V_H repertoire

```
                              H1                    H2
                 FR1         CDR1      FR2        CDR2              FR3
         1    10    20    30         40      50      60      70    80    90
         |    |     |     |          |        |       |       |    |     |
```

		FR1	CDR1	FR2	CDR2	FR3
V_H1	1-02 (DP-75)	QVQLVQSGAEVKKPGASVKVSCKASGYTFT	G--YYMH	WRQAPGQGLEWMG	WINP--NSGGTNYAQKFQG	RVTMTRDTSISTAYMELSRLRSDDTAVYYCAR
	1-03 (DP-25)	QVQLVQSGAEVKKPGASVKVSCKASGYTFT	S--YAMH	WRQAPGQRLEWMG	WINA--GNGNTKYSQKFQG	RVTITRDTSASTAYMELSRSEDTAVYYCAR
	1-08 (DP-15)	QVQLVQSGAEVKKPGASVKVSCKASGYTFT	S--YDIN	WRQAQTGQGLEWMG	WMNP--NSGNTGYAQKFQG	RVTMTRNTSISTAYMELSSLRSEDTAVYYCAR
	1-18 (DP-14)	QVQLVQSGAEVKKPGASVKVSCKASGYTFT	S--YGIS	WRQAPGQGLEWMG	WISA--YNGNTNYAQKLQG	RVTMTIDTSTSTAYMELSRLRSDDTAVYYCAR
	1-24 (DP-5)	QVQLVQSGAEVKKPGASVKVSCKVSGYILT	E--LSMH	WRQAPGKGLEWMG	GFDP--EDGETIYAQKFQG	RVTMTEDTSTDTAYMELSLSRSEDTAVYYCAT
	1-45 (DP-4)	QMQLVQSGAEVKKTGSSVKVSCKASGYTFT	Y--RYLH	WRQAPGQALEWMG	WITP--FNGNTNYAQKFQG	RVTITRDRSMSTAYMELSLSRSEDTAMYYCAR
	1-46 (DP-7)	QVQLVQSGAEVKKPGASVKVSCKASGYTFT	S--YYMH	WRQAPGQGLEWMG	IINP--SGGSTSYAQKFQG	RVTMTRDTSTSTVYMELSSLRSEDTAVYYCAR
	1-58 (DP-2)	QMQLVQSPEVKKPGTSVKVSCKASGFTFT	S--SAVQ	WRQARGQRLEWIG	WIVV--GGGNTNYAQKFQE	RVTITRDMSTSTAYMELSSLRSEDTAVYYCAA
	1-69 (DP-10)	QVQLVQSGAEVKKPGSSVKVSCKASGGTFS	S--YAIS	WRQAPGQGLEWMG	GIIP--IFGTANYAQKFQG	RVTITADESTSTAYMELSSLRSEDTAVYYCAR
	1-e (DP-88)	QVQLVQSGAEVKKPGSSVKVSCKASGGTFS	S--YAIS	WRQAPGQGLEWMG	GIIP--IFGTANYAQKFQG	RVTITADKSTSTAYMELSSLRSEDTAVYYCAR
	1-f (DP-3)	EVQLVQSGAEVKKPGATVKISCKVSGYTFT	D--YYMH	WVQQAPGKGLEWMG	LVDP--EDGETIYAEKFQG	RVTITADTSTDTAYMELSLSRSEDTAVYYCAT
V_H2	2-05 (DP-76)	QITLKESGPTLVKPTQTLTLTCTFSGFSLS	TSGVGVG	WIRQPPGKALEWLA	LIY---WNDDKRYSPSLKS	RLTITKDTSKNQVVLTMTNMDPVDTATYYCAHR
	2-26 (DP-26)	QVTLKESGPVLVKPTETLTLTCTVSGFSLS	NARMGVS	WIRQPPGKALEWLA	HIF---SNDEKSYSTSLKS	RLTISKDTSKSQVVLTMTNMDPVDTATYYCARI
	2-70 (DP-28)	QVTLKESGPALVKPTQTLTLTCTFSGFSLS	TSGMRVS	WIRQPPGKALEWLA	RID--WDDDKFYSTSLKT	RLTISKDTSKNQVVLTMTNMDPVDTATYYCARI
V_H3	3-07 (DP-54)	EVQLVESGGGLVQPGGSLRLSCAASGFTFS	S--YWMS	WVRQAPGKGLEWVA	NIKQ--DGSEKYYVDSVKG	RFTISRDNAKNSLYLQMNSLRAEDTAVYYCAR
	3-09 (DP-31)	EVQLVESGGGLVQPGRSLRLSCAASGFTFD	D--YAMH	WVRQAPGKGLEWVS	GISW--NSGSIGYADSVKG	RFTISRDNAKNSLYLQMNSLRAEDTALYYCAKD
	3-11 (DP-35)	QVQLVESGGGLVQPGGSLRLSCAASGFTFS	D--YYMS	WVRQAPGKGLEWVS	YISS--SGSTTYYADSVKG	RFTISRDNAKNSLYLQMNSLRAEDTAVYYCAR
	3-13 (DP-48)	EVQLVESGGGLVQPGGSLRLSCAASGFTFS	S--YDMH	WVRQATGKGLEWVS	AIG---TAGDTYYPGSVKG	RFTISRENAKNSLYLQMNSLRAGDTAVYYCAR
	3-15 (DP-38)	EVQLVESGGGLVKPGGSLRLSCAASGFTFS	N--AWMS	WVRQAPGKGLEWVG	RIKSKTDGGTTDYAAPVKG	RFTISRDDSKNTLYLQMNSLKTEDTAVYYCTT
	3-20 (DP-32)	EVQLVESGGGVVRPGGSLRLSCAASGFTFD	D--YGMS	WVRQAPGKGLEWVS	GINW--NGGSTGYADSVKG	RFTISRDNAKNSLYLQMNSLRAEDTALYHCAR
	3-21 (DP-77)	EVQLVESGGGLVKPGGSLRLSCAASGFTFS	S--YSMN	WVRQAPGKGLEWVS	SISS--SSSYTYYADSVKG	RFTISRDNAKNSLYLQMNSLRAEDTAVYYCAR
	3-23 (DP-47)	EVQLLESGGGLVQPGGSLRLSCAASGFTFS	S--YAMS	WVRQAPGKGLEWVS	AISG--SGGSTYYADSVKG	RFTISRDNSKNTLYLQMNSLRAEDTAVYYCAK
	3-30/3-30.5 (DP-49)	QVQLVESGGGVVQPGRSLRLSCAASGFTFS	S--YGMH	WVRQAPGKGLEWVA	VISY--DGSNKYYADSVKG	RFTISRDNSKNTLYLQMNSLRAEDTAVYYCAR
	3-30.3 (DP-46)	QVQLVESGGGVVQPGRSLRLSCAASGFTFS	S--YAMH	WVRQAPGKGLEWVA	VISY--DGSNKYYADSVKG	RFTISRDNSKNTLYLQMNSLRAEDTAVYYCAR
	3-33 (DP-50)	QVQLVESGGGVVQPGRSLRLSCAASGFTFS	S--YGMH	WVRQAPGKGLEWVS	VIWY--DGSNKYYADSVKG	RFTISRDNSKNTLYLQMNSLRAEDTAVYYCAR
	3-43 (DP-33)	EVQLVESGGVVVQPGRSLRLSCAASGFTFD	D--YTMH	WVRQAPGKGLEWVS	LISW--DGGSTYYADSVKG	RFTISRDNSKNSLYLQMNSLRTEDTALYYCAKD
	3-48 (DP-51)	EVQLVESGGGLVQPGGSLRLSCAASGFTFS	S--YSMN	WVRQAPGKGLEWVS	YISS--SSSTYYADSVKG	RFTISRDNAKNSLYLQMNSLRDEDTAVYYCAR
	3-49	EVQLVESGGGLVQPGRSLRLSCTASGFTFG	D--YAMS	WFRQAPGKGLEWVG	FIRSKAYGGTTEYTASVKG	RFTISRDGSKSIAYLQMNSLKTEDTAVYYCTR
	3-53 (DP-42)	EVQLVETGGGLIQPGGSLRLSCAASGFTVS	S--NYMS	WVRQAPGKGLEWVS	VIY---SGGSTYYADSVKG	RFTISRDNSKNTLYLQMNSLRAEDTAVYYCAR
	3-64 (DP-61)	EVQLVESGGGLVQPGGSLRLSCAASGFTFS	S--YAMH	WVRQAPGKGLEYVS	AISS--NGGSTYYANSVKG	RFTISRDNSKNTLYLQMNSLRAEDMAVYYCAR
	3-66 (DP-86)	EVQLVESGGGLVQPGGSLRLSCAASGFTVS	S--NYMS	WVRQAPGKGLEWVS	VIY---SGGSTYYADSVKG	RFTISRDNSKNTLYLQMNSLRAEDTAVYYCAR
	3-72 (DP-29)	EVQLVESGGGLVQPGGSLRLSCAASGFTFS	D--HYMD	WVRQAPGKGLEWVG	RTRNKANSYTTEYAASVKG	RFTISRDDSKNTLYLQMNSLKTEDTAVYYCAR
	3-73	EVQLVESGGGLVQPGGSLKLSCAASGFTFS	S--SAMH	WVRQASGKGLEWVG	RIRSKANSYATAYAASVKG	RFTISRDDSKNTAYLQMNSLKTEDTAVYYCTR
	3-74 (DP-53)	EVQLVESGGGLVQPGGSLRLSCAASGFTFS	S--YWMH	WVRQAPGKGLVWVS	RINS--DGSSTSYADSVKG	RFTISRDNAKNTLYLQMNSLRAEDTAVYYCAR
	3-d	EVQLVESRGVLVQPGGSLRLSCAASGPTVS	S--NEMS	WVRQAPGKGLEWVS	SI----SGGSTYYADSRKG	RFTISRDNSKNTLHLQMNSLRAEDTAVYYCKK
V_H4	4-04 (DP-70)	QVQLQESGPGLVKPSGTLSLTCAVSGGSIS	SS-NWWS	WRQPPGKGLEWIG	EIY---HSGSTNYNPSLKS	RVTISVDKSKNQFSLKLSSVTAADTAVYYCAR
	4-28 (DP-68)	QVQLQESGPGLVKPSDTLSLTCAVSGYSIS	SS-NWWG	WIRQPPGKGLEWIG	YIY---YSGSTYYNPSLKS	RVTMSVDTSKNQFSLKLSSVTAVDTAVYYCAR
	4-30.1/4-31 (DP-65)	QVQLQESGPGLVKPSQTLSLTCTVSGGSIS	SGGYSWS	WIRQHPGKGLEWIG	YIY---YSGSTYYNPSLKS	RVTISVDTSKNQFSLKLSSVTAADTAVYYCAR
	4-30.2 (DP-64)	QLQLQESGSGLVKPSQTLSLTCAVSGGSIS	SGGYSWS	WIRQPPGKGLEWIG	YIY---HSGSTYYNPSLKS	RVTISVDRSKNQFSLKLSSVTAADTAVYYCAR
	4-30.4 (DP-78)	QVQLQESGPGLVKPSQTLSLTCTVSGGSIS	SGDYYWS	WIRQPPGKGLEWIG	YIY---YSGSTYYNPSLKS	RVTISVDTSKNQFSLKLSSVTAADTAVYYCAR
	4-34 (DP-63)	QVQLQQWGAGLLKPSETLSLTCAVYGGSFS	G--YYWS	WIRQPPGKGLEWIG	EIN---HSGSTNYNPSLKS	RVTISVDTSKNQFSLKLSSVTAADTAVYYCAR
	4-39 (DP-79)	QLQLQESGPGLVKPSETLSLTCTVSGGSIS	SSSYYWG	WIRQPPGKGLEWIG	SIY---YSGSTYYNPSLKS	RVTISVDTSKNQFSLKLSSVTAADTAVYYCAR
	4-59 (DP-71)	QVQLQESGPGLVKPSETLSLTCTVSGGSIS	S--YYWS	WIRQPPGKGLEWIG	YIY---YSGSTNYNPSLKS	RVTISVDTSKNQFSLKLSSVTAADTAVYYCAR
	4-61 (DP-66)	QVQLQESGPGLVKPSETLSLTCTVSGGSVS	SGSYYWS	WIRQPPGKGLEWIG	YIY---YSGSTNYNPSLKS	RVTISVDTSKNQFSLKLSSVTAADTAVYYCAR
	4-b (DP-67)	QVQLQESGPGLVKPSETLSLTCAVSGYSIS	SG-YYWG	WIRQPPGKGLEWIG	SIY---HSGSTYYNPSLKS	RVTISVDTSKNQFSLKLSSVTAADTAVYYCAR
V_H5	5-51 (DP-73)	EVQLVQSGAEVKKPGESLKISCKGSGYSFT	S--YWIG	WVRQMPGKGLEWMG	IIYP--GDSDTRYSPSFQG	QVTISADKSISTAYLQWSSLKASDTAMYYCAR
	5-a	EVQLVQSGAEVKKPGESLRISCKGYSYSFT	S--YWIS	WVRQMPGKGLEWMG	RIDP--SDSYINYSPSFQG	HVTISADKSISTAYLQWSSLKASDTAMYYCAR
V_H6	6-1 (DP-74)	QVQLQQSGPGLVKPSQTLSLTCAISGDSVS	SNSAAWN	WIRQSPSRGLEWLG	RTYYR-SKWYNDYAVSVKS	RITINPDTSKNQFSLQLNSVTPEDTAVYYCAR
V_H7	7-4.1 (DP-21)	QVQLVQSGSELKKPGASVKVSCKASGYTFT	S--YAMN	WVRQAPGQGLEWMG	WINT--NTGNPTYAQGFTG	RFVFSLDTSVSTAYLQICSLKAEDTAVYYCAR

The Human Germline J_H repertoire

```
              H3
             CDR3
              100
               |
       J_H1   ---AEYFQHWGQGTLVTVS
       J_H2   ---YWYFDLWGRGTLVTVS
       J_H3   -----AFDIWGQGTMVTVS
       J_H4   -----YFDYWGQGTLVTVS
       J_H5   ----NWFDPWGQGTLVTVS
       J_H6   YYYYYGMDVWGQGTTVTVS
```

318

Index

ABI 373A automated sequencer 133–6, 137
ABx chromatography 293–4, 301, 302–3
acetate 232, 236, 244
affinity
 maturation
  mutagenesis of antibodies 41–3, 44–9
  screening altered antibodies 43–4, 54–7
  selection of altered antibodies 43, 49–54
  sequence/fingerprint analysis 57–8
 measurement in solution 77–96, 183
  ELISA- and RIA-based methods 80–1
  fluorescence-based methods 79–80
  indirect competition ELISA 81–90, 95–6
  overview 78–81
  RIA-based method 91–5, 96
 measurement using biosensors 99–115, 183
  anti-CEA antibodies 108–15
  immobilization, binding, and regeneration 103–8
  theoretical aspects 101–3
 screening phage antibodies for 43–4, 54–7
antibiotic resistance 212, 221, 222–3
antibody concentration, ELISA method 164–6
antibody-dependent cell-mediated cytotoxicity (ADCC) 189
 assay 192, 195–7
 IgG subclass comparisons 195–7
antibody fragments
 for production in *E. coli* 205, 213–15
 see also Fab fragments; Fab' fragments; Fv fragments; single chain Fv fragments
antibody gene repertoires
 human, transgenic approach 59–74
 production on phage 1–39, 169, 174, 205
  ELISA analysis of antibodies 30–2
  growth/expression of antibodies 23–8
  in-cell PCR assembly 32–8
  preparation/cloning of antibody DNA 6–22
  selection of antibody variants displayed 28–30
  vectors for antibody display 2–6
antibody genes
 databases of sequences 139–40
 expression in *E. coli* 203–48
 expression in mammalian cells 154, 269–88
 primary PCR 9–12, 14–16, 204
 synthesis 206
 see also V genes
anti-CEA antibodies 103, 104
 kinetic analysis 108–15
  anti-idiotypic antibody 115

 direct binding assays 108–9
 engineered fragments 112–15
 indirect binding assays 109–12
antigen (Ag)
 antibody selection and 43, 50
 density, cell-mediated lysis and 190–1
 guided antibody selection and 182–3
 immobilization on hydrogel 104–8
 regeneration from hydrogel 108
 soluble, solution capture on 51–3
 valency, K_d measurement and 78, 96
anti-idiotypic antibody (6G6.C4H) 103, 109, 111, 112, 115
anti-p185^{HER2} immunoliposomes 303
Apa I 65, 68
avidity 49–50

B cells, in-cell PCR 32–7
benchtop-cultivation flasks 231–5
BIAcore system (surface plasmon resonance) 78, 99–115
biosensors 99–115
biotinylated antigen 51
 capture of phage on 52–3
 off-rate selection 53–4
 preparation 52
BLAST algorithm 140
BSA (bovine serum albumin), NIP/NP labelling 191–2
*Bst*NI fingerprinting 44, 57–8, 183, 184

calcium phosphate co-precipitation procedure 275–8
carboxymethyl dextran hydrogel 100, 104–8
carcinoembryonic antigen (CEA) 103, 104, 115
 immobilization on biosensor 105–7
 periodate oxidation 107–8
 see also anti-CEA antibodies
CD3 ε chain, extracellular domain (ECD) 305, 306
cDNA preparation 8–9, 151, 206–7
CDRs, *see* complementarity-determining regions
CDw52 (Campath-1 antigen) 187, 191
CEA, *see* carcinoembryonic antigen
cell-mediated lysis (CML) 187, 188
 assay 192, 194–5
 IgG subclass comparisons 190–5
 measurement *in vitro* 190–5
 optimizing 200

Index

chain-shuffling 41–3, 44, 170–1
  parallel 171
  rodent V genes with human V genes 174–82
chaperones, molecular 228–9
Chinese hamster ovary (CHO) cells 269, 270
  dihydrofolate reductase selection 273–9
  glutamine synthetase selection 279–82
  growth media 273–4, 279
  transfection methods 275–8, 280–1
chloramphenicol (*Cam*) resistance 221, 222–3
chromium-51 (^{51}Cr) 192–3
complement 187
  activation 189, 197–8
    assay 193–4
    IgG subclass comparisons 190–5
    measurement *in vitro* 190–5
    structural requirements for 190
  mediation of lysis, *see* cell-mediated lysis
  receptors 197
complementarity-determining regions (CDRs) 41, 42
  antibody folding and 227
  anti-CEA antibodies 109
  grafting 228
    for humanizing rodent antibodies 147–67, 169
  heavy chain of third (V_HCDR3), spiking 47–9
  light chain of third (V_LCDR3), spiking 49
  loop conformation, important residues for 156
  modelling 155
  mutations 44, 47–9, 58, 143–4
COS cells
  expression of humanized antibodies 164
  transfection 272–3
  transient antibody expression 270–3
cytoplasm
  antibody inclusion bodies 216, 219, 226
  functional antibody expression 216, 218–19, 221–6

databases 139–40
diabodies 215
dihydrofolate reductase (DHFR) 269
  selection in CHO cells 273–9
disulfide bonds 216
disulfide-linked Fv fragments (dsFv) 205, 214, 215
5,5'-dithiobis(2-nitrobenzoic acid) (DTNB) 296–7
DNA sequencing, *see* sequencing
double-stranded DNA
  preparation of template 127
  sequencing 129, 132–3
doxorubicin 303, 304–5
drop-out media 62–3

Dulbecco modified Eagle medium (DMEM) 270, 272–3, 283, 284
DyeDeoxyTerminators, sequencing with 133–4
dye primers, sequencing with 133, 136

EDC/NHS (*N*-ethyl-*N*'-(dimethylaminopropyl) carbodiimide hydrochloride/*N*-hydroxysuccinimide) 104–6
effector functions 187–201
  choosing 187–8
  measurement *in vitro* 190–9
  mediation 188–9
  optimizing 200–1
electroporation
  Escherichia coli 22, 179–80
  NS0 myeloma cells 284–7
ELISA 44, 73
  affinity measurement 78, 80–1, 183
    indirect competition method 81–90, 95–6
    rapid screening method 54–7
  antibody concentration 164–6
  selection of humanized antibodies 183
  specificity screening 30–2, 54, 208–9
EMBL database 139
embryonic stem (ES) cells 60, 70–3
endotoxin removal 294, 296
Entrez database 139
Enzon expression vector 260
enzyme immunoassay (EIA) 109
equilibrium association constant (K_a) 102
equilibrium dissociation constant, *see* K_d
Escherichia coli
  electroporation/transformation 22, 179–80
  expression of antibodies in 203–48
    antibody purification 244–8
    in cytoplasm 216, 218–19, 221–6
    as cytoplasmic inclusion bodies 216, 219, 226
    growth and fermentation 229–44, 261–2
    improving 226–9
    as insoluble periplasmic protein 216, 218–19
    molecular biology 204–15
    overview 215–17
    by secretion 216, 217–21
    strategies 215–26
  Fab' fragment preparation/uses 291–307
  phage infection 1, 29–30, 50
  scFv production in 214–15, 223–4, 226, 259–65
  thioredoxin deficient strains 216

Fab fragments
  humanization by guided selection 171, 172

Index

NIP/NP labelling 191–2
  production in *E. coli* 205, 213, 214
Fab' fragments 291–307
  applications 297–307
  high-level production in *E. coli* 292
  immobilization for affinity purification 305–7
  immunoliposome attachment 303–5
  PEGylation 301–3
  recovery from *E. coli* 293–7
  targeting to periplasmic space 292–3
F(ab')$_2$ fragments
  bispecific 300–1
  monospecific 297–9
Fc fragments 213
FcγR 188, 189
  activities mediated by 189
  optimizing IgG binding 200–1
  phagocytosis/respiratory burst and 197
  structural requirements for binding 195
fd 1–2; *see also* phage, display
fingerprinting, *Bst*NI 44, 57–8, 183, 184
FLAG peptide 212, 215, 221
flow cytometry
  phagocytosis measurement 198–9
  respiratory burst measurement 199
flow injection analyser (FIA) 237, 239
fluorescence, for affinity measurements 79–80
fmol DNA sequencing system 129, 132–3
folding, protein 203–4, 216
  problems 214, 216–17
  solutions to problems 226–9
framework regions (FR) 47
  humanized rodent antibodies 155, 156–7
  mutations 143–4
fusion proteins 213
Fv fragments 205, 214, 253
  disulfide-linked (dsFv) 205, 214, 215
  single chain, *see* single chain Fv fragments

GenBank 139
gene 3 protein (g3p) 1
  antibody fusions 1, 2–5, 220, 221, 222–3
germline variability 141
glucose
  feeding during *E. coli* growth 240, 241–4
  on-line monitoring/control 237, 239, 240, 244
glucose–mineral salt medium (Glu–MM) 231, 232, 233–4, 235–6
glutamine synthetase (GS) 269
  selection in CHO cells 279–82
  selection in NS0 myeloma cells 282–8
glutaredoxin (Grx) 225–6
glycoproteins
  immobilization on hydrogel 106–7
  periodate oxidation 106, 107–8

GMEM-S medium 279, 280–2
guided selection 169–84
  cloning and expression of rodent scFv 173–4
  construction of large chain-shuffled repertoires 174–82
  display and enrichment of antibodies 182–4
  formats 171, 172
  principle 170
  strategy 171–2, 173

H6C8 anti-CEA antibody 103, 104, 111, 112, 113
hierarchical libraries 6
high cell-density cultivations (HCDCs) 235–44, 249, 291
  E. coli strains 236
  fed-batch strategies 237–44
  fermenter and accessories 236–7
  off-line analysis 244
high performance liquid chromatography (HPLC) 264–5, 266
histidine (his) tags 221, 245, 293
human antibodies 39, 59
  gene repertoires
    phage display 6, 7–8, 9–12, 17–21, 169
    transgenic approach 59–74
  sequence analysis 119–44
humanization of rodent antibodies 205
  by CDR grafting 147–67, 169
    chimeric antibody construction 154–5
    cloning and sequencing mouse variable regions 147–54
    design/construction of reshaped antibody 155–64, 167
    preliminary expression/analysis of reshaped antibodies 164–6
  by guided selection 169–84
hybridoma cells
  antibody gene amplification 204
  assembly of scFv from 206–9, 210–13, 255–9
  see also monoclonal antibodies
hydrazine 106, 107
hydrogel, carboxymethyl dextran 100, 104–8
hypoxanthine phosphoribosyl-transferase (*HPRT*) gene 64

IAsys system 99, 100
IgG
  antibody-dependent cell-mediated cytotoxicity 195–7
  complement activation/lysis 188, 189, 190–5
  Fc receptor activation 188, 189, 195
  mutations 141–4, 200–1
  optimizing function 200–1
  phagocytosis and 197–8

321

Index

IMDM medium 283, 284
iminodiacetic acid (IDA) 246
immobilized metal ion affinity chromatography (IMAC) 27, 221, 245–8, 259–60, 293
immunobeads 92–4
immunoglobulin genes, *see* antibody genes
immunoliposomes, Fab' attachment 303–5
inclusion bodies, cytoplasmic 216, 219, 226
integration
    homologous 63–4
    mapping 67–70
    by spheroplast transformation 65–7
ion-exchange chromatography 245–6, 264–5, 266
IPTG (isopropyl-β-D-thio-galactopyranoside) 5, 25, 26, 173, 220

K_a 102
k_a (kinetic rate constant for association)
    anti-CEA antibodies 109, 111
    K_d and 77–8
    measurement 99, 101–2
K_d (equilibrium dissociation constant) 77, 81
    anti-CEA antibodies 109–12
        anti-idiotypic antibodies 111, 115
        engineered fragments 111, 112–15
    association/dissociation kinetics and 77–8
    measurement 77–96
        biosensors for 99, 103, 108–15
        indirect competition ELISA 81–90, 95–6
        mAb/antigen valencies and 78
        overview of methods 79–81
        rapid screening assay 54–7
        RIA-based methods 80–1, 91–5, 96, 109–10
        in solid-phase assays 78
    see also affinity
k_d (kinetic rate constant for dissociation)
    anti-CEA antibodies 109, 111
    K_d and 77–8
    measurement 99, 101–2
Klotz equation 82, 90
Klotz plot 88, 90, 91

lac expression system 220, 222–3
law of mass action 77, 78
ligation 22, 178
linkers, scFv 17, 211, 214–15
    design 254–5
    preparation 17–18
liticase 66
lymphocytes, peripheral blood 7
LYS gene 62, 64
lysis
    cell-mediated, *see* cell-mediated lysis

cultured *E. coli* 230
yeast 66, 69, 71–2
lysozyme, hen egg-white 293, 294–5

M13K07 2, 23
macrophages 197
mammalian cells 154, 269–88
    dihydrofolate reductase selection in CHO cells 273–9
    glutamine synthase selection 279–88
    transient antibody expression 270–3
medium cell-density cultivation 231–5
methionine sulfoxime (MSX) 279, 280–2, 287–8
methotrexate (MTX) 278–9
microtitre plates 24–5, 27
miniantibodies 205, 215
minibodies, kinetic analysis 111, 112–14, 115
molecular modelling 155
monoclonal antibodies (mAb) 253
    measuring affinity in solution 77–96
    phage-display clones 7, 8, 17
    production in *E. coli* 204
    see also hybridoma cells
monocytes 197
mouse antibodies
    humanization 205
        by CDR grafting 147–67, 169
        by guided selection 169–84
    production in *E. coli* 204–5
    production of gene repertoires 6, 8, 9–12, 17–21
mouse myeloma cells 269; *see also* NS0 myeloma cells
mRNA
    DNA synthesis from 8–9, 125, 151, 206–7, 255–6
    preparation 7–8, 151
multimerization
    antibody selection and 43, 50
    scFv, *see* single chain Fv fragments (scFv), multimerization
mutagenesis
    in affinity maturation 41–3, 44–9
    oligonucleotide-directed 258
    site-directed 45–7
mutations
    replacement/silent (R/S) ratio 143–4
    somatic 141–4

neo (neomycin resistance) gene 64–5, 67–8
Ni-NTA (nitrilo-tri-acetic acid) 27, 246, 259–60
NIP (5-iodo-4-hydroxy-3-nitrophenyl-acetyl) 190
    labelling BSA and Fab fragments 191–2

Index

labelling of cells 192–3, 198
*Not*I 2, 4, 5–6, 20–2, 175, 181
NP 190
   labelling BSA and Fab fragments 191–2
   labelling of cells 192–3
NS0 myeloma cells 270, 282–8
   expression vectors 283–4
   glutamine synthetase selection 287–8
   media 282–3
   transfection 284–7

oligo (dT)-affinity purification 7–8
oligonucleotides
   primers, *see* primers
   quality 212–13
   spiked 47–8
ompA signal 217, 257, 260, 261
opsonins 197

pAK vectors 221, 222–3, 291–2
panning, selection of phage by 50–1
pCANTAB-5 2–5, 23, 46
pCANTAB-5E 4, 5, 7, 25, 30
pCANTAB-5myc 5, 25, 30, 173, 175, 178, 181
pCANTAB-6 5, 23, 25, 30, 181
pEE6hCMVgpt 271, 280, 283–4
pEE12 271, 274, 283–4
pEE13 271, 275, 284
pEE14 271, 272, 279–80
pelB signal 217, 222–3
periodate oxidation, glycoproteins 106, 107–8
periplasm
   production of insoluble antibodies 216, 218
   targeting Fab' fragments 292–3
permanence time, serum 298, 301
pH, Fab' fragment recovery and 293
phage
   display
      affinity maturation using 41–55
      antibody expression and 220, 221, 228
      antibody gene repertoires 1–39, 59, 169, 174
   helper 2, 23
   rescue 23–5
phagocytosis 189, 197–9
N,N'-1,2-phenylenedimaleimide (*o*-PDM) 297, 298
phenyl Sepharose 6 fast-flow chromatography 296, 303
phenyl Toyopearl chromatography 301, 302, 303
phylogenetic trees 143, 144
pIG vectors 221
pLNA 64, 65, 67–8

pLUNA 64, 65, 68
pNU 64, 65
polyethylene glycol (PEG) 23–4
   Fab' fragment modification (Fab'-PEG) 301–3
polymerase chain reaction (PCR)
   error-prone 41–3, 44–5
   humanized variable regions 162–3
   in-cell 7, 32–7
   mouse variable regions 150–2
   murine V_H and human V_L assembly 175–7
   mutagenesis methods 44–9
   primary, antibody genes 9–12, 14–16, 204
   products, as sequencing templates 127–9
   room, preparation and use 149–50
   scFV gene assembly 206–9, 210–13, 256–9
   screening of bacterial colonies 152–4
   V gene amplification/cloning 119–25
   YAC libraries 62, 73
polymorphonuclear leukocytes (PMN) 197–8
primers
   dye-labelled 133, 136
   human PCR reactions 9, 10, 14–16
   in-cell PCR 33–4, 37–8
   mouse PCR reactions 9–10, 12, 147, 148–9
   PCR screening 149
   scFv assembly 17, 210–11, 256–7, 258
   sequencing 126, 315
   'spiked' PCR, mutagenesis using 47–9
   V gene segment PCR 119–25
proteases 217, 246
Protein A/G affinity chromatography 293, 294–6
proteins
   folding, *see* folding, protein
   fusion 213
   immobilization on hydrogel 104–6
   sequencing 125
protoplasts, *see* spheroplasts
pSVM.*dhfr* 274–5, 276
pUC19-derived vectors 2–6, 46, 173, 221
pulse-field gel electrophoresis (PFGE) 67, 69–70
pUNA 64, 65
purification
   affinity, immobilized Fab'-SH for 305–7
   antibody 27–8, 244–8
   Fab' fragments 293–4
   F(ab')$_2$ fragments 299
   scFv 247–8, 259–60, 264–5, 266
pYAC vector 61
PyroBind ST filters 294, 296, 303

radioimmunoassay (RIA), affinity measurement 80–1, 91–5, 96, 109–10
random combinatorial libraries 6, 32, 37, 38
respiratory burst 197–9

Index

rodent antibodies
  humanization by CDR grafting 147–67, 169
  humanization by guided selection 169–84
RU (response units) 100

Scatchard equation 82, 89, 90
Scatchard plot 88, 89
scFv, *see* single chain Fv fragments
screening
  affinity 43–4, 54–7
  specificity 30–2, 43–4, 54
secretion, of antibodies by *E. coli* 216, 217–21
selection
  phage antibodies 28–30, 43, 49–54
    off-rate selection 53–4
    by panning 50–1
    solution capture on soluble antigen 51–3
  see also guided selection
Sequenase 129, 130, 133
sequence analysis 137–44
  databases 139–40
  editing/translating/comparing sequences 137–8
  germline variability 141
  mouse variable regions 155, 158
  multiple alignments 138–9
  software packages 137
  somatic mutations 141–4
  statistical methods 140–4
sequencing 44, 125–36
  affinity-matured antibodies 57–8
  automated 130–6
  chain termination (Sanger) method 125, 129
  immunoglobulin genes 125–36
  manual 129–30, 131–2
  techniques 129–36
  template preparation 126–9
*Sfi*I
  E. coli expression systems 222–3
  humanization by guided selection 175, 181
  phage-display libraries 2, 4, 5–6, 20–2
  scFV cloning 208, 212
shake-flask cultures 229–30, 291
sheep red blood cells (SRBC) 191, 192–3, 194
single chain Fv fragments (scFv) 205
  assembly from hybridoma cells 206–9, 210–13, 255–9
  from phage-display libraries 3, 7
    analysis by ELISA 30–2
    assembly 13, 17–20
    growth and soluble expression 25–8
    selection methods 28–30
  humanization by guided selection 171, 172, 173–4

linkers, *see* linkers, scFv
multimerization 50, 255, 263–4
K_d measurement and 112–15
mutagenesis 47–9
production 253–66
  in *E. coli* 214–15, 223–4, 226, 259–65
  fermentation/renaturation/purification 261–5
  protein renaturation 262–3
  purification 247–8, 259–60, 264–5, 266
  T84.66 anti-CEA antibody, kinetic analysis 111, 112–15
single-stranded DNA
  preparation of template 46–7, 126–7
  sequencing 129, 130
6G6.C4H anti-idiotypic antibody 103, 109, 111, 112, 115
sodium borohydride 106, 107
specificity
  affinity-matured antibodies 43, 54, 58
  screening for 30–2, 43–4, 54
spheroplasts
  fusion with ES cells 70–2
  preparation 66, 71–2
  transformation 66–7
'spiked' PCR primers 47–9
spleen, mouse 8, 204
stirred tank reactors 236–7
streptococcal Protein A/G affinity chromatography 293, 294–6
sulfur-35 (^{35}S) 129, 130, 132–3
super broth (SB) medium 230, 231, 232, 233, 234–5
surface plasmon resonance (SPR) 78, 99–115
SV40 origin of replication 270, 271

T84.12 anti-CEA antibody 104
  kinetic analysis 108–9, 110, 111
T84.66 anti-CEA antibody 104
  kinetic analysis 108, 109–12, 113
  scFV versions 111, 112–15
Taq polymerase 129, 133, 136
temperature
  antibody expression and 219–20, 260
  antibody folding and 229
templates, preparation 46–7, 126–9
tetracycline (*tet*) resistance 212, 223
thiol content, Fab' preparations 296–7
thiopropyl-Sepharose 6B resin 306–7
thioredoxin reductase (TrxB) 225–6
  deficient strains of *E. coli* (*trxB⁻*) 216, 226
transgenic mice, human antibody repertoires in 59–74

URA3 61, 64, 65, 68

Index

valency of antibodies/antigens, K_d and 78, 96
variable domain genes, *see* V genes
variable regions
  conserved residues 157
  humanized rodent antibodies 155–64
    construction 157–8
    design 155–6, 158–61, 167
    synthesis of DNA sequence coding for 161–4
  mouse, analysis 155, 158
V BASE 140
vectors
  E. coli expression 217–21, 222–3, 226, 260–1
  integration 63–4, 269
  mammalian cell expression 269–70, 271–2, 274–5, 279–80, 283–4
  phage display 2–6, 46
  replacement 63
  YAC 64–5
Vernier residues 41, 44, 58
V genes
  amplification and cloning 9–12, 119–25, 172
    germline segments 123–4
    rearranged V genes 125
  chain-shuffling 170–1, 174–82
  cloning and sequencing mouse 147–54
  databases of sequences 139–40
  germline variability 141
  humanization by guided selection 171
  mutagenesis 44–9
  replacement/silent mutation (R/S) ratio 143–4
  sequence analysis 138–9
  sequencing primers 315

somatic mutations 141–4
V_H genes
  chain-shuffling 170–1, 174, 175–81
  human, combination with human V_L genes 181–2
  human germline repertoire 316–17
  mutagenesis using 'spiked' PCR primers 47–9
  phylogenetic trees 143, 144
  somatic mutations 141–4
V_L genes
  chain-shuffling 170–1, 174, 175–81
  human, combination with human V_H genes 181–2
  mutagenesis using 'spiked' PCR primers 49

yeast
  spheroplasts, *see* spheroplasts
  strain AB1380 61–3
yeast artificial chromosomes (YACs) 59–74
  library screening 62
  maintenance 62–3
  modification 63–70
    mapping site-specific integration 67
    profile of single/multiple integrations 67–70
    site-directed introduction 65–7
    universal vectors 64–5
  transfer into ES cells 70–3
YPD plates/liquid medium 62, 63, 66

zymolyase 66, 69, 71–2